A

TREATISE

ON

ASTRONOMY,

DESCRIPTIVE, PHYSICAL, AND PRACTICAL.

DESIGNED FOR

SCHOOLS, COLLEGES, AND PRIVATE STUDENTS

BY H. N. ROBINSON, A. M.,
FORMERLY PROFESSOR OF MATHEMATICS IN THE U. S. NAVY; AUTHOR OF A TREATISE ON ARITHMETIC; ALGEBRA; NATURAL PHILOSOPHY; ETC.

ALBANY:
ERASTUS H. PEASE & CO., 82 STATE STREET.
CINCINNATI:
JACOB ERNST, NO. 183 MAIN STREET.
1850.

Entered, according to act of Congress, in the year 1849,

By HORATIO N. ROBINSON,

In the Clerk's Office of the District Court of the United States, for the District of Ohio.

J. MUNSELL,
STEREOTYPER AND PRINTER,
ALBANY.

PREFACE.

To give at once a clear explanation of the design and intended character of this work, it is important to state that its author, in early life, imbibed quite a passion for astronomy, and, of course, he naturally sought the aid of books; but, in this field of research, he was really astonished to find how little substantial aid he could procure from that source, and not even to this day have his desires been gratified.

Then, as now, books of great worth and high merit were to be found, but they did not meet the wants of a learner; the substantially good were too voluminous and mathematically abstruse to be much used by the humble pupil, and the less mathematical were too superficial and trifling to give satisfaction to the real aspirant after astronomical knowledge.

Of the less mathematical and more elaborate works on astronomy there are two classes—the pure and valuable, like the writings of Biot and Herschel; but, excellent as these are, they are not adapted to the purposes of instruction; and every effort to make class books of them has substantially failed. From the other class, which consists of essays and popular lectures, little substantial knowledge can be gathered, for they do not *teach* astronomy; as a general thing, they only *glorify* it; they may excite our wonder concerning the immensity or grandeur of the heavens, but they give us no additional power to investigate the science.

Another class of more brief and valuable productions were, and are always to be found, in which most of the important facts are recorded; such as the distances, magnitudes, and motions of the heavenly bodies; but how these facts became known is rarely explained: this is what the true searcher after science will always demand, and this book is designed expressly to meet that demand.

In the first part of the book we suppose the reader entirely unacquainted with the subject; but we suppose him competent to the task—to be, at least, sixteen years of age—to have a good knowledge of proportion, some knowledge of algebra, geometry, and trigonometry—and then, and not until then, can the study be pursued with any degree of success worth mentioning. Such a person, and with such acquirements as we have here designated, we believe, can take this book and learn astronomy in comparatively a short time; for the chief design of the work is, to teach whoever desires to learn: and it matters not where the learner may be, in a college, academy, school, or a solitary student at home, and alone in the pursuit.

The book is designed for two classes of students—the well prepared in the mathematics, and the less prepared; the former are expected to read the text notes, the latter should omit them. With the text notes, we conceive it, or rather designed it to be, a very suitable book to give sound elementary instruction in astronomy; but we do not offer the work as complete on practical astronomy; for whoever becomes a practical astronomer will, of course, seek the aid of complete and elaborate sets of tables, such as would be improper to insert in a school book.

We have inserted tables only for the purpose of carrying out a sound theoretical plan of instruction, and, therefore, we have given as few as possible, and those few in a very contracted form. The epochs for the sun and moon may be extended forward or backward, to any extent, by any one who understands the theory.

The chapters on comets, variable stars, &c., are compilations, and are printed in smaller type; and the works to which we are most indebted, are Herschel's Astronomy and the Cambridge Astronomy, originally the work of M. Biot.

Other parts of the work, we believe, will be admitted as mainly original, by all who take pains to examine it.

The chief merits claimed for this book are, brevity, clearness of illustration, anticipating the difficulties of the pupil, and removing them, and bringing out all the essential points of the science.

Some originality is claimed, also, in several of our illustrations, particularly that of showing the rationale of tides rising on the opposite sides of the earth from the moon; and in the general treatment of eclipses; but it is for others to determine how much merit should be awarded for such originalities; we have, however, used greater conciseness and perspicuity in general computations than is to be found in most of the books on this subject; and this last remark will apply to the whole work.

CONTENTS.

SECTION I.

SECTION II.

DESCRIPTIVE ASTRONOMY.

CHAPTER VIII.

CHAPTER IX.

CHAPTER X.

CHAPTER XI.

CHAPTER XII.

CHAPTER XIII.

SECTION III.

PHYSICAL ASTRONOMY

CHAPTER I.

CHAPTER II.

CHAPTER III.

SECTION IV.

PRACTICAL ASTRONOMY.

ASTRONOMY.

INTRODUCTION.

ASTRONOMY is the science which treats of the heavenly bodies, describes their appearances, determines their magnitudes, and discovers the laws which govern their motions. Astronomy defined.

When we merely state facts and describe appearances as they exist in the heavens, we call it *Descriptive Astronomy.* When we compute magnitudes, determine distances, record observations, and make any computations whatever, we call it *Practical Astronomy.* The divisions of astronomy.

The investigation of the laws which govern the celestial motions, and the explanation of the *causes* which bring about the known results, is called *Physical Astronomy.*

When the mariner makes use of the index of the heavens, to determine his position on the earth, such observations, and their corresponding computations, are called *Nautical Astronomy.* Nautical astronomy.

By nautical astronomy we determine positions on the earth, and subsequently, the magnitude of the earth; and thus, we perceive, that Geography and Astronomy must be linked together; and no one can fully understand the former science, without the aid of the latter. Geography and astronomy united.

Astronomy is the most ancient of all the sciences, for, in the earliest age, the people could not have avoided observing the successive returns of day and night, and summer and winter. They could not fail to perceive that short days corresponded to winter, and long days to summer; and it was thus, probably, that the attentions of men were first drawn to the study of astronomy. The antiquity of astronomy.

INTRODUC.

Facts alone not science.

In this work, we shall not take facts unless they are within the sphere of our own observations. We shall not peremptorily state that the earth is 7912 miles in diameter; that the moon is about 240,000 miles from the earth, and the sun 95,000,000 of miles; for such facts, alone, and of themselves, do not constitute knowledge, though often mistaken for knowledge. We shall direct the mind of the reader, step by step, through the observations and through the investigations, so that he can decide for himself that the earth must be of such a magnitude, and is thus far from the other heavenly bodies; and that will be knowledge of the most essential kind.

The foundation of astronomical knowledge.

All astronomical knowledge has its foundation in observation; and the first object of this book shall be to point out what observations must be taken, and what deductions must be made therefrom; but the great book which the pupil must study, if he would meet with success, is the one which spreads out its pages on the blue arch above; and he must place but secondary dependence on any book that is merely the work of human art.

As we disapprove of the practice of throwing to the reader astounding astronomical facts, whether he can digest them or not, and as we are to take the inductive method, and to lead the student by the hand, we must commence on the supposition that the reader is entirely unacquainted even with the common astronomical facts, and now for the first time seriously brings his mind to the study of the subject; but we shall suppose some maturity of mind, and some preparation, by the acquisition of at least respectable mathematical knowledge.

Conventional terms and definitions.

Every science has its technicalities and conventional terms; and astronomy is by no means an exception to the general rule; and as it will prepare the way for a clearer understanding of our subject, we now give a short list of some of the technical terms, which must be used in our composition.

Horizon.—Every person, wherever he may be, conceives himself to be in the center of a circle; and the circumference of that circle is where the earth and sky apparently meet. That circle is called the ***horizon.***

Altitude. — The perpendicular hight from the horizon, measured by degrees of a circle. INTRODUC.

Meridian. — An imaginary line, north and south from any point or place, whether it is conceived to run along the earth or through the heavens. If the meridian is conceived to divide both the earth and the heavens, it is then considered as a plane, and is spoken of as *the plane of the meridian.*

Poles. — The points where all meridians come together: poles of the earth — the extremities of the earth's axis.

Zenith. — The zenith of any place, is the point directly overhead; and the *Nadir* is directly opposite to the zenith, or under our feet. The *zenith* and *nadir* are the *poles* to the horizon. Poles of the horizon.

Verticals. — All lines passing from the zenith, perpendicular to the horizon, are called *Verticals*, or *Vertical Circles.* The one passing at right angles to the meridian, and striking the horizon at the east and west points, is called the *Prime Vertical.* Prime vertical.

Azimuth. — The angular position of a body from the meridian, measured on the circle of the horizon, is called its *Azimuth.*

The angular position, measured from its prime vertical, is called its *Amplitude.* Amplitude

The sum of the *azimuth* and *amplitude* must always make 90 degrees.

Equator. — The *Earth's Equator* is a great circle, east and west, and equidistant from the poles, dividing the earth into two hemispheres, a northern, and a southern.

The *Celestial Equator* is the plane of the earth's equator conceived to extend into the heavens. Celestial equator.

When the sun, or any other heavenly body, meets the celestial equator, it is said to be in the *Equinox*, and the equatorial line in the heavens is called the *Equinoctial.* Equinoctial.

Latitude. — The latitude of any place on the earth, is its distance from the equator, measured in degrees on the meridian, either north or south.

If the measure is toward the north, it is north latitude; if toward the south, south latitude.

INTRODUC.

The distance from the equator to the poles is 90 degrees — one-fourth of a circle; and we shall know the circumference of the whole earth, whenever we can find *the absolute length of one degree on its surface.*

Co-Latitude. — Co-latitude is the distance, in degrees, of any place from the nearest pole.

The latitude and co-latitude (complement of the latitude) must, of course, always make 90 degrees.

Parallels of latitude.

Parallels of latitude are small circles on the surface of the earth, parallel to the equator.

Every point, in such a circle, has the same latitude.

Longitude. — The *longitude* of a place, on the surface of the earth, is the inclination of its meridian to some other meridian which may be chosen to reckon from. English astronomers and geographers take the meridian which runs through Greenwich Observatory, as the zero meridian.

The first meridian arbitrary

Other nations generally take the meridian of their principal observatories, or that of the capital of their country, as the first meridian; but this is national vanity, and creates only trouble and confusion: it is important that the whole world should agree *on some one meridian*, from which to reckon longitude; but as nature has designated no particular one, it is not wonderful that different nations have chosen different lines.

We adopt the meridian of Greenwich; and why?

In this work, we shall adopt the meridian of Greenwich as the zero line of longitude, because most of the globes and maps, and all the important astronomical tables, are adapted to that meridian, and we see nothing to be gained by changing them.

Declination. — Declination refers only to the celestial equator, and is a leaning or *declining*, north or south of that line, and is similar to latitude on the earth.

Solstitial Points. — The points, in the heavens, north and south, where the sun has its greatest *declination.*

The northern point we call the *Summer Solstice*, and the southern point the *Winter Solstice;* the first is in longitude 90°, the other in longitude 270°.

As latitude is reckoned north and south, so longitude is

reckoned east and west; but it would add greatly to systematic regularity, and tend much to avoid confusion and ambiguity in computations, were this mode of expression abandoned, and longitude invariably reckoned *westward*, from 0 to 360 degrees.

Improvement suggested.

Latitude and longitude, on the earth, does not correspond to latitude and longitude in the heavens. Latitude, on the earth, corresponds with declination in the heavens; and longitude, on the earth, has a striking analogy to right ascension in the heavens, though not an exact correspondence. We shall more particularly explain latitude, longitude, and right ascension in the heavens, as we advance in this work; for it is only when we are forced to use these terms, that the nature and spirit of their import can be really understood.

Latitude, longitude, and right ascension.

There are other technicalities, and terms of frequent use, in astronomy, such as Conjunction, Opposition, Retrograde, Direct, Apogee, Perigee, &c., &c., all of which, for the sake of simplicity, had better not be explained until they fall into use; and, once for all, let us impress this fact on the minds of our readers, that we shall put far more stress on the substance and spirit of a thing, than on its name.

Other terms not explained.

SECTION I.

CHAPTER I.

PRELIMINARY OBSERVATIONS.

CHAP. I.

To commence the study of astronomy, we must observe and call to mind the real appearances of the heavens.

Take such a station, any clear night, as will command an extensive view of that apparent, concave hemisphere above us, which we call the sky, and fix well in the mind the directions of *north*, *south*, *east*, and *west*.

The apparent motion of the stars.

At first, let us suppose our observer to be somewhere in the United States, or somewhere in the northern hemisphere, about 40 degrees from the equator.

As yet, this imaginary person is not an astronomer, and neither has, nor knows how to use, any astronomical instrument; but we would have him mark with attention the *positions* of the heavenly bodies.

(1.) Soon he will perceive a variation in the position of the stars: those at the east of him will apparently rise; those at the west will appear to sink lower, or fall below the horizon; those at the south, and near his zenith, will apparently move westward; and those at the north of him, which he may see about half way between the horizon and zenith, will *appear stationary*.

Apparent revolution of the heavenly bodies.

Let such observations be continued during all the hours of the night, and for several nights, and the observer cannot fail to be convinced that not only all the stars, but the sun, moon, and planets, appear to perform revolutions, in about twenty-four hours, round a *fixed point;* and that fixed point, *as appears to us* (in the middle and northern part of the United States), is about midway between the northern horizon and the zenith.

Large and small circles.

It should always be borne in mind, that the sun, moon, and stars, have an apparent diurnal motion round a *fixed point*, and all those stars which are 90 degrees from that point, apparently describe a great circle. Those stars that are nearer to the fixed point than 90 degrees, describe smaller circles; and the circles are smaller and smaller as the objects are nearer and nearer the *fixed points*.

(2.) There is *one star* so near this fixed point, that the small circle it describes, in about 24 hours, is not apparent from mere inspection. To detect the apparent motion of this star, we must resort to nice observations, aided by mathematical instruments.

The North Star.

This *fixed point*, that we have several times mentioned, is the *North Pole of the heavens*, and this *one star* that we have just mentioned, is commonly called the *North Star*, or the *Pole Star*.

Position of the North Star.

(3.) This star, on the 1st of January, 1820, was 1° 39′ 6″ from the pole, and on 1st of January, 1847, its distance from the pole was 1° 30′ 8″; and it will gradually and more slowly approach within about half a degree of the pole, and afterward it will as gradually recede from the pole, and finally cease to be the *polar star*.

The pole in motion.

We here, and must generally, speak of the *star*, or the stars, as in motion; but this is not so. The fixed stars *are absolutely fixed;* it is the pole itself that has a slow motion among the stars, but the cause of this motion cannot now be explained; it is one of the most abstruse points in astronomy, and we only mention it as a fact.

As the North Star appears stationary, to the common observer, it has always been taken as the infallible guide to *direction;* and every sailor of the ocean, and every wanderer of the African and Arabian deserts, has held familiar acquaintance with it.

Changes of appearance on going southward.

(4.) If our observer now goes more to the southward, and makes the same observations on the apparent motions of the stars, he will find the same general results; each individual star will describe the same circle; but the pole, the *fixed point*, will be lower down, and nearer the northern horizon; and it will be lower and lower in proportion to the distance the observer goes to the south. After the observer has gone sufficiently far, the fixed point, the pole, will no longer be up

CHAP. I. in the heavens, but down in the northern horizon; and when the pole does appear in the horizon, the observer is at the equator, and from that line all the stars at or near the equator appear to rise up directly from the east, and go down directly to the west; and all other stars, situated out of the equator, describe their small circles parallel to this perpendicular equatorial circle.

Appearance from the equator.

South of the equator.

If the observer goes south of the equator, the apparent north pole of the heavens sinks below the northern horizon, and the south pole rises up into the heavens at the south.

Changes in appearance on going north.

(5.) If the observer should go north, from the first station, in place of going south, the north pole would rise nearer to the zenith; and, should he continue to go north, he would finally find the pole in his zenith, and all the stars would apparently make circles round the zenith, as a center, and parallel to the horizon; and the horizon itself would be the celestial equator.

(6.) When the north pole of the heavens appears at the zenith, the observer must then be at the north pole, on the earth, or at the latitude of 90 degrees.

Appearance from the north pole.

(7.) Any celestial body, which is north of the equator, is always visible from the north pole of the earth; hence the sun, which is north of the equator from the 20th of March to the 23d of September, must be constantly visible during that period, in a clear sky.

Just as the sun comes north of the equator, its diurnal progress, or rather, the progress of 24 hours, is around the horizon. When the sun's declination is 10 degrees north of the equator, the progress of 24 hours is around the horizon at the altitude of 10 degrees; and so for any other degree.

From the north pole, all directions, on the surface of the earth, are south. North would be in a vertical direction toward the zenith.

How to find the circumference and diameter of the earth.

We have observed that the pole of the heavens rises as we go north, and sinks toward the horizon as we go south; and when we observe that the pole has changed its position one degree, in relation to the horizon, we know that we must have changed place one degree on the surface of the earth.

(8.) Now we know by observation, that if we go north about 69¼ English miles on the earth, the north pole will be *one degree* higher above the horizon. Therefore 69¼ miles corresponds to one degree, on the earth; and hence the whole circumference of the earth must be 69¼ multiplied by 360: for there are 360 degrees to every circle. This gives 24,930 miles for the circumference of the earth, and 7,930 miles for its diameter, which is not far from the truth.

Circumpolar stars.

(9.) Here, in the United States, or anywhere either in Europe, Asia, or America, north of the equator, say in latitude 40°, the north pole of the heavens must appear at an altitude of 40° above the horizon; and as all the stars and heavenly bodies apparently circulate round this point as a center, it follows that all those stars which are within 40° of the pole can never go below the horizon, but circulate round and round the pole. All those stars which never go below the horizon, are called *circumpolar* stars.

The sun a circumpolar body, as seen from the north of latitude 66 degrees

At the north, and very near the north pole, the sun is a *circumpolar body* while it is north of the equator, and it is a circumpolar body as seen from the south pole, while it is south of the equator; this gives six months day and six months night, at the poles.

(10.) North of latitude 66°, and when the sun's declination is more than 23° north (as it is on and about the 20th of June in each year), then the sun comes at, or very near, the northern horizon, at midnight; it is nearly east, at 6 o'clock in the morning; it is south, at noon, and about 23° in altitude; and is nearly west at 6 in the afternoon.

(11.) In the southern hemisphere, there is no prominent star near the south pole; that is, no southern polar star; but, of course, there are circumpolar stars, and more and more as one goes south; and if it were possible to go to the south pole, the whole southern hemisphere would consist of circumpolar stars, and the pole, or fixed point of the heavens, would be directly overhead; and the sun himself, when south of the equator, would be a circumpolar body, going round and round every 24 hours, nearly parallel with the horizon.

(12.) In all latitudes, and from all places, the sun is

CHAP. I. observed to circulate round the nearest pole, as a center; and when the sun is on the same side of the equator as the observer, more than half of the sun's diurnal circle is above the horizon, and the observer will have more than 12 hours sunlight.

The nearest pole is the center of the sun's diurnal motion.

When the sun is on the equator, the horizon, of every latitude, cuts the sun's diurnal circle into two equal parts, and gives 12 hours day, and 12 hours night, the world over. When the sun is on the opposite side of the equator from the observer, the smaller segment of the sun's diurnal circle is above the horizon, and, of course, gives shorter days than nights.

We have, thus far, made but rude and very imperfect observations on the apparent motion of the heavenly bodies, and have satisfied ourselves only of two facts:

Facts settled.

1. That all the stars, sun, moon, and planets included, apparently circulate round the pole, and round the earth, in a day, or in *about* 24 hours.

2. That the sun comes to the meridian, at different altitudes above the horizon, at different seasons of the year, giving long days in June, and short days in December.

(13.) Let us now pay attention to some other particulars. Let us look at the different groups of stars, and individual stars, so that we can recognize them night after night.

Necessity of having a measure of time.

We should now have some means of measuring time; but, in early days, when astronomy was no further advanced than it is supposed to be in this work, a clock could hardly have had existence; and the advancement of timepieces has been nearly as gradual as the advancement of astronomy itself.

But we will not dwell on the history, and difficulties, of clockmaking: whatever these difficulties may have been, or whatever niceties modern science and art may have attained, there never was a period when people had not a good *general idea* of time, and some means to measure it. For instance, sunrise and sunset could be always noted as distinct points of time; and the interval of a day and a night, or an astronomical day, which we now call 24 hours, was soon observed to be a constant quantity.

At first, only rude timepieces could be made, designed to mark off equal intervals of time; but we will suppose, at once, that the reader of this work, or *our imaginary observer*, can have the use of a common clock, which measures mean solar time of 24 hours in a natural day, which is marked by the sun.

The particular position of stars in relation to time.

(14.) Now, having power to recognize certain stars, or groups of stars, such as the *Seven Stars*, the *Belt of Orion*, *Aldebaran*, *Sirius*, and the like, and having likewise the use of a clock, he can observe *when any particular star comes to any definite position.*

Let a person place himself at any particular point, to the north of any perpendicular line, as the edge of a wall or building, and let him observe the stars as they pass behind the building, in their diurnal motions from the east to the west. For example, let us suppose that the observer is watching the star *Aldebaran*, and that, when the eye is placed in a particular definite position, the star passes behind the building at exactly 8 o'clock.

The next evening, the same star will come to the same point about 4 minutes before 8 o'clock; and it will not come to the same point again, at 8 o'clock in the evening, until after the expiration of *one year.*

(15.) But in any year, on the same day of the month, and at the same hour of the day, the same star will be at, or very near, the same position, as seen from the same point.

On stars coming to the meridian.

For instance, if certain stars come on the meridian at a particular time in the evening, on the first day of December, the same stars will not come on the meridian again, at the same time of the night, until the first day of the next December.

Index to the length of a year.

On the first of January, certain stars come to the meridian at midnight; and (speaking loosely) every first of January the same stars come to the meridian at the same time; and there will be no other day during the whole year, when the same stars will come to the meridian at midnight.

Thus, the same day of every year is observed to have the same position of the stars at the same hour of the night; *and this is the most definite index for the expiration of a year.*

CHAP. I.

Another index of the length of the year.

(16.) The year is also indicated by the change of the sun's declination, which the most careless observer cannot fail to notice. On the 21st of June, the sun declines about 23½ degrees from the equator toward the north; and, of course, to us in the northern hemisphere, its meridian altitude is so much greater, and the horizontal shadows it casts from the same fixed objects will be shorter; and the same meridian altitude and short shadow will not occur again until the following June, or after the expiration of one year.

Thus, we see, that the time of the stars coming on to the meridian, and the declination of the sun, have a close correspondence, *in relation to time.*

Fixed stars; why this term is applied.

In all our observations on the stars, we notice that their apparent relative situations are not changed by their diurnal motions. In whatever parts of their circles they are observed, or at whatever hour of the night they are seen, the same configuration is recognized, although the same group, in the different parts of its course, will stand differently, in respect to the horizon. For instance, a configuration of stars resembling the letter A, when east of the meridian, will resemble the letter V, when west of the meridian.

Wandering stars.

As the stars, in general, do not change their positions in respect to each other, they are called *fixed stars;* but there are a few important stars that do change, in respect to other stars; and for that reason they become especial objects of attention, and form the most interesting portion of astronomy.

Planets.

In the earliest ages, those stars that changed their places, were called *wandering stars;* and they were subsequently found to be the planetary bodies of the solar system, like the earth on which we live.

CHAPTER II.

APPEARANCES IN THE HEAVENS.

CHAP. II.

IN the preceding chapter we have only called to mind the most obvious and preliminary observations, which force themselves on every one who pays the least attention to the subject.

We shall now consider the observer at one place, making more minute and scientific observations.

How to find the latitude of the place of observation.

(17.) We have already remarked, that if the observer was on the equator, the poles, to him, would be in his horizon. If he were at one of the poles, for instance, the north pole, the equator would then bound the horizon. If he were half way between the equator and one of the poles, that pole would appear half way between the horizon and the zenith.

Therefore, *by observing the altitude of the pole above the horizon,* we determine the number of degrees we are from the equator, which is called *the latitude of the place.*

(18.) To carry the mind of the reader progressively along, in astronomy, we must now suppose that he not only has the use of a good clock, but *has also some instrument to measure angles.*

Clocks and astronomical instruments progressed toward perfection in about the same ratio as astronomy itself; but, as we are investigating or leading the young mind to the investigation of astronomy, and not making clocks or mathematical instruments, we therefore suppose that the observer has all the necessary instruments at his command, and we may now require him to make a *correct map* of the visible heavens; but to accomplish it, we must allow him at least one year's time, and even then he cannot arrive at anything like accuracy, as several incidental difficulties, instrumental errors, and practical inaccuracies, must be met and overcome.

Sources of error in making observation.

(19.) There are three principal sources of error, which must be taken into consideration, in making astronomical observations. 1. Uncertainty as to the exact time. 2. Inex-

CHAP. II. pertness and want of tact in the observer; and 3. Imperfection in the instruments. Everything done by man is necessarily imperfect.

Practical difficulties and causes of error. "It may be thought an easy thing," says Sir John Herschel, "by one unacquainted with the niceties required, to turn a circle in metal, to divide its circumference into 360 equal parts, and these again into smaller subdivisions,—to place it accurately on its center, and to adjust it in a given position; but practically it is found to be one of the most difficult. Nor will this appear extraordinary, when it is considered that, owing to the application of telescopes to the purposes of angular measurement, every imperfection of structure or division becomes magnified by the whole optical power of that instrument; and that thus, not only direct errors of workmanship, arising from unsteadiness of hand or imperfection of tools, but those inaccuracies which originate in far more uncontrollable causes, such as the unequal expansion and contraction of metallic masses, by a change of temperature, and their unavoidable flexure or bending by their own weight, become perceptible and measurable."

Necessary instruments. (20.) The most important instruments, in an observatory, aside from the clock, are a *circle*, or *sector*, *for altitudes*; and a *transit instrument*.

The former consists of a circle, or a portion of a circle, of firm and durable material, divided into degrees, at the rate of 360 to the whole circle. Each degree is divided into equal parts; and, by a very ingenious mechanical adjustment of an index, called a *Vernier* scale, the division of the degree is practically (though not really) subdivided into seconds, or 3600 equal parts.

The whole instrument must now be firmly placed and adjusted to the *true horizontal* (which is exactly at right angles to a plumb line), and so made as to turn in any direction. With this instrument we can measure angles of altitude.

The transit instrument. (21.) The transit instrument is but *a telescope*, firmly fastened on a horizontal axis, *east* and *west*, so that the telescope itself moves up and down in the *plane of the meridian*, but can never be turned aside from the meridian to the east or west.

To place the instrument in this position, is a very difficult matter; but it is a difficulty which, at present, should not come under consideration: we simply conceive it so placed, ready for observations.

Transit Instrument.

A line in the transit instrument a visible meridian.

"In the focus of the eyepiece, and at right angles to the length of the telescope, is placed a system of one horizontal and five equidistant vertical threads or wires, as represented in the annexed figure, which always appear in the *field of view*, when properly illuminated, by day by the light of the sky, by night by that of a lamp, introduced by a contrivance not necessary here to explain. The place of this system of wires may be altered by adjusting screws, giving it a lateral (horizontal) motion; and it is by this means brought to such a position, that the middle one of the vertical wires shall intersect *the line of collimation* of the telescope, where it is arrested and permanently fastened. In this situation it is evident that the middle thread will be a visible representation of that portion of the celestial meridian to which the telescope is pointed; and when a star is seen to cross this wire in the telescope, it is in the act of culminating, or passing the celestial meridian. The instant of this event is noted by the clock or chronometer, which forms an indispensable accompaniment of the transit instrument. For greater precision, the moment of its crossing each of the vertical threads is noted, and a mean taken, which (since the threads are equidistant) would give exactly the same result, were all the observations perfect, and will, of course, tend to subdivide and destroy their errors in an average of the whole."

Meridian Wires.

Practical artifices, to attain accuracy.

Intervals between the fixed stars passing the meridian always constant.

(22.) Thus, all prepared with a transit instrument and a clock, we fix on some bright star, and mark when it comes to the meridian, or appears to pass behind the central wire of the instrument. By noting the same event the next evening, the next, and the next, we find the interval to be very sensi-

CHAP. II. bly less than 24 hours; but the intervals are equal to each other; *and all the fixed stars are unanimous in giving equal intervals of time between two successive transits of the same star,* if measured by the same clock.

The following observations were actually taken by M. Arago and Lacroix, in the small island of Formentera, in the Mediterranean, in December, 1807.

Date of Observations.	Time of transit of the Star α Arietis.			Intervals between successive Transits.		
	h.	m.	s.	h.	m.	s.
1807, Dec. 24,	9	42	32.36			
" " 25,	9	41	29.70	23	58	57.34
" " 26,	9	40	26.72	23	58	57.02
" " 27,	9	39	23.90	23	58	57.18
" " 28,	9	38	21.38	23	58	57.48

These intervals between the transits agree so nearly, that it is very natural to suppose them *exactly* equal, and the small difference of the fraction of a second to arise from some slight irregularities of the clock, or imperfection in making the observations.

The equality of these intervals is not only the same for all the fixed stars, in passing the meridian, but they are the same in *passing all other planes.*

Standard of measure for time.

Now as this has been the universal experience of astronomers in all ages, it completely establishes the fact, that all the fixed stars come to the meridian in exactly equal intervals of time; and this gives us a *standard measure* for time, and the only standard measure, for all other motions are variable and unequal.

Time of the earth's revolution on its axis.

Again, this interval must be the time that the earth employs in turning on its axis; for if the star is *fixed,* it is a mark for the time that the meridian is in exactly the same position in relation to *absolute space.*

M. Arago's clock.

(23.) That the reader may not imbibe erroneous impressions, we remark, that the clock used for the preceding observations, made by M. Arago and Lacroix, ran too fast, if it was a *common clock,* and too slow, if it was an *astronomical clock*. It was not mentioned which clock was used, nor was it material simply to establish the fact of *equal intervals*; nor was it essential that the clock should run 24 hours, in a mean solar day: it was only essential that it ran uniformly, and marked off equal hours in equal times.

If it had been a common clock, and ran at a *perfect rate*, the interval would have been 23 h. 56 m. 4.09 s.

An astronomical clock.

(24.) In the preceding section we have spoken of an *astronomical clock*. Soon after the fact was established that the fixed stars came to the meridian in *equal times*, and that interval less than 24 hours, astronomers conceived the idea of *graduating a clock* to that interval, and dividing it into 24 hours. Thus graduating a clock to the *stars*, and not to the sun, is called a *sidereal*, and not a solar, or common clock; and as it was suggested by astronomers, and used only for the purposes of astronomy, it is also very appropriately called an astronomical clock; but save its graduation, and the nicety of its construction, it does not differ from a common clock.

To determine the rate of an astronomical clock.

*With a perfect astronomical clock, the same star will pass the meridian at exactly the same time, from one year's end to another.** If the time is not the same, the clock does not run

* Sidereal time has been slightly modified since the discovery of the *precession of the equinoxes*, though such modification has never been *distinctly* noticed in any astronomical work.

At first, *it was designed* to graduate the interval between two successive transits of the same star over the meridian, to 24 hours, and to call this a sidereal day; *which, in fact, it is.*

But it was necessary, in some way, to connect sidereal with solar time; and, to secure this end, it was determined to commence the sidereal day (not from the passage of any particular star across the meridian, but from the passage of the *imaginary point* in the heavens, where the sun's path crosses the vernal equinox, called the first point of Aries), thus making the *sidereal day* and the *equinoctial year* commence at the same moment of absolute time.

For some time, it was supposed that the interval between two successive transits of the first point of Aries, over the meridian, was the same as two successive transits of a star; but the two intervals are not *identical;* the first point of Aries has a very slow motion westward among the stars, which is called the *precession of the equinox*, and

CHAP. II. to sidereal time; and the variation of time, or the difference between the time when the star passes the meridian, and the time which ought to be shown by the clock, will determine the rate of the clock. And with the *rate* of the clock, and its *error*, we can readily deduce the true time from the time shown by the face of the clock.

Solar days not equal. (25.) When we examine the sun's passage across the meridian, and compare the elapsed intervals with the sidereal clock, we find regular and progressive variations, above and below a mean period, that cannot be accounted for by errors of observation.

The mean interval, from one transit of the sun to another, or from noon to noon, when we take the average of the whole year, is 24 hours of solar time, or 24 h. 3 m. 56.5554 s. of sidereal time; but, as we have just observed, these intervals are not uniform; for instance, about the 20th of December, they are about half a minute *longer*, and about the 20th of September, they are as much shorter, than the mean period.

The sun must have real or apparent motion. From this fact, we are compelled to regard the sun, not as a fixed point; it must have motions, real or apparent, independent of the rotation of the earth on its axis.

(26.) When we compare the times of the moon passing the meridian, with the astronomical clock, we are very forcibly struck with the *irregularity* of the interval.

General motion of the moon. The least interval between two successive transits of the moon (which may be called a *lunar day*), is observed to be about 24 h. 42 m.; the greatest, 25 h. 2 m.; and the mean, or average, 24 h. 54 m., of mean *solar time*.

These facts show, conclusively, that the moon is not a

---

which makes its transits across the meridian a *fraction of a second shorter* than the transits of a star.

The time required for 366 transits of a star across the meridian, is (3″.34), *three seconds and thirty-four hundredths of a second* of sidereal time, greater than for 366 transits of the equinox.

This difference would make a day in about 25000 years. The time elapsed between two successive transits of the equinox being now called a sidereal day of - - - - - - 24 h. 0 m. 0 s., the time between the transits of the same star, is - 24 h. 0 m. 0.00916 s.

Every astronomer understands Art. (24) with this modification.

fixed body, like a fixed star, for then the interval would be 24 hours of sidereal time.

But as the interval is always more than 24 hours, it shows that the general motion of the moon is eastward among the stars, with a daily motion varying from $10\frac{1}{2}$ to 16 degrees,* traveling, or appearing to travel, through the whole circle of the heavens (360°) in a little more than 27 days.

Chief object of astronomy.

Thus, these observations, however imperfectly and rudely taken, at once disclose the important fact, that the sun and moon are in constant change of position, in relation to the stars, and to each other; and, we may add, that the chief object and study of astronomy, is, to discover the reality, the causes, the nature, and extent of such motions.

Other movable and wandering bodies.

(27.) Besides the sun and moon, several other bodies were noticed as coming to the meridian at very unequal intervals of time — intervals not differing so much from 24 sidereal hours as the moon, but, unlike the sun and moon, the intervals were sometimes more, sometimes less, and sometimes equal to 24 sidereal hours.

These facts show that these bodies have a real, or apparent motion, *among the stars*, which is sometimes westward, sometimes eastward, and sometimes stationary; but, on the whole, the eastward motion preponderates; and, like the sun and moon, they finally perform revolutions through the heavens from west to east.

Wandering stars known to the ancients.

Only *four* such bodies (stars) were known to the ancients, namely, *Venus*, *Mars*, *Jupiter*, and *Saturn*.

Modern discoveries

These stars are a portion of the *planets* belonging to our solar system, and, by subsequent research, it was found that the Earth was also one of the number. As we come down to more modern times, several other planets have been discovered, namely, *Mercury*, *Uranus*, *Vesta*, *Juno*, *Ceres*, *Pallas*, and, very recently (1846), the planet *Neptune*.†

* Four minutes above 24 hours corresponds to one degree of arc.

† We have not mentioned the names of these *planets* in the order in which they stand in the system, but rather in the order of their discovery. As yet, we have really no idea of a planet, or a planetary system.

CHAP. II.

We shall again examine the meridian passages of the sun, moon, and planets, and deduce other important facts concerning them, besides that of their apparent, or real motions among the fixed stars.

Observations which determine the meridian distances of the stars.

(28.) But let us return to the fixed stars. We have several times mentioned the fact, that the same star returns to the same meridian again and again, after every interval of 24 sidereal hours. So two different stars come to the meridian at constant and invariable intervals of time from each other; and by *such intervals* we decide how far, or how many degrees, one star is east or west of another. For instance, if a certain fixed star was observed to pass the meridian when the sidereal clock marked 8 hours, and another star was observed to pass at 9, just one sidereal hour after, then we know that the latter star is on a celestial meridian, just 15 degrees eastward of the meridian of the first mentioned star.

Correspondence between hours and degrees.

As 24 hours corresponds to the whole circle, 360 degrees, therefore one hour corresponds to 15 degrees; and 4 minutes, in time, to one degree of arc. Hence, whatever be the observed interval of time between the passing of two stars over the meridian, that interval will determine the actual difference of the meridians running through the stars; and when we know *the position* of any one, in relation to any *celestial* meridian, we know the *positions of all* whose meridian observations have been thus compared.

Right ascension.

The position of a star, in relation to a *particular celestial meridian*, is called *Right Ascension*, and may be expressed either in time or degrees. Astronomers have chosen that

---

It is true, we might mention every fact, and every particular respecting each planet; such as its period of revolution, size, distance from the sun, &c.; but such facts, arbitrarily stated, would not convey the science of astronomy to the reader, for they can be told alike to the man and to the child — to the intellectual and to the dull — to the learned and to the unlearned.

To constitute true knowledge — to acquire true science — the pupil must not only know the fact, *but how that fact was discovered*, or deduced from other facts. Hence we shall *mainly* construct our theories from observations, as we pass along, and teach the pupil to decide the case from the facts, evidences, and circumstances presented.

meridian, for the *first meridian*, which passes through the sun's center at the instant the sun crosses the celestial equator in the spring, on the 20th of March.

CHAP. II. First meridian.

Right ascension is measured from the first meridian, eastward, on the equator, all the way round the circle, from 0 to 360 degrees, or from 0 h. to 24 h.

The reason why right ascension is not called *longitude* will be explained hereafter.

To find the right ascensions of the sun, moon, and planets.

(29.) If we observe and note the difference of sidereal time between the coming of a star to the meridian, and the coming of any other celestial body, as the *sun*, *moon*, *planet*, or *comet*, such difference, applied to the right ascension of the star, will give the right ascension of the body.

But every astronomer regulates, or aims to regulate, his sidereal clock, so that it shall show 0 h. 0 m. 0 s. when the equinox is on the meridian; and, if it does so, and runs regularly, then the time that any body passes the meridian by the clock, will give the right ascension of the body in time, without any correction or calculation; but, practically, this is never the case: *a clock is never exact*, nor can it ever run exactly to any given rate or graduation.

We have thus shown how to determine the right ascensions of the heavenly bodies. We shall explain how to find their positions in *declination*, in the next chapter.

CHAPTER III.

REFRACTION. — POSITION OF THE EQUINOX, AND OBLIQUITY OF THE ECLIPTIC — HOW FOUND BY OBSERVATION.

CHAP. III.

(30.) To determine the angular distance of the stars from the *pole*, the observer must first know the distance of his zenith from the same point.

As any zenith is 90 degrees from the true horizon, if the observer can find the altitude of the pole above the horizon (which is the latitude of the place of observation), he, of course, knows the distance between the *zenith* and the *pole*.

Preparations for determining the latitude by original observations.

As the north *pole* is but an imaginary point, *no star being there*, we cannot directly observe its altitude. But there is a very bright star near the pole, called the Polar Star, which, as all other stars in the same region, apparently revolves round the pole, and comes to the meridian twice in 24 sidereal hours; once above the pole, and once below it; and it is evident that the altitude of the pole itself must be midway between the greatest and least altitudes of the same star, *provided the apparent motion of the star round the pole is really in a circle;* but before we examine this fact, we will show how altitudes can be taken by the *mural circle*.

The mural circle.

Fig. 2.

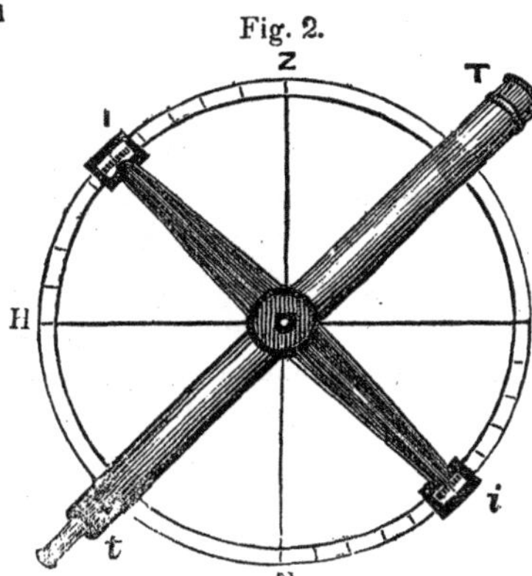

(31.) *The mural*, or *wall circle*, is a large metallic circle, firmly fastened to a wall, so that its plane shall coincide with the plane of the meridian.

A perpendicular line through the center, *Z N*, (Fig. 2), represents the zenith and nadir points; and at right angles to this, through the center, is the horizontal line, *H h*.

How to observe meridian altitudes.

A telescope, *T t*, and an index bar, *I i*, at right angles to the telescope, are firmly fixed together, and made to revolve on the center of the mural circle.

The circle is graduated from the zenith and nadir points, each way, to the horizon, from 0 to 90 degrees.

When the telescope is directed to the horizon, the index points, *I* and *i*, will be at *Z* and *N*, and, of course, show 0° of altitude. When the telescope is turned perpendicular to *Z*, the index bar will be horizontal, and indicate 90 degrees of altitude.

When the telescope is pointed toward any star, as in the

figure, the index points, I and i, will show the position of the telescope, or its angle from the horizon, *which is the altitude of the star.*

Mural circle also a transit instrument.

As the telescope, and index of this instrument, can revolve freely round the whole circle, we can measure altitudes with it equally well from the north or the south; but as it turns only in the plane of the meridian, we can observe *only* meridian altitudes with it.

This instrument has been called a *transit circle*, and, says Sir John Herschel, "The mural circle is, in fact, at the same time, a transit instrument; and, if furnished with a proper system of vertical wires in the focus of its telescope, may be used as such. As the axis, however, is only supported at one end, it has not the strength and permanence necessary for the more delicate purposes of a transit; nor can it be verified, as a transit may, by the *reversal* of the two ends of its axis, east for west. Nothing, however, prevents a divided circle being permanently fastened on the axis of a transit instrument, near to one of its extremities, so as to revolve with it, the reading off being performed by a microscope fixed on one of its piers. Such an instrument is called a *transit circle*, or a *meridian circle*, and serves for the simultaneous determination of the right ascensions and polar distances of objects observed with it; the time of transit being noted by the clock, and the circle being read off by the lateral microscope."

Altitude and azimuth instrument.

(32.) To measure altitudes in all directions, we must have another instrument, or *a modification of this.*

Conceive this instrument to turn on a perpendicular axis, parallel to $Z N$, in place of being fixed against a wall; and conceive, also, that the perpendicular axis rests on the center of a horizontal circle, and on that circle carries a horizontal index, to measure *azimuth angles.*

This instrument, so modified, is called an altitude and azimuth instrument, because it can measure altitudes and azimuths at the same time.

(33.) After astronomy is a little advanced, and the *angular distance* of each particular *star*, *sun*, *moon*, and *planet*,

3

CHAP. III.

The latitude taken by the altitude of the pole.

from the pole is known, then we can determine the latitude by observing the meridian altitude of any known celestial body; but before their positions are established (as is now supposed to be the case with the reader of this work), the only way to observe the latitude is by the altitudes of some *circumpolar star*, as mentioned in Art. 30.

To settle this *very important* element, the observer turns the telescope of his mural circle to the pole star, and observes its greatest and least altitudes, and takes the half sum for his latitude. But is this really his latitude? Does it require any correction, and if so, what, and for what reason?

A difficulty.

At first, it was very natural to suppose that this gave the exact latitude; but astronomers, ever suspicious, chose to verify it, by taking the same observations on other circumpolar stars; and if the theory was correct, and the observations correctly taken, all circumpolar stars would give the same, or very nearly the same, result. Such observations were made, and stars at the same distance from the pole gave the *same* latitude, and stars at different distances from the pole gave *different* latitudes; and the greater the distance of any star from the pole, the greater the latitude deduced from it. A star 30 or 35 degrees from the pole, observed from about the latitude of 40 degrees, will give the latitude 12 or 15 minutes of a degree greater than the pole star.

New and important truths.

Astronomers were now troubled and perplexed. These great and manifest discrepancies could not be accounted for by imperfection of instruments, or errors of observations, and some unconsidered natural cause was sought for as a solution.

Curves described by circumpolar stars.

To bring more evidence to bear on the case, astronomers examined the apparent paths of the stars round the pole, by means of the *altitude and azimuth instrument*, and they were found to be not *exact circles;* but departed more and more from a circle, as the star was a greater and greater distance from the pole.

These curves were found to be somewhat like *ovals* — the longer diameter passing horizontally through the pole — the upper segments *very nearly semicircles*, and the lower segments *flattened* on their under sides.

With such evidences before the mind, men were not long in deciding that these discrepancies were owing to

ASTRONOMICAL REFRACTION.

(34.) It is shown, in every treatise on natural philosophy, that light, passing obliquely from a rarer medium into a denser, is bent toward a perpendicular to the new medium. General effect of refraction.

Now, when rays of light pass, or are conceived to pass, from any celestial object, through the earth's atmosphere to an observer, the rays must be *bent downward*, unless they pass perpendicularly through the atmosphere; that is, come from the zenith.

Let *AB*, *CD*, *EF*, &c. (Fig. 3), represent different strata of the earth's atmosphere. Let *s* be a star, and conceive a line of light to pass from the star through the various strata of air, to the observer, at O.

Fig. 3.

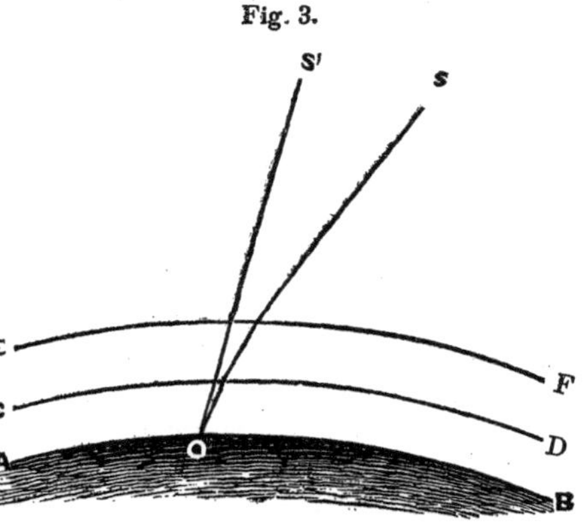

When it meets the first strata, as *EF*, it is slightly bent downward; and as the air becomes more and more dense, its refracting power becomes greater and greater, which more and more bends the ray. But the direction of the ray, at the point where it meets the eye of the observer, will determine the position of the star as seen by him. Hence the observer at O will see the star at *s'*, when its real position is at *s*. Refraction increases altitudes.

As a ray of light, from any celestial object, is bent down-

CHAP. III. ward, therefore, as we may see by inspecting the figure, *the altitude of all the heavenly bodies is increased by refraction.*

This shows that all the altitudes, taken as described in Art. 33, must be *apparent* altitudes — greater than *true* altitudes — and the resulting latitudes, deduced from them, all too great.

The object is now to obtain the amount of the refraction corresponding to the different altitudes, in order to *correct* or *allow* for it.

To determine the amount of refraction, we must resort to observations of some kind. But what sort of observations will meet the case?

How to find the amount of refraction corresponding to every degree of altitude.

Conceive an observer at the equator, and when the sun or a star passes through, or very near his zenith, it has *no refraction.* But, at the equator, the diurnal circles are perpendicular to the horizon; and those stars which are very near the equator, *really change their altitudes in proportion to the time.*

Now a star may be observed to pass the zenith, at the equator, at a particular moment: *four hours* afterward (sidereal time), the zenith distance of this star must be 4 times 15, or 60 degrees, and its altitude just 30 degrees. But, by *observation,* the altitude will be found to be 30° 1′ 38″. From this, we perceive, that 1′ 38″ is the amount of refraction corresponding to 30 degrees of altitude.

In six sidereal hours from the time the star passed the zenith, the true position of the star would be in the horizon; but, by observation, the altitude would be 33′ 0″, or a little more than the angular diameter of the sun.

Amount of horizontal refraction.

From this, we perceive, that 33′ 0″ is the amount of refraction at the horizon.

Thus, by taking observations at all intervals of time, between the zenith and the horizon, we can determine the refraction corresponding to every degree of altitude.

(35.) In the last article, we carried the observer to the equator, to make the case clear; but the mathematician need not go to the equator, for he can manage the case wherever he may be — he takes into consideration the curves, as mentioned in Art. 33. CHAP. III.

The mathematician's method of finding the amount of refraction.

If it were not for refraction, the curves round the pole would be *perfect circles*, and the mathematician, by means of the altitude and azimuth, which can be taken at any and every point of a curve, can determine how much it deviates from a circle, and from thence the amount of refraction, or *nearly* the amount of refraction, at the several points.

By using the refraction thus imperfectly obtained, he can correct his altitudes, and obtain his latitude, to considerable accuracy. Then, by repeating his observations, he can further approximate to the refraction.

In this way, by a multitude of observations and computations, the table of refraction (which appears among the tables of every astronomical work) was established and drawn out.

Refraction increases the time of sun-light.

(36.) The effect of refraction, as we have already seen, is to increase the altitude of all the heavenly bodies. Therefore, by the aid of refraction, the sun rises before it otherwise would, and does not set as soon as it would if it were not for refraction; and thus the apparent length of every day is increased by refraction, and *more than half of the earth's surface is constantly illuminated.* The extra illumination is equal to a zone, entirely round the earth, of about 40 miles in breadth.

As the refraction in the horizon is about 33′ of a degree, the length of a day, at the equator, is *more than four minutes* longer than it otherwise would be, and the *nights four minutes less.*

At all other places, where the diurnal circles are oblique to the horizon, the difference is still greater, especially if we take the average of the whole year.

Effects in high latitudes.

In high northern latitudes, the long days of summer are very materially increased, in length, by the effects of refraction; and near the pole, the sun rises, and is kept above the horizon, even for days, longer than it otherwise would be, owing to the same cause.

Refraction varies very rapidly, in its amount, near the hori-

CHAP. III. zon; and this causes a visible distortion of both sun and moon, just as they rise or set.

Distortion of the sun and moon in the horizon. For instance, when the lower limb of the sun is just in the horizon, it is elevated, by refraction, 33′.

But the altitude of the upper limb is then 32′, and the refraction, at this altitude, is 27′ 50″, elevating the upper limb by this quantity. Hence, we perceive, that the lower limb is elevated more than the upper; and the perpendicular diameter of the sun is apparently shortened by 5′ 10″, and the sun is distinctly seen of an oval form, which deviates more from a circle below than above.

An optical illusion. The apparently dilated size of the sun and moon, when near the horizon, has nothing to do with refraction: *it is a mere illusion*, and has no reality, as may be known by applying the following means of measurement.

Roll up a tube of paper, of such a size and dimensions as just to take in the rising moon, at one end of the tube, when the eye is at the other. After the moon rises some distance in the sky, observe again with this tube, and it will be found that the apparent size of the moon will even more than fill it.

The reason of this illusion is well understood by the student of philosophy; but we are now too much engaged with realities to be drawn aside to explain illusions, *phantoms*, or any *Will-o'-the-wisp*.

When small stars are near the horizon, they become invisible; either the refraction enfeebles and dissipates their light, or the vapors, which are always floating in the atmosphere, serve as a cloud to obscure them.

Application of refraction. (37.) Having shown the possibility of making a table of refraction corresponding to all apparent altitudes, we can now, by applying its effects to the observed altitudes of the circumpolar stars, obtain the true latitude of the place of observation.

Let it be borne in mind, that the latitude of any place on the earth, *is the inclination of its zenith to the plane of the equator;* which inclination *is equal to the altitude of the pole above the horizon.*

We demonstrate this as follows. Let E (Fig. 4) repre-

A demonstration.

sent the earth. Now, as an observer always conceives himself to be on the topmost part of the earth, the vertical point, Z, truly and naturally represents his zenith.

Fig. 4.

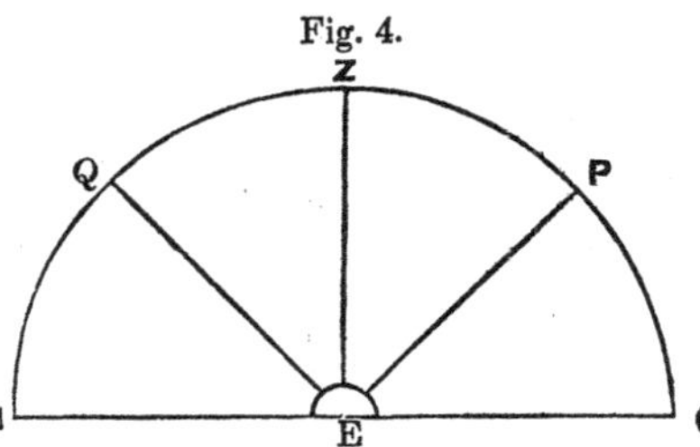

Through E, draw $H E O$, at right angles to $E Z$; then $H E O$ will represent the horizon (for the horizon is always at right angles to the zenith).

Let $E Q$ represent the plane of the equator, and at right angles to it, from the center of the earth, *must be the earth's axis;* therefore, $E P$, at right angles to $E Q$, is the direction of the pole.

Now the arcs, - - $ZP+PO=90°$,

Also, - - - $ZP+ZQ=90°$,

By subtraction, - $PO-ZQ=0$;

Or, by transposition, the arc $PO=ZQ$; that is, the altitude of the pole is equal to the latitude of the place; which was to be demonstrated.

In the same manner, we may demonstrate that the arc $H Q$ is equal to the arc $Z P$; that is, *the polar distance of the zenith is equal to the meridian altitude of the celestial equator.* Now, we perceive, that by knowing the latitude, we know the several divisions of the celestial meridian, from the northern to the southern horizon, namely, $O P$, $P Z$, $Z Q$, and $Q H$.

(38.) We are now prepared to observe and determine the *declinations* of the stars.

Declination defined.

The declination of a star, or any celestial object, is its meridian distance from the celestial equator.

This corresponds with latitude on the earth, and declination might have been called latitude.

The term latitude, as applied in astronomy, is to be defined hereafter.

CHAP. III.

How to find the declination of a star.

To determine the declination of a star, we must observe its meridian altitude (by some instrument, say the mural circle, Fig. 2), and correct the altitude for refraction (see table of refraction); the difference will be the star's *true altitude.*

If the true meridian altitude of the star is less than the meridian altitude of the celestial equator, then the declination of the star is south. If the meridian altitude of the star is greater than the meridian altitude of the equator, then the declination of the star is north.

These truths will be apparent by merely inspecting Fig. 4.

EXAMPLES.

Examples of the method pursued to find any star's declination.

1. Suppose an observer in the latitude of 40° 12′ 18″ north, observes the meridian altitude of a star, from the southern horizon, to be 31° 36′ 37″; what is the declination of that star?

From - - - - - -				90°	0′	00″
Take the latitude, - - -				40	12	18
Diff. is the meridian alt. of the equator,				49°	47′	42″
Alt. of star,	31°	36′	37″			
Refraction,		1	32			
True altitude,	31°	35′	5″ - -	31°	35′	5″
Declination of the star, south, - -				18°	12′	37″

2. The same observer finds the meridian altitude of another star, from the southern horizon, to be 79° 31′ 42″; what is the declination of that star?

Observed altitude, - - -	79°	31′	42″
Refraction, - - - -			11
True altitude, - - - -	79	31	31
Altitude of equator, - - -	49	47	42
Star's declination, north, - - -	29°	43′	49″

3. The same observer, and from the same place, finds the meridian altitude of a star, from the *northern horizon,* to be 51° 29′ 53″; what is the declination of that star?

Observed altitude, - - -	51°	29′	53″
Refraction, - - - -			46
True altitude of star, - - -	51	29	7
Altitude of pole (or latitude), -	40	12	18
Star from the pole (or polar dist.),	11	16	49
Polar dist., from 90°, gives decl., north,	78°	43′	11″

In this way the *declination* of every star in the visible heavens can be determined.

Elements for a chart of the stars.

(39.) In Art. 28 we have explained how to obtain the difference of the *right ascensions* of the stars; and in the last article we have shown how to obtain their *declinations*.

With the declinations and differences of right ascensions, we may mark down the positions of all the stars on a globe or sphere — the true representation of the appearance of the heavens.

Quite a region of stars exists around the south pole, which are never seen from these northern latitudes; and to observe them, and define their positions, Dr. Halley, Sir John Herschel, and several other English and French astronomers, have, at different periods, visited the southern hemisphere. Thus, by the accumulated labors of the many astronomers, we at length have correct catalogues of all the stars in both hemispheres, even down to many that are never seen by the naked eye.

The zero meridian of right ascension.

(40.) In Art. 28, we have explained how to find the *differences* of the right ascensions of the stars; but we have not yet found the *absolute right ascension* of any star, for the want of the first meridian, or *zero line*, from which to reckon. But astronomers have agreed to take that meridian for the *zero meridian*, which passes through the sun's center the instant the sun comes to the celestial equator, in the spring (which point on the equator is called the equinoctial point); *but the difficulty is to find exactly where (near what stars) this meridian line is.* Before we can define this line, we must take observations on the sun, and determine *where* it crosses the equator, and from the *time* we can determine the *place*. But before we can place much reliance on solar observations, we must ask ourselves this question. *Has the sun any parallax?*

CHAP. III. that is, is the position of the sun *just where* it appears to be? Is it really in the plane of the equator, when it appears to be there?

Parallax.

To all *northern* observers, is not the sun *thrown back* on the face of the sky, to a more southern position than the one it really occupies? Undoubtedly it is; and this change of position, caused by the *locality* of the observer, is called *parallax;* but, in respect to the sun, it is *too small* to be considered in these primary observations.

The early astronomers asked themselves these questions, and based their conclusions on the following consideration:

Sun's parallax insensible, in common observations.

If the sun is *materially projected out of its true place;* if it is thrown to the *southward*, as seen by a *northern observer*, it will cross the equator in the spring sooner than it appears to cross.

But let an observer be in the southern hemisphere, and, to him, the sun would be apparently thrown over to the north, and it would *appear* to cross the equator before it really did cross. Hence, if the sun is thrown out of place by parallax, an observer in the southern hemisphere would decide that the sun crossed the equator quicker, in absolute time, than that which would correspond to northern observations.

Northern and southern observations compared.

But, in bringing observations to the test, it was found that both northern and southern observers fixed on the same, or *very nearly* the same, absolute time for the sun crossing the equator. This proves that the position of the sun was not sensibly affected by parallax.

We will now suppose (for the sake of simplicity) that a sidereal clock has been so regulated as to run to the rate of sidereal time; that is, measure 24 hours between any two successive transits of the same star, over the same meridian, but the sidereal time *not known.*

Also, suppose that, at the Observatory of Greenwich, in the year 1846, the following observations were made:*

* In early times, such observations were often made. We took these results from the Nautical Almanac, and called them observations; but, for the purpose of showing principles, it is immaterial whether observations are real or imaginary.

Date.	Face of the Sidereal Clock.			Declination by Observa. (Art. 38.)			
	h.	m.	s.	°	′	″	
March 18,	1	3	20.00	0	58	53.4	south,
" 19,	1	6	58.62	0	35	11.3	"
" 20,	1	10	37.10	0	11	29.4	"
" 21,	1	14	15.47	0	12	12.0	north,
" 22,	1	17	54.07	0	35	52.0	"

Observations to find the equinox, and the sidereal time.

From these observations, it is required to determine the sidereal time, or the error of the clock; the time that the sun crossed the equator; the sun's right ascension; its longitude, and the obliquity of the ecliptic.

It is understood that the observations for declinations must have been meridian observations, and, of course, must have been made at the instant of *apparent noon*, local solar time.

By merely inspecting these observations, it will be perceived that the sun must have crossed the equator between the 20th and 21st; for at the apparent noon of the 20th, the declination was 11′ 29″.4 south; and on the 21st, at apparent noon, it was 12′ 12″ north. Between these two observations, the clock *measured out* 24 h. 3 m. 38.37 s., of sidereal time.

If the sun had not changed its meridian among the stars, the time would have been just 24 hours. The excess (3 m. 38.37 s.) must be changed into *arc*, at the rate of four minutes to one degree. Hence, to find the arc, we have this *proportion:*

As 4^{m} : $3^{m.}\ 38.37^{s.}$: : 1° : to the required result.

The result is 54′ 35″.4; the extent of arc which the sun changed right ascension during the interval between *noon* and *noon* of the 20th and 21st of March.

To examine this matter understandingly, draw a line EQ (Fig. 5), and make it equal to 54′ 35″.4.

Computations to find the equinox.

From E, draw ES at right angles to EQ, and make it equal to 11′ 29″.4. From Q, draw QN at right angles to EQ, and make it equal to 12′ 12″. Then S will represent the sun at *apparent noon*, March 20th, and N the position of the sun at apparent noon on the 21st, and SN is the line of

CHAP. III.

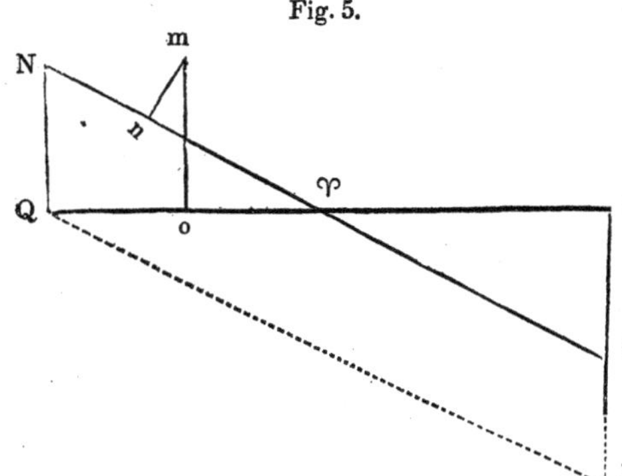

the sun among the stars, and the point ♈ (called the first point of Aries), and it is where the sun crosses the equator.

Now we wish to find where the line EQ is crossed by the line SN; or, the object is, to find E ♈, *expressed in time.*

To facilitate the computation, continue ES to P, making $SP=QN$, and draw the dotted line PQ. Then $SPQN$ is a parallelogram. $EP=11'\ 29''.4+12'\ 12''=23'\ 41''.4$; and the two triangles, PEQ and SE♈, are similar; therefore we have

$$PE : EQ :: SE : E♈.$$

To have the value of E ♈, *in time*, EQ must be taken in time; which is 3 m. 38.37 s.

Hence, $(23'\,41''.4) : (3^{m.}\ 38.37^{s.}) : 11'\,29''.4 : E♈$.

The result gives, $E♈=1^{m.}\ 45.91^{s.}$

But the clock time that the point E passed the meridian,

was - - - - -	1 h.	10 m.	37.10 s.
Add, - - - - - -		1	45.91
The equi. passed merid. (by clock) at	1 h.	12 m.	23.01

Error of the clock.

But, at the instant that the equinox is on the meridian, the sidereal clock ought to show 0 h. 0 m. 0 s.

The error of the clock was, therefore, 1 h. 12 m. 23.01 s. (subtractive).

Sun's right ascension.

As the whole line, EQ (in time), is -	3 m.	38.37 s.
And the part E ♈ is - - -	1	45.91
Therefore, ♈ Q is - - - -	1 m.	52.46

But ♈ Q is the *right ascension* of the sun at apparent noon, at Greenwich, on the 21st of March, 1846; a very important element. CHAP. III.

How to find the absolute right ascension of the stars, sun, moon, and planets.

The right ascension of any heavenly body, whether it be *sun*, *moon*, *star*, or *planet*, is the true sidereal time that it passes the meridian; and now, as we have the error of the clock, we can determine the true sidereal time that any star passes the meridian, and, of course, its *right ascension;* thus, for example,

If a star passed the meridian at -	10 h.	15 m.	47 s.
Error of the clock is (subtractive)	1	12	23
Right ascension of the star is -	9 h.	3 m.	24 s.

(42.) To find the *Greenwich apparent time*, when the sun crossed the equinox, we refer to Fig. 5; and as the point E corresponds to apparent noon of March 20th, and the Q to apparent noon of March 21st, and supposing the motion of the sun uniform (*as it is nearly*) *for that short interval*, we have the following proportion:

$$EQ : E\Upsilon :: 24\text{h.} : x.$$

Giving to EQ and $E\Upsilon$ their numeral values in seconds of sidereal time, the proportion becomes:

$$218''.37 : 105''.91 :: 24\text{h.} : x.$$

Time of the equinox.

The result of this proportion gives 11 h. 38 m. 24 s. for the interval, after the noon of the 20th of March, when the sun crossed the equator.

This result is in apparent time. The difference between apparent time, and mean clock time, will be explained hereafter. At this period, the difference between the sun and the common clock was 7 m. 36 s., to be added to apparent time.

Equinox of 1846, March - -	20 d.	11 h.	38m.	24s.
Equation of time (add), -			7	36
Equinox, clock time (Greenwich),	20 d.	11 h.	46 m.	0

Obliquity of the ecliptic, how found.

(43.) The two triangles, $ES\Upsilon$ and ΥQN, are really spherical triangles; but triangles on a sphere whose sides are less than a degree may be regarded as plane triangles, without any appreciable error. In the triangle $ES\Upsilon$,

$$E\Upsilon = 1588''.65, \quad ES = 689''.4;$$

CHAP. III. and, if we regard these seconds of *arc* as mere numerals, and calculate the angle E ♈ S, we find it 23° 27′ 43″; which is the *obliquity of the ecliptic.*

Sun's longitude. *By computing the length of the line SN, we find it* 59′ 30″; *which was the variation in the sun's longitude, between the noon of the* 20*th and* 21*st.*

Both longitude and right ascension are reckoned from the equinoctial point ♈: longitude along the line ♈ N (which line is called the ecliptic), and right ascension along the celestial equator ♈ Q.

Computing the length of the line ♈ N, we find it equal to 30′ 36″.6; which was the sun's longitude at the instant of apparent noon, at Greenwich, March 21st, 1846.

Latitude, in astronomy, from what line reckoned. Meridians of right ascension are at right angles to the celestial equator (at right angles to ♈ Q). The *first meridian* runs through the point ♈. Meridians of latitude are at right angles to the ecliptic (at right angles to the line SN). *Latitude, in astronomy, is reckoned north and south of the ecliptic.*

Thus a star at m (Fig. 5), ♈ n would be its longitude, nm its north latitude, ♈ o its right ascension, and om its north declination.

Path of the sun. (44.) Thus, it may be perceived, that these observations are very fruitful in giving important results; but, as yet, we have used only two of them — those made on the 20th and 21st. By bringing the other observations into computation, and extending Fig. 5, we can find the points where the sun was on the other days mentioned; and then, by taking observations every day in the year, the sun's *right ascension and longitude* can be determined for every day, and its exact *pathway* through the apparent celestial sphere. The same kind

Length of a year, how observed. of observations taken on the 20th, 21st, 22d, 23d, and 24th days of September, will show when the sun crosses the equator from *north* to *south*; and how long it remains north of the equator, and how long south of it. In March, 1847, the same observations might have been made, and the exact length of an *equinoctial* year determined: and in this way that *important interval* has been decided, *even to seconds.*

The true length of an equinoctial year was early a very

interesting problem to astronomers; and, before they had good clocks and refined instruments, it was one of some difficulty to settle. But the more the difficulty, the greater the zeal and perseverance; and we are often astonished at the accuracy which the ancients attained. CHAP. III.

The length of the equinoctial year, as stated in the tables of

	Days.	hours.	min.	secs
Ptolomée, is - - - -	365	5	55	12
Tycho Brahe, made it - -	365	5	48	45
Kepler, in his tables, - - -	365	5	48	57
M. Cassini, in his tables, - -	365	5	48	52
M. De Lalande, - - - -	365	5	48	45
Sir John Herschel, - - -	365	5	48	49.7

Solar and sidereal year.

The last cannot differ from the truth more than *one or two seconds.* Let the reader notice that this is the equinoctial year — the one that must ever regulate the change of seasons. There is another year — the *sidereal* year — which is about 20 *minutes* longer than the *equinoctial* year. *The sidereal year is the time elapsed from the departure of the sun from the meridian of* ANY STAR, *until it arrives at the same meridian again, and consists of* 365 *d.* 6 *h.* 9 *m.* 9 *s.*

Cause of difference.

As the stars are really the fixed points in space, this latter period is the apparent revolution of the sun; and the shorter period, for the equinoctial year, is caused by the motion of the equinoctial points to the westward, called the *precession of the equinoxes.* Since astronomers first began to record observations, the fixed stars have increased, in *right ascension,* about 2 hours in time, or 30 degrees of arc.

The mean annual precession of the equinoxes is 50″.1 of arc; which will make a revolution, among the stars, in 25868 years.*

---

* The computation is thus: As 50″.1 is to the number of seconds in 360 degrees; so is one year to the number of years. Which gives 25868 years, nearly.

We say, the stars increase in right ascension; and this is true; but the stars do not move — they are fixed; the meridian moves from the stars.

CHAPTER IV.

GEOGRAPHY OF THE HEAVENS.

CHAP. IV.

Groups of stars.

(45.) The fixed stars are the only *landmarks* in astronomy, in respect to both time and space. They seem to have been thrown about in irregular and ill-defined groups and clusters, called *constellations*. The individuals of these groups and clusters differ greatly as to brightness, hue, and color; but they all agree in one attribute — a high degree of permanence, as to their relative positions in the group; and the groups are as permanent in respect to each other. This has procured them the title of *fixed stars;* an expression which must be understood in a comparative, and not in an absolute, sense; for, after long investigation, it is ascertained that some of them, if not all, are in motion; although too slow to be perceptible, except by very delicate observations, continued through a long series of years.

Magnitudes of the stars.

The stars are also divided into different classes, according to their degree of brilliancy, called *magnitudes*. There are *six* magnitudes, visible to the naked eye; and ten telescopic magnitudes — in all, *sixteen.*

The brightest are said to be of the first magnitude; those less bright, of the second *magnitude*, etc.; the sixth magnitude is just visible to the naked eye.

One star of the first magnitude.

The stars are very unequally distributed among these classes; nor do all astronomers agree as to the number belonging to each; for it is impossible to tell where one class ends, and another begins; nor is it important, for all this is but a matter of fancy, involving no principle. In the first magnitude there is really but one star (*Sirius*); for this is manifestly brighter than any other; but most astronomers put 15 or 20 into this class.

The second magnitude includes from 50 to 60; the third, about 200, the numbers increasing very rapidly, as we descend in the scale of brightness.

From some experiments on the intensity of light, it has

been determined, that if we put the light of a star, of the average 1st magnitude, 100, we shall have: **Chap. IV.**

1st magnitude	= 100	4th magnitude	= 6
2d "	= 25	5th "	= 2
3d "	= 12	6th "	= 1

On this scale, Sir William Herschel placed the brightness of Sirius at 320.

Ancient astronomy has come down to us much tarnished with superstition, and heathen *mythology*. Every constellation bears the name of some pagan deity, and is associated with some absurd and ridiculous fable; yet, strange as it may appear, these masses of rubbish and ignorance — these clouds and fogs, intercepting the true light of knowledge, are still not only retained, but cherished, in many modern works, and dignified with the name of astronomy.

Ancient names must be continued.

Merely as names, either to constellations or to individual stars, we shall make no objections; and it would be useless, if we did; for names long known, will be retained, however improper or objectionable; hence, when we speak of *Orion*, the *Little Dog*, or the *Great Bear*, it must not be understood that we have any *great respect* for mythology.

It is not our purpose now to describe the starry heavens — to point out the *variable, double*, and *multiple* stars — the *Milky Way* and *nebulæ;* these will receive special attention in some future chapter: at present, our only aim is to point out the method of obtaining a knowledge of the mere appearance of the sky, to the common observer, which may be called *the geography of the heavens.*

To give a person an idea of locality, on the earth, we refer to points and places supposed to be known. Thus, when we say that a certain town is 15 miles north-west of Boston, a ship is 100 miles east of the Cape of Good Hope, or a certain mountain 10 miles north of Calcutta, we have a pretty definite idea of the localities of the town, the ship, and the mountain, on the face of the earth, provided we have a clear idea of the face of the earth, and know the position of Boston, the Cape of Good Hope, and Calcutta.

So it is with the geography of the heavens; the apparent

CHAP. IV.

surface of the whole heavens must be in the mind, and then the localities of certain bright stars must be known, as *land-marks*, like Boston, the Cape of Good Hope, and Calcutta.

Stars about the pole.

We shall now make some effort to point out these *land-marks*. The *North Star* is the first, and most important to be recognized; and it can always be known to an observer, in any northern latitude, from its stationary appearance and altitude, equal to the latitude of the observer. At the distance of about 32 degrees from the pole, are seven bright stars, between the 1st and 2d magnitudes, forming a figure resembling a *dipper*, four of them forming the cup, and three the handle. The two forming the sides of the cup, opposite to the handle, are always in a line with the North Star; and are therefore called *pointers: they always point to the North Star*. The line joining the equinoxes, or the first meridian of right ascension, runs from the pole, between the other two stars forming the cup. The first star in the handle, nearest the cup, is called *Alioth*, the next *Mizar*, near which is a small star, of the 4th magnitude; the last one is *Benetnasch*. The stars in the handle are said to be in the *tail of the Great Bear*.

About four degrees from the pole star, is a star of the 3d magnitude, δ *Ursæ Minoris*. A line drawn through the pole (not pole star) and this star, will pass through, or very near, the poles of the *ecliptic* and the *tropics*. A small constellation, near the pole, is called *Ursa Minor*, or the Little Bear. An irregular semicircle of bright stars, between the dipper and the pole, is called the *Serpent*.

Imaginary lines from star to star.

If a line be drawn from δ *Ursæ Minoris*, through the pole star, and continued about 45 degrees, it will strike a very beautiful star, of the 1st magnitude, called *Capella*. Within five degrees of Capella are three stars, of about the 4th magnitude, forming a very exact isosceles triangle, the vertical angle about 28 degrees. A line drawn from *Alioth*, through the *pole star*, and continued about the same distance on the other side, passes through a cluster of stars called *Cassiopia in her chair*. The principal star in *Cassiopea*, with the pole star and Capella, form an isosceles triangle, Capella at the vertex.

CHAP. IV.

(46.) More attention has been paid to the constellations along the equator and ecliptic, than to others in remoter regions of the heavens, because the *sun*, *moon*, and *planets*, traverse through them. The ecliptic is the sun's apparent annual path among the stars (so called because all eclipses, of both sun and moon, can take place only when the moon is either in or near this line).

Ecliptic defined.

Eight degrees on each side of the ecliptic is called the *zodiac;* and this space the ancients divided into 12 equal parts (to correspond with the 12 months of the year), and each part (30°) is called a *sign*—and the whole, the *signs of the zodiac.* These divisions are useless; and, of late years, astronomers have laid them aside; yet custom and superstition will long demand a place for them in the common almanacs.

Signs of the zodiac.

The signs of the zodiac, with their symbolic characters, are as follows: *Aries* ♈, *Taurus* ♉, *Gemini* ♊, *Cancer* ♋, *Leo* ♌, *Virgo* ♍, *Libra* ♎, *Scorpio* ♏, *Sagittarius* ♐, *Capricornus* ♑, *Aquarius* ♒, *Pisces* ♓.

Owing to the precession of the equinoxes, these signs do not correspond with the constellations, as originally placed: the variation is now about 30 degrees; the stars remain in their places; and the first meridian, or first point of *Aries*, has drawn back, which has given to the stars the appearance of moving forward.

Beginning with the first point of *Aries* as it now stands, no prominent star is near it; and, going along the ecliptic to the eastward, there is nothing to arrest special attention, until we come to the *Pleiades*, or *Seven Stars*, though only six are visible to the naked eye. This little cluster is so well known, and so remarkable, that it needs no description. South-east of the *Seven Stars*, at the distance of about 18 degrees, is a remarkable cluster of stars, said to be in the *Bull's Head;* the largest star in this cluster is of the 1st magnitude, of a red color, called *Aldebaran*. It is one of the eight stars selected as points from which to compute the moon's distance, for the assistance of navigators.

Method of tracing the stars.

This cluster resembles an A when east of the meridian, and

CHAP. IV. a V when west of it. The *Seven Stars, Aldebaran,* and *Capella,* form a triangle very nearly isosceles — Capella at the vertex. A line drawn from the *Seven Stars,* a little to the west of *Aldebaran,* will strike the most remarkable constellation in the heavens, *Orion* (it is out of the zodiac, however): some call it the Ell and Yard. The figure is mainly distinguished by three stars, in one direction, within two degrees of each other; and two other stars, forming, with one of the three first mentioned, another line, at right angles with the first line.

The five stars, thus in lines, are of the 1st or 2d magnitude. A line from the *Seven Stars,* passing near *Aldebaran* and through *Orion,* will pass very near to *Sirius,* the most brilliant star in the heavens. The ecliptic passes about midway between the Seven Stars and Aldebaran, in nearly an eastern direction. Nearly due east from the northernmost and brightest star in *Orion,* and at the distance of about 25 degrees, is the star *Procyon;* a bright, lone star.

The northernmost star in *Orion,* with *Sirius* and *Procyon,* form an equilateral triangle.

The constellations are above the horizon, and visible every evening during the winter season.

Directly north of *Procyon,* at the distances of 25 and 30 degrees, are two bright stars, *Castor* and *Pollux.* Castor is the most northern. *Pollux* is one of the eight *lunar stars.* Thus we might run over that portion of the heavens which is ever visible to us, and by this method every student of astronomy can render himself familiar with the aspect of the sky; but it is not sufficiently *definite* and *scientific* to satisfy a mathematical mind.

(47.) The only scientific method of defining the position of a place on the earth, is to mention its *latitude* and *longitude;* and this method fully defines any and every place, however unimportant and unfrequented it may be: so in astronomy, the only scientific methods of *defining the position* of a star, is to mention its *latitude* and *longitude,* or, more conveniently, its *right ascension* and *declination.*

General and indefinite descriptions not satisfactory.

It is not sufficient to tell the navigator that a coast makes off in such a direction from a certain point, and that it is so far to a certain cape; and, from one cape to another, it is *about* 40 miles south-west — he would place very little reliance on any such directions. To secure his respect, and command his confidence, the *latitude* and *longitude* of every point, promontory, river, and harbor, along the coast, must be given; and then he can shape his course to any point, or strike in upon it from the indefinite expanse of a pathless sea. So with an astronomer; while he understands and appreciates the *rough and general descriptions*, such as we have just given, he requires the *certain description*, comprised in *right ascension* and *declination*.

CHAP. IV.

What constitutes a definite description.

Accordingly, astronomers have given the *right ascensions* and declinations of every visible star in the heavens (and of very many that are invisible), and arranged them in tables, in the order of right ascension.

There are far too many stars, for each to have a proper name; and, for the sake of reference, Mr. John Bayer, of Augsburg, in Suabia, about the year 1603, proposed to denote the stars by the letters of the Greek and Roman alphabets; by placing the first Greek letter α to the principal star in the constellation, β to the second in magnitude, γ to the third, and so on; and if the Greek alphabet shall become exhausted, then begin with the Roman, *a*, *b*, *c*, etc.

John Bayer's method of reference.

"*Catalogues* of particular stars, in sections of the heavens, have been published by different astronomers, each author numbering the individual stars embraced in his list, according to the places they respectively occupy in the catalogue." These references to particular catalogues are sometimes marked on celestial globes, thus: 79 H, meaning that the star is the 79th in Herschel's catalogue; 37 M, signifies the 37th number in the catalogue of Mayer, etc.

Particular catalogues.

Among our tables will be found a catalogue of a *hundred* of the principal stars, *inserted for the purpose of teaching a definite and scientific method of making a learner acquainted with the geography of the heavens*.

To have a clear understanding of the method we are about to explain, we again consider that right ascension is reckoned from the equinox, eastward along the equator, from 0 h. to 24 hours. When the sun comes to the equator, in March, its

CHAP. IV. right ascension is 0; and from that time its right ascension increases about four minutes in a day, throughout the year, to 24 hours; and then it is again at the equinox, *and the* 24 *hours are dropped.*

When it is apparent noon. But whatever be the right ascension of the sun, it is apparent noon when it comes to the meridian; and the more eastward a body is, the later it is in coming to the meridian. Thus, *if a star comes to the meridian at two o'clock in the afternoon* (*apparent time*), *it is because its right ascension* IS TWO HOURS GREATER *than the right ascension of the sun.*

Therefore, if from the right ascension of a star we subtract the right ascension of the sun, the remainder will be the time for that star to come to the meridian.

Connection between R. A. and meridian passage. If we put ($R\ast$) to represent the star's right ascension, and ($R\odot$) to represent that of the sun, and T to represent the *apparent time* that the star passes the meridian, then we shall have the following equation:

$$R\ast - R\odot = T;$$

By transposition . . $R\ast = R\odot + T$:

That is, *the right ascension of a star* (*or any celestial body*), *is equal to the right ascension of the sun, increased by the time that the star* (*or body*) *comes to the meridian.*

The right ascension of the sun is given, in the Nautical Almanac (and in many other almanacs), for every day in the year, when the sun is on the meridian of Greenwich; but many of the readers of this work may not have such an almanac at hand, and, for their benefit, we give the right ascension for every fifth day of the year 1846 (Table III): the local time is the apparent noon at Greenwich.

We take the year 1846, because it is the second year after leap year; and the sun's right ascension for any day in that year, will not differ more than *two minutes* from its right ascension, on the same day, of any other year; and will correspond with the right ascension of the same day in 1850, by adding $7\frac{3}{10}$ seconds; and so on for each succeeding *period of four years.*

To apply the preceding equation, the observer should adjust his watch to *apparent time;* that is, apply the equation of time, and know the direction of his meridian, at least approximately. In short, by the range of definite objects, he must be able to decide, within *two or three minutes*, when a celestial body is on his meridian. CHAP. IV.

Thus, all prepared, we will give a few

EXAMPLES.

Examples to find stars.

1. *On the 20th of May* (no matter what year, if not many years from 1850), *in the latitude of 40° N., and longitude of 80° W., at 9 h. 24 m. in the evening, clock time, I observed a lone, bright star, of about the 2d magnitude, on the meridian. It had a bland, white light; and, as I had no instrument to measure its altitude, I simply judged it to be 42°. What star was it?*

We decide the question thus:

Time per watch, - - -	9 h.	24 m.	00 s.
Equation of time (see Table), add		3	46
Apparent time, - - -	9	27	46
Lon. 80° W., equal, in time, to	5	20	00
Apparent time, at Greenwich, -	14	47	46

Correction of the sun's R. A.

The right ascension of the sun, on the 20th of May (noon, Greenwich time), is 3 h. 47 m. 15 s. (see Table III). The increase, estimated at the rate of 4 minutes in 24 hours, will give 1 minute in 6 hours, or 10 seconds to 1 hour; this, for 14 h. 47 m., gives 2 m. 27 s.

Hence, the right ascension of the sun, *at the time of observation*, was - - - -	3 h.	49 m.	42 s.
Apparent time of observation, -	9	27	46
Right ascension of the star, - -	13 h.	17 m.	28 s.

By inspecting the catalogue of the stars (Table II), we find the right ascension of *Spica* to be 13 h. 17 m. 08 s., and its declination, 10° 21′ 35″.

But, in the latitude of 40° N., the meridian altitude of the celestial equator must be 50°; and any stars south of that must be of a less altitude. Therefore, the meridian altitude of *Spica* must be 50°, less 10° 21′, or 39° 39′; but the star I observed, I simply judged to have had an altitude of 42°.

CHAP. IV.

It is very possible that I should err, in altitude, two or three degrees;* *but, it is not possible that the star I observed should be any other star than Spica;* for there is no other bright star near it. This is one of the lunar stars.

Personal observations recommended.

Being now certain that this star is Spica, I can observe it in relation to its appearance — the small stars that are near it, and the clusters of stars that are about it — or the fact, that no remarkable constellation is near it. In short, I can so make its acquaintance as to know it ever after; but I am unable to convey such acquaintance to others, by language: true knowledge, in this particular, demands personal observation.

Continuation of examples to find stars.

2. *On the 3d day of July,* 1846, *at* 9 *h.* 34 *m., P. M., mean time per watch, a star of the 1st magnitude came to the meridian. I was in latitude* 39° *N., and about* 75° *W. The star was of a deep red color, and, as near as my judgment could decide, its altitude was between* 25° *and* 30°. *Two small stars were near it, and a remarkable cluster of smaller stars were west and northwest of it, at the distances of* 5°, 6°, *or* 7°. *What star was this?*

Time per watch, - - - - -	9 h.	34 m.	00 s.
Equa. of time (subtr. from mean time)		3	48
Apparent time, - - - - -	9	30	12
Longitude, 75°, equal to - -	5		
Apparent time, at Greenwich, - -	14 h.	30 m.	00 s.

By examining the table for the sun's R. A., I find that,

On the 1st of July, it is - -	6 h.	40 m.	00 s.
On the 5th, - - - - -	6	56	30
Variation, for 4 days, - - -		16 m.	30 s.

At this rate, the variation for 2 days, 14½ hours, cannot be

* Ten or twenty degrees, near the horizon, is apparently a much larger space than the same number of degrees near the zenith. Two stars, when near the horizon, appear to be at a greater distance asunder than when their altitudes are greater. The variation is a mere optical illusion; for, by applying instruments, to measure the angle in the different situations, we find it the same. Unless this fact is taken into consideration, an observer will always conceive the altitude of any object to be greater than it really is, especially if the altitude is less than 45 degrees.

far from 10 m. 10 s.; and the right ascension of the sun, at the time of observation, must have been

An example of finding Antares

Nearly - - - - -	6 h.	50 m.	10 s.
To which add, apparent time, - -	9	30	12
Right ascension of the star, - -	16 h.	20 m.	22 s.

By inspecting the catalogue of stars, I find *Antares* to have a right ascension of 16h. 20m. 2s. and a declination of 26° 4′, south.

In the latitude mentioned, the meridian altitude of the celestial equator must be - - -	50°	0′
Objects *south of that plane must be less, hence* (sub.)	26	4
Meridian altitude of Antares, in lat. 50°,	23°	56

As the observation corresponds to the right ascension of *Antares* (as near as possible, considering errors in observation, and probably in the watch), and as the altitudes do not differ many degrees (within the limits of guess work), it is certain that the star observed was ANTARES. By its peculiar *red color*, and the remarkable clusters of stars surrounding it, I shall be able to recognize this star again, without the trouble of direct observation.

To find Altair.

3. *On the night of the* 20*th of June*, 1846, *latitude* 40° *N., and longitude* 75° *W., at* 1 *h.* 48 *m. past midnight, clock time, I observed a star of the* 1*st magnitude nearly on the meridian; two other stars, of about the* 3*d magnitude, within* 3° *of it; the three stars forming nearly a right line, north and south; the altitude of the principal star about* 60°. *What star was it?*

In these examples, the time must be reckoned on from noon to noon again; therefore 1 h. 48 m. after midnight must be

written, - - - - - -	13 h.	48 m.	00 s.
Equation of time, to subtract, - -		1	12
Apparent time, - - - - -	13	46	48
Longitude, - - - - - -	5		
Greenwich apparent time, June 20,	18 h.	46 m.	48 s.
Sun's right ascension, at this time, -	5 h.	57 m.	40 s.
Time, - - - - - -	13	46	48
Star's right ascension, - - -	19 h.	44 m.	28 s.

CHAP. IV.

By inspecting the catalogue of stars, we find the right ascension of *Altair* 19 h. 43 m. 15 s., and its declination 8° 27′ N. In latitude 40° N., the declination of 8° 27′ N. will give a meridian altitude of 58° 27′; and, in short, I know the star observed must be *Altair*, and the two other stars, near it, I recognize in the catalogue.

By taking these observations, any person may become acquainted with all the principal stars, and the general aspect of the heavens; but no efforts, confined merely to the study of books, will accomplish this end.

The equation in Art. 47 is not confined to a star; it may be any heavenly body, *moon*, *comet*, or *planet*. The time of passing the meridian is but another term for right ascension. If observations are made on any bright star, and no corresponding star is found in the catalogue, such a star would probably be a planet; and if a planet, its right ascension will change.

The Southern Cross, and Magellan Clouds.

(48.) The whole region of stars south of declination 50°, is never seen in latitude 40° north, nor from any place north of that parallel; and, to register these stars in a catalogue, it has been necessary for astronomers to visit the southern hemisphere, as we have before mentioned; but these stars are mostly excluded from our catalogues. There are several constellations, in the southern region, worthy of notice — the *Southern Cross* and the *Magellan Clouds*. The Southern Cross very much resembles a cross; so much so, that any person would give the constellation that appellation. Its principal star is, in right ascension, 12 h. 20 m., and south declination 33°.

The Magellan Clouds were at first supposed to be clouds by the navigator Magellan, who first observed them. They are four in number; two are white, like the Milky Way, and have just the appearance of little white clouds. They are *nebulæ*. The other two are black — extremely so — and are supposed to be places entirely devoid of all stars; yet they are in a very bright part of the Milky Way: right ascension 10 h. 40 m., declination 62° south.

SECTION II.

DESCRIPTIVE ASTRONOMY.

CHAPTER I.

FIRST CONSIDERATIONS AS TO THE DISTANCES OF THE HEAVENLY BODIES.—SIZE AND EXACT FIGURE OF THE EARTH.

CHAP. I.

Distance is but relative.

(49.) Hitherto we have considered only appearances, and have not made the least inquiry as to the nature, magnitude, or distances of the celestial objects.

Abstractly, there is no such thing as great and small, near and remote; *relatively* speaking, however, we may apply the terms great, and very great, as regards both magnitude and distance. Thus an error of *ten* feet, in the measure of the length of a building, is very great—when an error of ten rods, in the measure of one hundred miles, would be too trifling to mention.

Fig. 6.

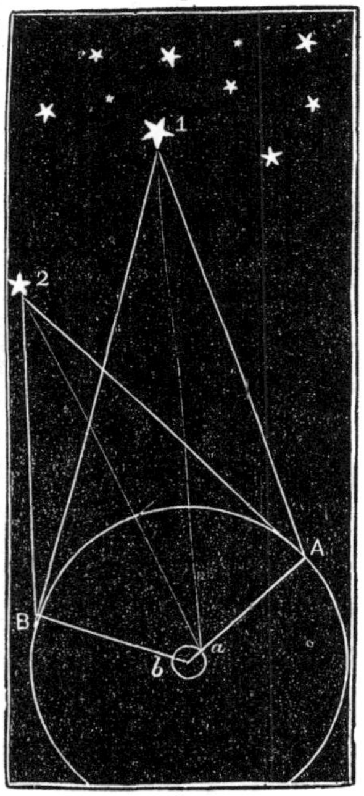

Are the heavenly bodies remote?

Now if we consider the distance to the stars, it must be relative to some measure taken as a standard, or our inquiries will not be definite, or even intelligible. We now make this general inquiry: *Are the heavenly bodies near to, or remote from, the earth?* Here, the earth itself seems to be the natural standard for measure; and if any body were but two, three, or even ten times the diameter of the earth, in distance, we

CHAP. I. should call it *near;* if 100, 200, or 2000 times the diameter of the earth, we should call it remote. To answer the inquiry, *Are the heavenly bodies near or remote?* we must put them to all possible mathematical tests; a mere opinion is of no value, without the foundation of some positive knowledge. Let 1, 2 (Fig. 6), represent the absolute position of two stars; and then, if *A B C* represents the circumference of the earth, these stars may be said to be *near;* but if *a b c* represents the circumference of the earth, the stars are many times the diameter of the earth, in distance, and therefore may be said to be remote. If *A B C* is the circumference of the earth, in *relation* to these stars, the apparent distance of the two stars asunder, as seen from A, is measured by the angle 1 A 2; and their apparent distance asunder, as seen from the point B, is measured by the angle 1 B 2; and when the circumference *A B C* is very large, as represented in our figure, the angle *A*, between the two stars, is manifestly greater than *B*. But if *a b c* is the circumference of the earth, the points *a* and *b* are relatively the same as *A* and *B*. And, it is *an ocular demonstration* that the angle under which the two stars would appear at *a* is the same, or nearly the same, as that under which they would appear at *b*; or, at least, we can conceive the earth so small, in relation to the distance to the stars, that the angle under which two stars would appear, would be the same seen from any point on the earth.

The means of deciding this question pointed out.

The conclusion.

Conversely, then, if the angle under which two stars appear is the same as seen from all parts of the earth's surface, it is certain that the diameter of the earth is very small, compared with the distance to the stars; or, which is the same thing, *the distance to the stars is many times the diameter of the earth.* Therefore observation has long since decided this important point. Sir John Herschel says: "The nicest measurements of the apparent angular distance of any two stars, *inter se*, taken in any parts of their diurnal course (after allowing for the unequal effects of refraction, or when taken at such times that this cause of distortion shall act equally on both), manifest *not the slightest* perceptible variation. Not only this, but at whatever point of the earth's surface the measurement is performed, the results are *absolutely identical.* No instruments ever yet invented by man are delicate enough to indicate, by an increase or diminution of the angle subtended, that one point of the earth is nearer to or farther from the stars than another."

Another illustration of the great distance to the stars.

(50.) Perhaps the following view of this subject will be more intelligible to the general reader.

Fig. 7.

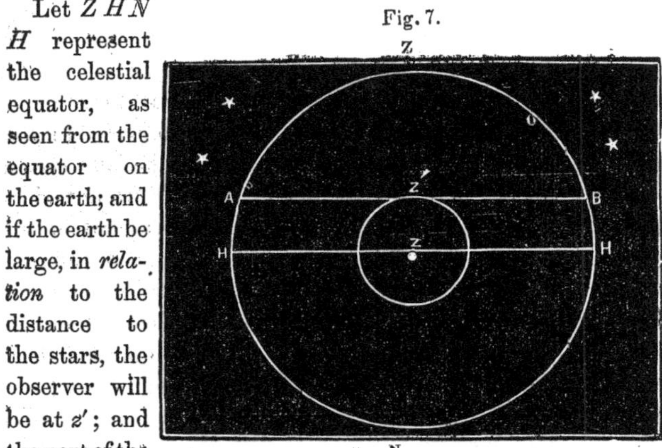

Let *Z H N H* represent the celestial equator, as seen from the equator on the earth; and if the earth be large, in *relation* to the distance to the stars, the observer will be at *z'*; and the part of the celestial arc above his horizon would be represented by *A Z B*, and the part below his horizon by *A N B*, and these arcs are obviously *unequal;* and their relation would be measured by the time a star or heavenly body remains above the horizon, compared with the time below it; but by observation (refraction being allowed for), we know that the stars are as long above the horizon as they are below; which shows that the observer is not at *z'*, but at *z*, and even more near the center; so that the arc *A Z B*, is imperceptibly unequal to the arc *H N H*; that is, they are equal to each other; and the earth is comparatively but a point, in relation to the distance to the stars.

The moon an exception.

This fact is well established, as applied to the fixed *stars*, *sun*, and *planets; but with the moon it is different:* that body

CHAP. I.

is longer below the horizon than above it; which shows that its distance from the earth is at least measurable.

(51.) It is improper, at present, or rather, it is too advanced an age, to pay any respect to the ancient notion, that the earth is an extended plane, bounded by an unknown space, inaccessible to men. Common intelligence must convince even the child, that the earth must be a large ball, of a regular, or an irregular shape; for every one knows the fact, that the earth has been *many times* circumnavigated; which settles the question.

Earth's surface convex.

In addition to this, any observer may convince himself, that the surface of the sea, or a lake, is not a plane, but *everywhere convex;* for, in coming in from sea, the high land, back in the country, is seen before the shore, which is nearer the observer; the tops of trees, and the tops of towers, are seen before their bases. If the observer is on shore, viewing an approaching vessel, he sees the topmast first; and from the top, downward, the vessel gradually comes in view. This being the case on every sea, and on every portion of the earth, proves that the surface of the earth is convex on every part — hence it must be a globe, or nearly a globe. These facts, last mentioned, are sufficiently illustrated by

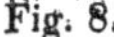
Fig. 8.

(52.) On the supposition that the earth is a sphere, there are several methods of measuring it, without the labor of applying the measure to every part of it. The first, and most natural method (which we have already mentioned), is that of measuring any definite portion of the meridian, and from thence computing the value of the whole circumference.

How to find the circumference of the earth.

Thus, if we can know the number of degrees, and parts of a degree, in the *arc* AB (Fig. 9), and then measure the distance in miles, we in fact virtually know the whole circumference; for whatever part the arc AB is of 360 degrees, the same part, the number of miles in AB, is of the miles in the whole circumference.

To find the arc AB, the latitudes of the two points, A and B, must be very accurately taken, and their difference will give the *arc* in *degrees*, *minutes*, and *seconds*. Now AB must be measured simply in distance, as *miles*, *yards*, or *feet;* but this is a laborious operation, requiring great care and perseverance. To measure *directly* any considerable portion of a meridian, is indeed impossible, for local obstructions would soon compel a deviation from any definite line; but still the measure can be continued, by keeping an account of the deviations, and *reducing* the measure to a meridian line.

Let m be the miles or feet in AB; then the whole circumference will be expressed by $\left(\frac{360\,m}{\text{arc}\,AB}\right)$.

How to find the diameter.

(53.) When we know the hight of a mountain, as represented in Fig. 9, and at the same time know the distance of its visibility from the surface of the earth; that is, know the line MA; then we can compute the line MC, by a simple theorem in geometry; thus,

Fig. 9.

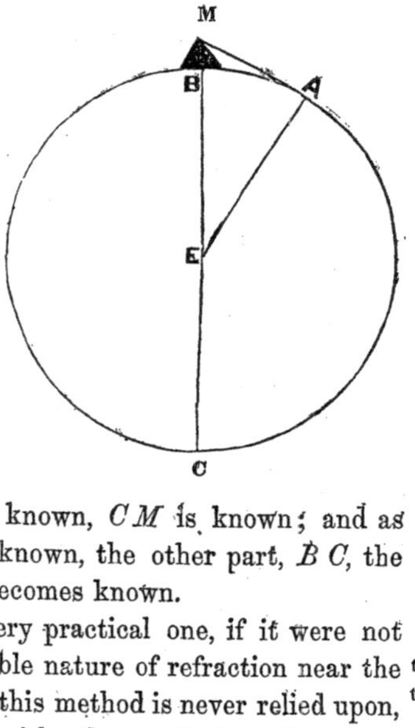

$$CM\times MB=(AM)^2;$$

Or, $CM=\frac{(AM)^2}{MB}$.

Now as the right hand member of this equation is known, CM is known; and as part of it (MB) is already known, the other part, BC, the diameter of the earth, thus becomes known.

Objection to this method.

This method would be a very practical one, if it were not for the uncertainty and variable nature of refraction near the horizon; and for this reason, this method is never relied upon, although it often well agrees with other methods. As an example under this method, we give the following:

CHAP. I.

A mountain, two miles in perpendicular hight, was seen from sea at a distance of 126 miles. If these data are correct, what then is the diameter of the earth?

Solution: $MC = \frac{(126)^2}{2} = 63 \times 126 = 7938.$ $BC = 7936.$

Dip of the horizon.

(54.) This same geometrical theorem serves to compute the *dip of the horizon.* The true horizon is a right angle from the zenith; but the navigator, in consequence of the motion of his vessel, can never use the true horizon; he must use the sea offing, making allowance for its dip. If the navigator's eye were on a level with the sea, and the sea perfectly stable, the true and apparent horizon would be the same. But the observer's eye must always be above the sea; and the higher it is, the greater the dip; and the amount of dip will depend on the hight of the eye, and the diameter of the earth. The difference between the angle AMC (Fig. 9), and a right angle (which is the same as the angle AEM), is the measure of the dip corresponding to the hight BM.

For the benefit of navigators, a table has been formed, showing the dip for all common elevations.*

The angle at the center is equal to the dip.

* The dip is computed thus:

Put BC (Fig. 9) $= D, BM = h$;

Then $EM = \left(\frac{D}{2} + h\right)$; and $(MA)^2 = CM \times MB = (D+h)h.$

By trigonometry, $(EA)^2 : (MA)^2 :: R^2 : \tan.^2 AEM$;

That is, - - - $\frac{D^2}{4} : (D+h)h :: R^2 : \tan.^2 AEM.$

For very moderate elevations, h is extremely small, in relation to D; and the second term of the proportion may be Dh. (R represents the radius of the tables.) Making this consideration, we have

$$\frac{D^2}{4} : Dh :: R^2 : \tan.^2 AEM;$$

Or, - - $D : h :: 4R^2 : \tan.^2 AEM$;

Or, - - $\sqrt{D} : \sqrt{h} :: 2R : \tan. AEM.$

(55.) All such computations are made on the supposition that the earth is exactly spherical; and it is, in fact, so nearly spherical, that no corrections are required in consequence of its deviation from that figure.

The earth not exactly spherical.

After correct views began to be entertained, as to the magnitude of the earth, and its revolution on an axis, philosophers concluded that its equatorial diameter might be greater than its polar diameter; and investigations have been made to decide the fact.

If the earth were exactly spherical, it is plain that the curvature over its surface would be the same in every latitude; but if not of that figure, a degree would be longer on one part of the earth than on another. "But," says Herschel, "when we come to compare the measures of meridional arcs made in various parts of the globe, the results obtained, although they agree sufficiently to show that the supposition of a spherical figure is not *very* remote from the truth, yet exhibit discordances far greater than what we have shown to be attributable to error of observation; and which render it evident that the hypothesis, in strictness of its wording, is untenable. The following table exhibits the lengths of a degree of the meridian (astronomically determined as above described), ex-

---

By inspecting this last proportion, it will be perceived that the tangent of the dip varies as the *square root* of the elevation. To apply this proportion, we adduce the following problem:

The diameter of the earth is 7912 miles; the elevation of the eye, above the surface, is *ten feet*. *What is the dip?*

$2R$ . . log. - - - - - - - - -			10.301030
$\sqrt{h}$ . . log. - - - - - - - - -			.500000
Product of the means (log.), - - - -			10.801030
D miles, 7912, - -	log. -	3.898286	
Feet, - 5280, - -	log. -	3.722634	
	2)	7.620920	
$\sqrt{D}$ in feet, - -	(log.)	3.810460 . .	3.810460
	tan. 3′ 22″	- - -	6.990570

CHAP. I. pressed in British standard feet, as resulting from actual measurement, made with all possible care and precision, by commissioners of various nations, men of the first eminence, supplied by their respective governments with the best instruments, and furnished with every facility which could tend to insure a successful result of their important labors.

Country.	Latitude of Middle of the Arc.	Arc measured.	Length of Degree concluded	Observers.
Sweden	66 20 10	1°37′ 19″	365782	Svanberg.
Russia	58 17 37	3 35 5	365368	Struve.
England.......	52 35 45	3 57 13	364971	Roy, Kater.
France	46 52 2	8 20 0	364872	Lacaille, Cassini.
France	44 51 2	12 22 13	364535	Delambre, Mechain.
Rome	42 59 0	2 9 47	364262	Boscovich.
America, U. S...	39 12 0	1 28 45	363786	Mason, Dixon.
Cape of G. Hope	33 18 30	1 13 17½	364713	Lacaille.
India..........	16 8 22	15 57 40	363044	Lambton, Everest.
India..........	12 32 21	1 34 56	363013	Lambton.
Peru	1 31 0	3 7 3	362808	Condamine, etc.

The earth less curved at the poles than at the equator

"It is evident, from a mere inspection of the second and fourth columns of this table, that *the measured length of a degree increases with the latitude*, being greatest near the poles, and least near the equator."

"Assuming," continues Herschel, "that the earth is an ellipse, the geometrical properties of that figure enable us to assign the proportion between the lengths of its axes which shall correspond to any proposed rate of variation in its curvature, as well as to fix upon their absolute lengths, corresponding to any assigned length of the degree in a given latitude. Without troubling the reader with the investigation (which may be found in any work on the conic sections), it will be sufficient to state that the lengths, which agree on the whole best with the entire series of meridional arcs, which have been satisfactorily measured, are as follow:—

		Feet.	Miles.
Greater, or equatorial diam.,	=	41,847,426 =	7925.648
Lesser, or polar diam., - -	=	41,707,620 =	7899.170
Difference of diameters, or polar compression, - - -	=	139,806 =	26.478

The proportion of the diameters is very nearly that of

298 : 299, and their difference $\frac{1}{299}$ of the greater, or a very little greater than $\frac{1}{300}$." CHAP. I.

(56.) The shape of the earth, thus ascertained by actual measurement, is just what theory would give to a body of water equal to our globe, and revolving on an axis in 24 hours; and this has caused many philosophers to suppose that the earth was formerly in a fluid state.

Explanation of radius of curvature.

If the earth were a sphere, a plumb line at any point on its surface would tend directly toward the center of gravity of the body; but the earth being an ellipsoid, or an *oblate spheroid*, and the plumb lines, being perpendicular to the surface at any point, do not tend to the center of gravity of the figure, but to points as represented in Fig. 10.

Fig. 10.

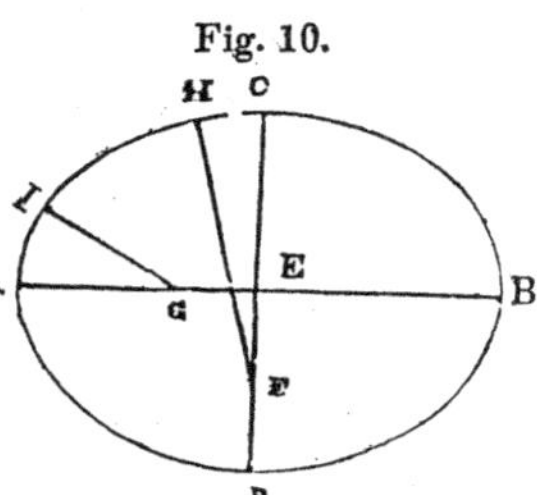

The plumb line at *H* tends to *F*, yet the mathematical center, and center of gravity of the figure, is at *E*. So at *I*, the plumb line tends to the point *G*; and as the length of a degree at *A*, is to the length of a degree at *H*, so is *IG* to *HF*. If, however, a passage were made through the earth, and a body let drop through it, the body would not pass from *I* to *G*: its *first tendency* at *I* would be toward the point *G*; but after it passed below the surface at *I*, its tendency would be *more and more* toward the point *E*, the center of gravity; but it would not pass exactly through that point, *unless dropped* from the point *A*, or the point *C*.

Force of gravity different on different parts of the earth; and why?

(57.) If the earth were a perfect and stationary sphere, the force of gravity, on its surface, would be everywhere the same; but, it being neither stationary, nor a perfect sphere, the force of gravity, on the different parts of its surface, must be different. The points on its surface nearest its center of gravity, must have more attraction than other points more remote from the center of gravity; and if those points which are more remote from the center of gravity have also a rotary motion, there will be a diminution of gravity on that account.

Let *AB* (Fig. 10) represent the equatorial diameter of

Gravity diminished by rotation.

the earth, and CD the polar diameter; and it is obvious that E will be the center of gravity, of the whole figure, and that the force of gravity at C and D will be greater than at any other points on the surface, because $E\,C$, or $E\,D$, are less than any other lines from the point E to the surface. The force of gravity will be greatest on the points C and D, also, because they are stationary: all other points are in a circular motion; and circular motion has a tendency to depart from the center of motion, and, of course, to diminish gravity. The diminution of the earth's gravity by the rotation on its axis, amounts to its $\frac{1}{289}$ part,* at the equator. By this frac-

Computation of the amount of diminution.

Fig. 11

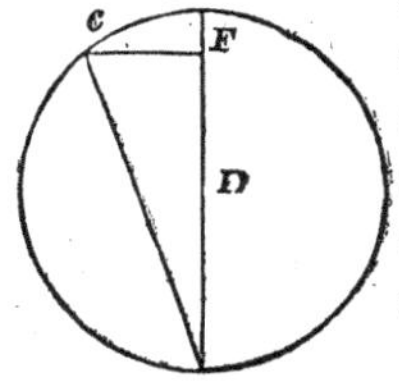

* Let D be the equatorial diameter of the earth, F the versed sine of an arc corresponding to the motion in a second of time, and c the chord or arc (for the chord and arc of so small a portion of the circumference will coincide, *practically* speaking).

A portion of the earth's gravity, equal to F, is destroyed by the rotation of the earth, and we are now to compute its value.

By proportional triangles, $F : c :: c : D$;

Or - - - - - - - - - $F = \frac{c^2}{D}$ . . . (1)

The value of c is found by dividing the whole circumference into as many equal parts as there are *seconds* in the time of revolution. But the time of revolution is 23 h. 56 m. 4 s., = 86164 seconds.

The whole circumference is $(3.1416)D$;

Therefore, - - - - - - - $c = \frac{(3.1416)D}{(86164)}$. (2)

By this value of c, we have $F = \frac{(3.1416)^2 D}{(86164)^2}$.

The visible force of gravity, at the equator, is the distance a body will fall the first second of time, expressed in feet. Let us call this distance g. Now the part of gravity des-tion, then, is the weight of the sea about the equator *lightened*, and thereby rendered susceptible of being supported at a higher level than at the poles, where no such *counteracting* force exists.

---

troyed by rotation, as we have just seen, is $\frac{c^2}{D}$; therefore the whole force of gravity is $\left(g+\frac{c^2}{D}\right)$

Ratio of the diminution computed.

Our next inquiry is: *what part of the whole is the part destroyed?* Or what part of $\left(g+\frac{c^2}{D}\right)$ is $\frac{c^2}{D}$?

Which, by common arithmetic, is,

$$\frac{\frac{c^2}{D}}{g+\frac{c^2}{D}}=\frac{c^2}{gD+c^2}=\frac{1}{\frac{gD}{c^2}+1}$$

From (2) - $D^2=\frac{(86164)^2c^2}{(3.1416)^2}$ or, $\frac{D}{c^2}=\frac{(86164)^2}{(3.1416)^2D}$:

Hence,

$$\frac{gD}{c^2}=\frac{(86164)^2g}{(3.1416)^2D}=\frac{(86164)^2(16.07)}{(3.1416)^2(7925)(5280)}.$$

By the application of logarithms, we soon find the value of this expression to be 288.4. Therefore, $\frac{1}{\frac{gD}{c^2}+1}=\frac{1}{289.4}$.

We may now inquire, how rapidly the earth must revolve on its axis, so that the whole of gravity would be destroyed on the equator. That is, so that F shall equal g. Equation (1) then becomes, $g=\frac{c^2}{D}$, or $c=\sqrt{gD}$.

But as often as c is contained in the whole circumference, is the corresponding number of seconds in a revolution; that is, the time in seconds must correspond to the expression,

$$\frac{(3.1416)D}{\sqrt{gD}}, \quad \text{or} \quad (3.1416)\sqrt{\frac{D}{g}}.$$

CHAP. I.

(58.) It is this centrifugal force itself that changed the shape of the earth, and made the equatorial diameter greater than the polar. Here, then, we have the same cause, exercising at once a direct and an indirect influence. The amount of the former (as we may see by the note) is easily calculated; that of the latter is far more difficult, and requires a knowledge of the integral calculus, "But it has been clearly treated by Newton, Maclaurin, Clairaut, and many other eminent geometers; and the result of their investigations is to show, that owing to the elliptic form of the earth alone, and independently of the centrifugal force, its attraction ought to increase the weight of a body, in going from the equator to the pole, by nearly its $\frac{1}{590}$th part; which, together with the $\frac{1}{289}$th part, due from centrifugal force, make the whole quantity $\frac{1}{194}$th part; which corresponds with observations as deduced from the vibrations of pendulums."— *See Natural Philosophy.*

Rotation has a direct and indirect effect on gravity.

Fig. 12.

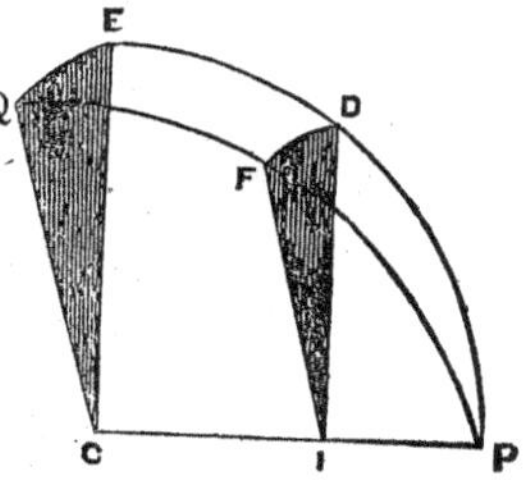

(59.) The form of the earth is so nearly a sphere, that it is considered such, in geography, navigation, and in the general problems of astronomy.

English and geographical miles.

The average length of a degree is $69\frac{1}{4}$ English miles; and, as this number is fractional, and inconvenient, navigators have tacitly agreed to retain the ancient, rough estimate of sixty miles to a degree; calling the mile a *geographical* mile. Therefore, the geographical mile is longer than the English mile.

D, in feet, $=(7925)(5280)$; $g=16.076$. By the application of logarithms, we find this expression to be 5069 seconds, or 1 h. 24 m. 29 s.; which is about 17 times the rapidity of its present rotation.

In a subsequent portion of this work, we shall show how to arrive at this result by another principle, and through another operation.

As all meridians come together at the pole, it follows that a degree, between the meridians, will become less and less as we approach the pole; and it is an interesting problem to trace the *law* of decrease.* CHAP. I.

* This law of decrease will become apparent, by inspecting Fig. 12. Let EQ represent a degree, on the equator, and EQC a sector on the plane of the equator, and of course EC is at right angles to the axis CP. Let DFI be any plane parallel to EQC; then we shall have the following proportion:

$$EC : DI :: EQ : DF.$$

In trigonometry, EC is known as the radius of the sphere; DI as the cosine of the latitude of the point D (the numerical values of sines and cosines, of all arcs, are given in trigonometrical tables): therefore we have the following rule, to compute the length of a degree between two meridians, on any parallel of latitude.

RULE.—*As radius is to the cosine of the latitude; so is the length of a degree on the equator, to the length of a parallel degree in that latitude.*

Calling a degree, on the equator, 60 miles, what is the length of a degree of longitude, in latitude 42°? Example

SOLUTION BY LOGARITHMS.

As radius (see tables), - - - -	10.000000
Is to cosine 42° (see tables), - - -	9.871073
So is 60 miles (log.), - - - - -	1.778151
To $44\frac{589}{1000}$ miles, - - - - - -	1.649224

At the latitude of 60°, the degree of longitude is 30 miles; the diminution is very slow near the equator, and very rapid near the poles.

In navigation, the DF's are the known quantities obtained by the estimations from the log line, etc.; and the navigator wishes to convert them into longitude, or, what is the same thing, he wishes to find their values projected on the equator, and he states the proportion thus: To reduce departure to longitude.

$$DI : EC :: DF : EQ;$$

That is, *as cosine of latitude is to radius, so is departure to difference of longitude.*

CHAPTER II.

PARALLAX, GENERAL AND HORIZONTAL. — RELATION BETWEEN PARALLAX AND DISTANCE. — REAL DIAMETER AND MAGNITUDE OF THE MOON.

Chap. II.

(60.) Parallax is a subject of very great importance in astronomy: it is the key to the measure of the planets — to their distances from the earth — and to the magnitude of the whole solar system.

Parallax in general.

Parallax is the difference in position, of any body, as seen from the center of the earth, and from its surface.

When a body is in the zenith of any observer, *to him* it has no parallax; for he sees it in the same place in the heavens, as though he viewed it from the center of the earth. The greatest possible parallax that a body can have, takes place when the body is in the horizon of the observer; and this parallax is called *horizontal parallax.* Hereafter, when we speak of the parallax of a body, *horizontal parallax* is to be understood, unless otherwise expressed.

A clear and summary illustration of parallax in general, is given by Fig. 13.

Horizontal parallax.

Fig. 13.

Let C be the center of the earth, Z the observer, and P, or p, the position of a body. From the center of the earth, the body is seen in the direction of the line CP, or Cp; from the observer at Z, it is seen in the direction of ZP, or Zp; and the *difference in direction*, of these two lines, is parallax. When P is in the zenith, there is no parallax; when P is in the horizon, the angle ZPC is then greatest, and is the *horizontal parallax*.

Relation between parallax and distance.

We now perceive that the horizontal parallax of any body is equal to the *apparent semidiameter of the earth*, as seen from the body. The greater the distance to the body, the less the horizontal parallax; and when the distance is so great that the semidiameter of the earth would appear only as a point, then the body has no parallax. Conversely, if we can detect no sensible parallax, we know that the body must be at a vast distance from the earth, and the earth itself appear as a point from such a body, if, in fact, it were even visible.

Trigonometry gives the relation between the angles and sides of every conceivable triangle; therefore we know all about the horizontal triangle ZCP, when we know CZ and the angles. Calling the horizontal parallax of any body p, and the radius of the earth r, and the distance of the body from the center of the earth x (the radius of the table always R, or *unity*), then, by trigonometry, we have,

$$R \ : \ x \ :: \ \sin. p \ : \ r:$$

Therefore, - - - $x = \left(\frac{R}{\sin. p}\right) r.$

From this equation we have the following general rule, to find the distance to any celestial body:

Rule to find the distances to the heavenly bodies.

RULE. — *Divide the radius of the tables by the sine of the horizontal parallax. Multiply that quotient by the semidiameter of the earth, and the product will be the result.*

This result will, of course, be in the same terms of *linear* measure as the semidiameter of the earth: that is, if r is in feet, the result will be in feet; if r is in miles, the result will be in miles, etc.: but, for astronomy, our terrestrial measures are too diminutive, to be convenient (not to say inappropriate); and, for this reason, it is customary to *call the semidiameter of the earth unity*, and then the distance of any body from the earth is simply the quotient arising from *dividing the radius by the sine of the horizontal parallax pertaining to*

CHAP. II. *the body;* and it is obvious, that the less the parallax, the greater this quotient; that is, the greater the distance to the body; and the difficulty, and the *only difficulty*, is to obtain the *horizontal parallax*.

Horizontal parallax cannot be observed.

(61.) The horizontal parallax cannot be directly observed, by reason of the great amount and irregularity of horizontal refraction; but if we can obtain a parallax at any considerable altitude, we can compute the horizontal parallax therefrom.*

The fixed stars have no sensible horizontal parallax, as we have frequently mentioned; and the parallax of the sun is so small, that it cannot be directly observed (see 40); the moon is the only celestial body that comes forward and presents its parallax; and from thence we know that the moon is the only body that is within a moderate distance of the earth.

That the moon had a sensible parallax, was known to the earliest observers, even before mathematical instruments were at all refined; but, to decide upon its exact amount, and detect its variations, required the combined knowledge and observations of modern astronomers.

Deduction of horizontal parallax.

* In the two triangles ZpC and ZPC (Fig. 13), call the angle p the parallax in altitude, and the angle $ZPC = x$, and Cp and CP each equal D. Then, by trigonometry, we have

$$\sin. pZC : \sin. p :: D : r;$$

$$\text{And} \quad R : \sin. x :: D : r.$$

Therefore, by equality of ratios (see algebra),

$$\sin. pZC : \sin. p :: R : \sin. x.$$

But the sine pZC is the sine of the apparent zenith distance. Therefore,

$$\sin. x = \frac{R \sin. p}{\sin. \text{zenith distance}};$$

That is, *the sine of the horizontal parallax is equal to the sine of the parallax in altitude, into the radius, and divided by the sine of the apparent zenith distance.*

By what observations the lunar parallax was first indicated.

The lunar parallax was first recognized in European and northern countries, *by its appearing to describe more than a semicircle south of the equator, and less than a semicircle north of that line;* and, on an average, it was observed to be a longer time south, than north of the equator; *but no such inequality could be observed from the region of the equator.*

Observers at the south of the equator, observing the position of the moon, see it for a longer time north of the equator than south of it; *and, to them, it appears to describe more than a semicircle north of the equator.*

Here, then, we have observation against observation, unless we can reconcile them. But the only reconciliation that can be made, is to conclude that the moon is really as long in one hemisphere as the other, and the observed discrepancy must arise from the *positions* of the observers; and when we reflect that parallax must always depress the object (see Fig. 13), and throw it farther from the observer, it is therefore perfectly clear that a northern observer should see the moon farther to the south than it really is, and a southern observer see the same body farther north than its true position.

(62.) To find the amount of the lunar parallax, requires the concurrence of two observers. They should be near the same meridian, and as far apart, in respect to latitude, as possible; and every circumstance, that could affect the result, must be known.

Observations to obtain the amount of parallax

The two most favorable stations are Greenwich (England) and the Cape of Good Hope. They would be more favorable if they were on the same meridian; but the small change in declination, while the moon is passing from one meridian to the other, can be allowed for; and thus the two observations are reduced to the same meridian, and equivalent to being made at the same time.

The most favorable times for such observations, are when the moon is near her greatest declinations, for then the change of declination is *extremely slow.*

Let A (Fig. 14) represent the place of the Greenwich observatory, and B the station at the Cape of Good Hope. C is the center of the earth, and Z and Z' are the zenith

CHAP. II.

Fig. 14.

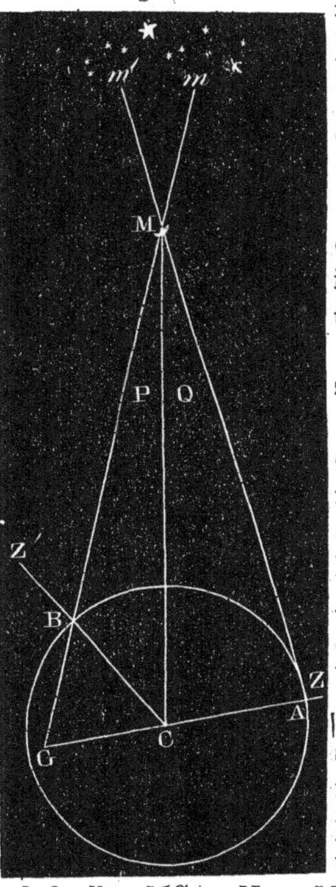

points of the observers. Let M be the position of the moon, and the observer at A will see it projected on the sky at m', and the observer at B will see it projected on the sky at m.

Illustration of primary observations.

Now the figure $A\,C\,B\,M$ is a quadrilateral; the angle $A\,CB$ is known by the latitudes of the two observers; the angles $MA\,C$ and $MB\,C$ are the respective zenith distances, taken from 180°.

But the sum of all the angles of any quadrilateral is equal to four right angles; and hence the angles at A, C, and B, being known, the parallactic angle at M is known.

In this quadrilateral, then, we have two sides, $A\,C$ and CB, and all the angles; and this is sufficient for the most ordinary mathematician to decide every particular in connection with it; that is, we can find AM, MB, and finally MC.* Now MC being known, the horizontal

---

A mathematical deduction.

* The direct and analytical method of obtaining MC, will be very acceptable to the young mathematician; and, for that reason, we give it.

Put $AC=CB=r$, $CM=x$, and the two parts of the observed parallactic angle, M, represented by P and Q, as in the figure. Also, let a represent the *natural sine* of the angle $MA\,C$, and b the *natural sine* of the angle $MB\,C$:

Then, by trigonometry, - $x : a :: r : \sin. Q$;

Also, - - - - - - - $x : b :: r : \sin. P$;

Hence, - - - - $\sin. P+\sin. Q=\frac{(a+b)r}{x}$. . . (1)

parallax can be computed, for it is but a *function* of the distance (see 60). CHAP. II.

By the equation (Art. 60), $x=\left(\frac{R}{\sin. p}\right)r$

By changing, - - - $\sin. p=\left(\frac{R}{x}\right)r$; and when x, the distance, is known, sin. p, or sine of the horizontal parallax, is known.

(63.) The result of such observations, taken at different times, show all values to MC, between $55\frac{92}{100}$, and $63\frac{84}{100}$; taking the value of r as unity. Variable distance to the moon.

These variations are regular and systematic, both as to time and place, in the heavens; and they show, without further investigation, that the moon does not go round the earth in a circle, or, if it does, the earth is not in the center of that circle.

The parallaxes corresponding to these extreme distances, are 61′ 29″ and 53′ 50″.

When the moon moves round to that part of her orbit which is most remote from the earth, it is said to be in *apogee;* and, when nearest to the earth, it is said to be in *perigee.* The points apogee and perigee, mainly opposite to each other, do not keep the same places in the heavens, but gradually move forward in the same direction as the motion of the moon, and perform a revolution in a little less than nine years. Apogee and perigee.

But, by a general theorem in trigonometry,

$$\sin. P+\sin. Q=2\sin.\frac{P+Q}{2}\cos.\frac{P-Q}{2}. \quad . \quad (2)$$

Now by equating (1) and (2), and observing that $P+Q=M$, and that $\left(\cos.\frac{P-Q}{2}\right)$ must be extremely near *unity;* and, therefore, as a factor, may disappear; we then have,

$$2\sin.\frac{M}{2}=\frac{(a+b)r}{x}, \quad \text{or} \quad x=\frac{(a+b)r}{2\sin.\frac{1}{2}M}$$

A more ancient method is to compute the value of the little triangle BCG, and then of the whole triangle AMG, and then of a part AMC or MGC.

CHAP. II.

(64.) Many times, when the moon comes round to its perigee, we find its parallax less than 61′ 29″, and, at the opposite apogee, more than 53′ 50″. It is only when the sun is in, or near a line with the lunar perigee and apogee, that these greatest extremes are observed to happen; and when the sun is near a right angle to the perigee and apogee, then the moon moves round the earth in an orbit nearer a circle; and thus, by observing with care the variation of the moon's parallax, we find that its orbit is a *revolving* ellipse, of *variable eccentricity*.

(65.) Because the moon's distance from the earth is variable, therefore there must be a *mean* distance: we shall show, hereafter, that her motion is variable; therefore there is a *mean* motion; and, as the eccentricity is variable, there is a *mean eccentricity*.

MEAN parallax and parallax at MEAN distance.

The extreme parallaxes, at *mean eccentricity*, are 60′ 20″ and 54′ 05″, and the corresponding distances from the earth are 56.93 and 63.64, the radius of the earth being *unity*. The mean parallax, or mean between 60′ 20″ and 54′ 05″, is 57′ 12″.5; but the parallax, at mean distance, is 57′ 03″*.

* It may seem paradoxical that the *mean* parallax, and the parallax at *mean* distance are different quantities; but the following investigation will set the matter at rest. Let d and D be extreme distances, and M the mean distance.

Then, - - - - $d+D=2M$. . . . (1)

Also, let p and P be the parallaxes corresponding to the distances d and D; and put x to represent the parallax at mean distance. Then, by Art. 60 (if we call the radius of the tables unity), we have

$$d=\frac{1}{\sin. p},\quad D=\frac{1}{\sin. P},\quad \text{and } M=\frac{1}{\sin. x}.$$

Substituting these values of d, D, and M, in equation (1) we have, - -

$$\frac{1}{\sin. p}+\frac{1}{\sin. P}=\frac{2}{\sin. x};$$

Or, - - -

$$\sin. P+\sin. p=\frac{2\sin. p\sin. P}{\sin. x} \quad (2)$$

The mean between *extreme distances* is $\frac{55.92+63 \cdot 84}{2}$ or 59.88; but the *true mean* distance is 60.26, corresponding to the parallax 57′ 3″. The mean, between extremes, is a variable quantity; but the true *mean distance* is ever the same, a little more than $60\frac{1}{4}$ *times* the semidiameter of the earth.

Mean distance to the moon.

(66.) *The variations in the moon's real distance must correspond to apparent variations in the moon's diameter;* and if the moon, or any other body, should have no variation in apparent diameter, we should then conclude that the body was always at the same distance from us.

The change, in apparent diameter, of any heavenly body, is numerically *proportioned to its real change in distance;* as appears from the demonstration in the note below.*

But by a well known, and general theorem in trigonometry, we have, $\sin. P + \sin. p = 2 \sin. \left(\frac{P+p}{2}\right) \cos. \left(\frac{P-p}{2}\right)$ (3)

Mean parallax.

By equating (3) and (2), and observing that the cosines of very small arcs may be practically taken as unity, or radius; therefore,

$$\sin. \left(\frac{P+p}{2}\right) = \frac{\sin. P \sin. p}{\sin. x};$$

Or, - - - - - $$\sin. x = \frac{\sin. P \sin. p}{\sin. \frac{1}{2}(P+p)}.$$

On applying this equation, we find $x = 57' \ 3''$.

* Let A be the point of vision, and d the diameter of any body at different distances, AB, AC.

Fig. 15.

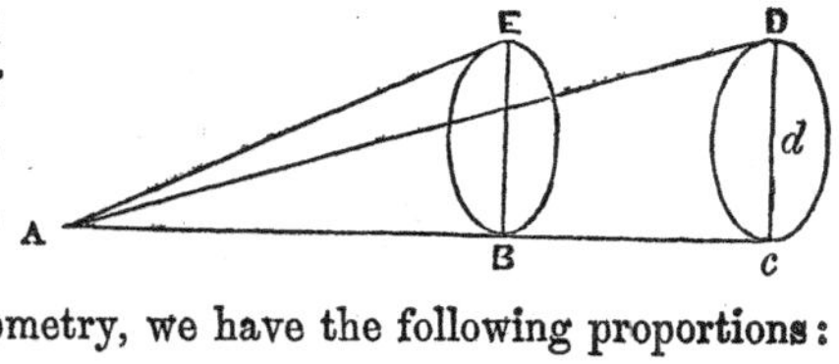

Now, by trigonometry, we have the following proportions:

$$AC : d :: R : \tan. CAD$$
$$AB : d :: R : \tan. BAE.$$

CHAP. II.

Now if the moon has a real change in distance, as observations show, such change *must be* accompanied with apparent changes in the moon's diameter; and, by directing observations to this particular, we find a perfect correspondence; showing the harmony of truth, and the beauties of real science.

Connection between semidiameter and horizontal parallax.

We have several times mentioned that the moon's horizontal parallax is the semidiameter of the earth, as seen from the moon; and now we further say, that what we call the moon's semidiameter, an observer at the moon would call the earth's horizontal parallax; and the variation of these two angles depends on the *same circumstance* — the variation of the distance between the earth and moon; and, depending on one and the same cause, they must vary in just the same proportion.

When the moon's horizontal parallax is greatest, the moon's semidiameter is greatest; and, when least, the semidiameter is the least; and if we divide the tangent of the semidiameter by the tangent of its horizontal parallax, we shall always find the *same quotient* (the decimal 0.27293); and that quotient is the ratio between the real diameter of the earth and the diameter of the moon.* Having this ratio, and the diameter of the earth, 7912 miles, we can compute the diameter of the moon thus:

$$7912 \times 0.27293 = 2169.4 \text{ miles.}$$

From the first proportion, - - - $AC \tan. CAD = dR$;

From the second, - - - - - $AB \tan. BAE = dR$:

By equality, - - - - $AC \tan. CAD = AB \tan. BAE$.

This last equation, put into an equivalent proportion, gives:

$$AC : AB : \tan. BAE :: \tan. CAD.$$

But tangents of very small arcs (such as those under which the heavenly bodies appear) are to each other as the arcs themselves. Therefore,

$$AC : AB :: \text{angle } BAE : \text{angle } CAD;$$

That is; *the angular measures of the same body are inversely proportional to the corresponding distances.*

* This requires demonstration. Let E be the *real* semi-

As spheres are to each other in proportion to the *cubes* of their diameters, therefore the bulk (not mass) of the earth, is to that of the moon, as 1 to $\frac{1}{49}$, nearly.

Augmentation of the moon's semidiameter: its cause.

As the moon's distance is $60\frac{1}{4}$ times the radius of the earth, it follows that it is about $\frac{1}{60}$th nearer to us, when at the zenith, than when in the horizon. Making allowance for this (in proportion to the cosine of the altitude), is called the *augmentation of the semidiameter*.

The earth a moon to the moon.

(68.) It may be remarked, by every one, that we always see the same face of the moon; which shows that she must roll on an axis in the same time as her mean revolution about the earth; for, if she kept her surface toward the same part of the heavens, it could not be constantly presented to the earth, because, to her view, the earth revolves round the moon, the same as to us the moon revolves round the earth; and the earth presents *phases* to the moon, as the moon does to us, except opposite in time, because the two bodies are opposite in position. When we have new moon, the lunarians have full earth; and when we have first quarter, they have last quarter, etc. The moon appears, to us, about half a degree in diameter; the earth appears, to them, a moon, about

---

diameter of the earth (Fig. 16), m that of the moon, D the distance between the

Fig. 16.

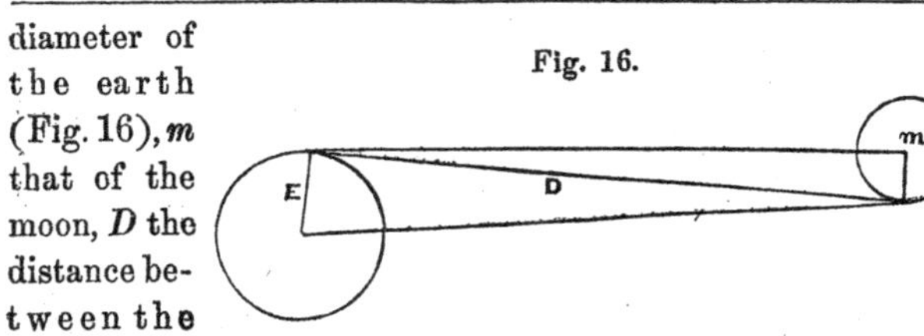

two bodies; and let the radius of the tables be unity. Put P to represent the moon's horizontal parallax, and s its apparent semidiameter. Then, by trigonometry,

$$D : E :: 1 : \tan. P; \text{ and } D : m :: 1 : \tan. s.$$

From the first, $D = \dfrac{E}{\tan. P}$; from the 2d, $D = \dfrac{m}{\tan. s}$:

Therefore, $\dfrac{E}{\tan. P} = \dfrac{m}{\tan. s}$, or $\dfrac{\tan. s}{\tan. P} = \dfrac{m}{E}$. Q. E. D.

CHAP. II. two degrees in diameter, *invariably fixed in their sky*, and the stars passing slowly behind it.

The moon revolves on an axis.

"But," says Sir John Herschel, "the moon's rotation on her axis is uniform; and since her motion in her orbit is not so, we are enabled to look a few degrees round the equatorial parts of her visible border, on the eastern or western side, according to circumstances; or, in other words, the line joining the centers of the earth and moon fluctuates a little in its position, from its mean or average intersection with her surface, to the east, or westward. And, moreover, since the axis about which she revolves is not exactly perpendicular to her orbit, her poles come alternately into view for a small space at the edges of her disc. These phenomena are known by the name of *librations*. In consequence of these two distinct kinds of libration, the same identical point of the moon's surface is not always the center of her disc; and we therefore get sight of a zone of a few degrees in breadth on all sides of the border, beyond an exact hemisphere."

CHAPTER III.

THE EARTH'S ORBIT ECCENTRIC. — THE APPARENT ANGULAR MOTION OF THE SUN NOT UNIFORM. — LAWS BETWEEN DISTANCE, REAL, AND ANGULAR MOTION. — ECCENTRICITY OF THE ORBIT.

CHAP. III

The sun larger than the earth.

(69.) THE sun's parallax is too small to be detected by any common means of observation; hence it remained unknown, for a long series of years, although many ingenious methods were proposed to discover it. The only decision that ancient astronomers could make concerning it was, that it must be less than 20″ or 15″ of arc; for, were it as much as that quantity, it could not escape observation.

Now let us suppose that the sun's horizontal parallax is less than 20″; that is, the apparent semidiameter of the earth, as seen from the sun, must be less than 20″; but the semidiameter of the sun is 15′ 56″, or 956″; therefore the sun must be vastly larger than the earth — by at least 48 times its diameter; and the bulk of the earth must be, to that of the sun, in as high a ratio as 1 to the cube of 48. But as we do not suffer ourselves to know the true horizontal parallax of the sun, all the decision we can make on this subject is, that the sun is *vastly larger* than the earth.

Does the sun go round the earth, or the earth round the sun?

(70.) Previous observations, as we explained in the first section of this work, clearly show, or give the *appearance* of the sun going round the earth once in a year; but the *appearance* would be the same, whether the earth revolves round the sun, or the sun round the earth, or both bodies revolve round a point between them. We are now to consider which is the most probable: *that a large body should circulate round a much smaller one; or, the smaller one round a large one.* The last suggestion corresponds with our knowledge and experience in mechanical philosophy; the first is opposed to it.

(71.) We have seen, in the last chapter, that the semidiameter and horizontal parallax of a body have a constant relation to each other; and, while we cannot discover the one, we will examine all the *variations* of the other (*if it have variations*), and thereby determine whether the earth and sun always remain at the same distance from each other.

Methods of measuring apparent diameters.

Here it is very important that the reader should clearly understand, how the apparent diameter of a heavenly body can be determined to great precision.

As an example, we shall take the diameter of the sun; but the same principles are to be followed, and the same deductions are to be made, whatever body, moon, or planet, may be under observation.

The micrometer.

An instrument to measure the apparent diameter of a planet is called a *micrometer*. It is an eyepiece to a telescope, with opening and closing parallel wires; the amount of the opening is measured by a mathematical contrivance. For the measure of all small objects, the micrometer is exclusively used; and since it is impossible that any one observation can be relied upon as accurate (on account of the angular space eclipsed by the wires), a great number of observations are taken, and

CHAP. III. the mean result is regarded as a single observation. Generally speaking, the following method is more to be relied upon, when large angles are measured, and to it we commend *special attention.*

The method by time in passing the meridian.

The method *depends on the time employed by the body in passing the perpendicular wires of the transit instrument.*

All bodies (by the revolution of the earth) come to the meridian at right angles, and 15 degrees pass by the meridian in one hour of sidereal time; and, in four minutes, one degree will pass; and, in two minutes of time, 30 minutes of arc will pass the meridian wire.

Now if the sun is on the equator, and *stationary* there, and employs two minutes of sidereal time in passing the meridian, then it is evident that its apparent diameter is just 30′ of arc; if the time is more than two minutes, the diameter is more; if less, less.

But we have just made a supposition that is *not true;* we have supposed the sun stationary, in respect to the stars; but it is not so: it apparently moves eastward; therefore it will not get past the meridian wire as soon as it would if stationary. *Hence we must have a correction, for the sun's motion, applied to the time of its passing the meridian.*

Corrections to be made.

We have also supposed the sun on the equator, and for a moment continue the supposition, and also conceive its diameter to be *just* 30′ *of arc.* Now suppose it brought up to the 20th degree of declination, on *that parallel,* it will extend *over more than* 30′ *of arc,* because meridians converge toward the pole; *therefore the farther the sun, or any other body is from the equator, the longer it will be in passing the meridian on that account;* the increase of time depending on the *cosine of the declination.* (See 59.)

Hence two corrections must be made to the actual time that the sun occupies in crossing the meridian wire, before we can proportion it into an arc: one for the progressive motion of the sun in right ascension; and one for the existing declination. We give an example.

Method of deciding the

On the first day of June, 1846, the sidereal time (time measured by the sidereal clock) of the sun passing the meridian wire, was observed to be 2 m. 16.64 s.; the declination was 22° 2′ 45″, and the hourly increase of right ascension was 10.235 s. What was the sun's semidiameter?

CHAP. III. exact apparent diameter of the sun, moon, or planets.

3600 s. : 10.235 s. : : 136.64 : 0.39 s.

Observed dura. of tran., in secs.,	136.64	
Reduction for solar motion, -	.39	
	136.25 . . log.	2.134337
Dec. 22° 2′ 45″; cosine, - - -		9.967021
Duration, if stationary on equa.,	126.3 s. . . log.	2.101358

Minutes or seconds of time can be changed into minutes or seconds of arc, by multiplying by 15; therefore the diameter of the sun, at this time, subtended an arc of 1894″.5, and its semidiameter 947″.2, or 15′ 47″.2; which is the result given in the Nautical Almanac, from which any number of examples of this kind can be taken. We give one more example, for the benefit of those who may not have a Nautical Almanac.

On the 30th day of December (not material what year), the sidereal time of the sun's diameter passing the meridian was observed to be 2 m. 22.2 s., or 142.2 s. The sun's hourly motion in right ascension, at that time, was 11.06 s., and the declination was 22° 11′. What was the sun's semidiameter? * Ans. 16′ 17″.3.

Extreme values of the sun's apparent semidiameter.

These observations may be made every clear day throughout the year; and they have been made at many places, and for many years; and the combined results show that the

* The following is the formula for these reductions:

$$\frac{15(t-c)\cos. D}{R}=s.$$

Here t is the observed interval in seconds, c is the correction for the increase in right ascension, D is the declination, R the radius of the tables, and s is the result in *seconds of arc*. c is always very small; for one hour, or 3600 s., the variation is never less than 8.976 s., nor more than 11.11 s. The former happens about the middle of September; the latter about the 20th of December. For the meridian passage of the moon, the correction c is considerable; because the moon's increase of right ascension is comparatively very rapid. For the planets, c may be disregarded.

CHAP. III. apparent diameter of the sun is the same, on the same day of the year, from whatever station observed.

The least semidiameter is 15′ 45″.1; which corresponds, in time, to the first or second day of July; and the greatest is 16′ 17″.3, which takes place on the 1st or 2d of January.

Now as we cannot suppose that there is any *real* change in the diameter of the sun, we must impute this apparent change to real change in the distance of the body, as explained in Art. 66.

Variation of the distance from the earth to the sun. Therefore the distance to the sun on the 30th of December, must be to its distance on the first day of July, as the number 15′ 45″.1 is to the number 16′ 17″.3, or as the number 945.1 to 977.3; and all other days in the year, the *proportional distance* must be represented by intermediate numbers.

From this, we perceive that the sun must go round the earth, or the earth round the sun, in very nearly a circle; for were a representation of the curve drawn, corresponding to the apparent semidiameter in different parts of the orbit, and placed before us, the eye could scarcely detect its departure from a circle.

(72.) It should be observed that the time elapsed between the greatest and least apparent diameter of the sun, or the reverse, is just half a year; and the change in the sun's longitude is 180°.

Eccentricity of the earth's orbit, how known. If we would consider the mean distance between the earth and sun as *unity* (as is customary with astronomers), and then put x to represent the least distance, and y the greatest distance, we shall have

$$x+y=2.$$

$$\text{And,} \quad x : y :: 9451 : 9773.$$

A solution gives $x=0.98326$, nearly, and $y=1.01674$, nearly; showing that the *least, mean,* and *greatest* distance to the sun, must be very nearly as the numbers .98326, 1., and 1.01674.

The fractional part, .01674, or the difference between the extremes and mean (*when the mean is unity*), is called the *eccentricity* of the orbit.

The *eccentricity*, as just mentioned, must not be regarded as accurate. It is only a first approximation, deduced from the first and most simple view of the subject; but we shall, hereafter, give other expositions that will lead to far more accurate results.

Eccentricity from apparent diameters only approximate.

In theory, the apparent diameters are sufficient to determine the eccentricity, could we really observe them to rigorous exactness; but all luminous bodies are more or less affected by *irradiation*, which dilates a little their apparent diameters; and the exact quantity of this dilatation is not yet well ascertained.

(73.) The sun's right ascension and declination can be observed from any observatory, any clear day; and from thence we can trace its path along the celestial concave sphere above us, and determine its change from day to day; and we find it runs along a great circle called the *ecliptic*, which crosses the equator at opposite points in the heavens; and the ecliptic inclines to the equator with an angle of about 23° 27′ 40″.

The plane of the ecliptic passes through the center of the earth, showing it to be a great circle, or, what is the same thing, showing that the apparent motion of the sun has its center in the line which joins the earth and sun.

Variations in the distance of the sun, compared with its variations in longitude.

The apparent motion of the sun along the ecliptic is called longitude; and this is its most regular motion.

When we compare the sun's motion, in longitude, with its semidiameter, we find a correspondence—at least, an apparent connection.

When the semidiameter is greatest, the motion in longitude is greatest; and, when the semidiameter is least the motion in longitude is least; *but the two variations have not the same ratio.*

When the sun is nearest to the earth, on or about the 30th of December, it changes its longitude, in a mean solar day, 1° 1′ 9″.95. When farthest from the earth, on the 1st of July, its change of longitude, in 24 hours, is only 57′ 11″.48. A uniform motion, for the whole year, is found to be 59′ 8″.33.

The ancient philosophers contended that the sun moved

CHAP. III.

about the earth in a *circular orbit,* and its real velocity uniform; but the earth not being in the center of the circle, the same portions of the circle would appear under *different angles;* and hence the variation in its apparent angular motion.

The result shows that the angular motion is in the inverse proportion to the square of the distance.

Now if this is a true view of the subject, the variation in angular motion must be in *exact proportion* to the variation in distance, as explained in the note to Art. 66; that is, 945″.1 should be to 977″.3, as 57′ 11″.48 to 61′ 9″.95, if the supposition of the first observers were true. But these numbers have *not* the same ratio; therefore this supposition is not satisfactory; and it was probably abandoned for the want of this mathematical support. The ratio between 945″.1, and 977″.3 is - - - - - - - - - $\frac{9773}{9451}=1.0341$, nearly;

between 57′ 11″.48, and 61′9″.95, $\frac{3669''.95}{3431''.48}=1.0694$, nearly.

If we square (1.0341) the *first* ratio, we shall have 1.06936, a number so near in value to the *second* ratio, that we conclude it ought to be the same, and would be the same, provided we had perfect accuracy in the observations.

Law between motion and distance.

Thus we compare the angular motion of the sun in different parts of its orbit; and we always find, *that the inverse square of its distance is proportional to its angular motion;* and this incontestible *fact* is so exact and so regular, that we lay it down as a *law;* and if solitary observations do not correspond with it, we must condemn the observations, and not the law.

(74.) To investigate this subject thoroughly, we cannot avoid making use of a little geometry.

Let Fig. 17 represent the solar orbit,* the sun apparently revolving about the observer at O. The distance from O to

* We say solar orbit, when it is really the earth's orbit; so we speak of the sun's motion, when it is really the motion of the earth; and it is customary, with astronomers, to speak of apparent motions as real · and none object to this manner of speaking, who have a clear or enlarged view of the science—for to depart from it would lead to oft-repeated and troublesome technicalities, if not to confusion of ideas. Clearness does not always correspond with exactness of expression.

any point in the orbit is called the *radius vector;* and it is a varying quantity, conceived to sweep round the point O.

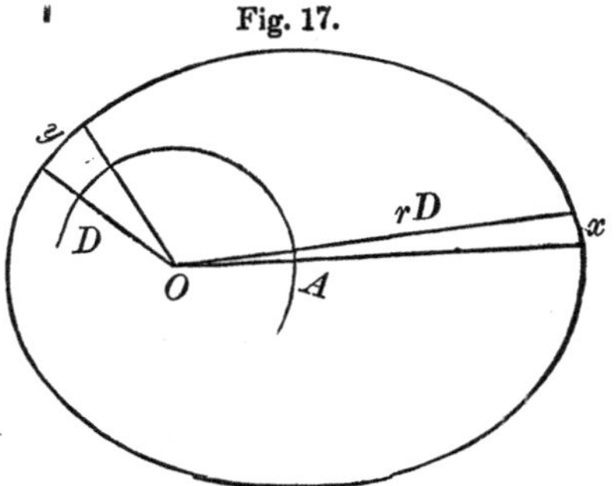

Fig. 17.

Let D be the value of the radius vector at any point, and rD its value at some other point, as represented in the figure. Let y represent the real motion of the sun, for a *very* short interval of time, at the extremity of the radius vector D; and x represent the real motion, at the extremity of the radius vector rD, in the same time.

Variations in real and angular motion.

From O, as a center, at the distance of unity, describe a circle. Put A to represent the angle under which x appears from O; then, *by observation*, r^2A is the angle under which y appears from the same point.

Now, considering the sectors as triangles, we have the following proportions:

$$1 : A :: rD : x;$$
$$1 : r^2A :: D : y.$$

From the first, - - $x=rAD$,

From the second, - $y=r^2AD$.

Multiply the first of these equations by r, and we perceive that - - - - - - - $y=rx$.

This last equation shows that the *real* velocity of the earth in its orbit varies in the inverse ratio as the radius vector; *or it varies directly as the apparent diameter of the sun.*

The real velocity of the earth in its orbit varies as the sun's apparent diameter.

(75.) If we multiply rD by x, the product will express the double of an *area* passed over by the *radius vector* in a certain interval of time; and if we multiply D by y, we shall have the double of another *area* passed over by the *radius vector* in the same time. But the first product is rDx, and the second is the same, as we shall see by taking the value of y (rx); that is $rDx=rDx$; hence we announce this general law:

CHAP. III.

That the solar radius vector describes equal areas in equal times.

The radius vector describes equal areas in equal times.

When expressed in more general terms, this is one of the three laws of Kepler, which will be fully brought into notice in a subsequent part of this work.

If we draw lines from any point in a plane, reciprocally proportional to the sun's apparent diameter, and at angles differing as the change of the sun's longitude, and then connect the extremities of such lines made all round the point, the connecting lines will form a curve, corresponding with an ellipse (see Fig. 18), which represents the apparent solar orbit; and, from a review of the whole subject, we give the following summary:

Laws of motion in an ellipse.

1. *The eccentricity of the solar ellipse, as determined from the apparent diameter of the sun, is* .01674.*

2. *The sun's angular velocity varies inversely as the square of its distance from the earth.*

3. *The real velocity is inversely as the distance.*

4. *The areas described by the radius vector are proportional to the times of description.*

(76.) We have several times mentioned, that, as far as appearances are concerned, it is immaterial whether we consider the sun moving round the earth, or the earth round the sun; for, if the earth is in one position of the heavens, the

---

* By making use of the 2d principle, above cited, we can compute the eccentricity of the orbit to greater precision than by the apparent diameters, because the same error of observation on longitude would not be as proportionally great as on apparent diameter.

Let E be the eccentricity of the orbit; then $(1-E)$ is the least distance to the sun, and $(1+E)$ the greatest distance. Then, by observation, we have

$$(1-E)^2 : (1+E)^2 :: 57' \, 11''.48 : 61' \, 9''.95;$$

$$\text{Or, } (1-E)^2 : (1+E)^2 :: 343148 : 366995;$$

$$\text{Or, } 1-E : 1+E :: \sqrt{343148} : \sqrt{366995}.$$

Whence $E=.016788+$. We shall give a still more accurate method of computing this important element.

sun appears exactly in the opposite position, and every motion made by the earth must correspond to an apparent motion made by the sun.

Fig. 18.

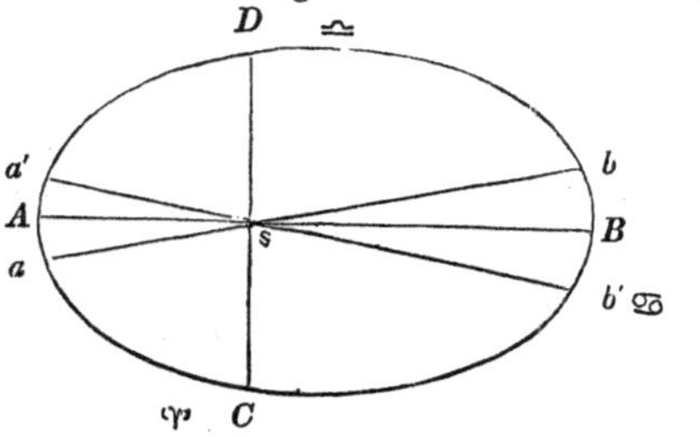

D ♎ a' A a S b B b' ♋ ♈ C

But, for the purpose of getting nearer to fact, we will now suppose the earth revolves round the sun in an elliptical orbit, as represented by Fig. 18.

We have very much exaggerated the eccentricity of the orbit, for the purpose of bringing principles clearer to view.

The *greatest* and *least* distances, from the sun to the earth, make a *straight line* through the sun, and cut the orbit into two *equal* parts. When the earth is at *B*, the greatest distance from the sun, it is said to be in *apogee*, and when at *A*, the least distance, it is in *perigee;* and the line joining the apogee and perigee is the major, or greater diameter of the orbit; *and it is the only diameter passing through the sun, that cuts the orbit into two equal parts.*

Observations to determine the positions of the solar apogee and perigee.

Now, as *equal areas* are described in *equal times*, it follows that the earth must be just *half a year* in passing from apogee to perigee, and from perigee to apogee; provided that these points are stationary in the heavens, and they are so, very nearly.*

If we suppose the earth moves along the orbit from *D* to *A*, and we observe the sun from *D*, and continue observations upon it until the earth comes to *C*, then the longitude of the sun has changed 180°; and if the time is less than

* The longer axis of the orbit, or apogee point, changes position by a very slow motion of about 12″ per annum, to the eastward: but this motion must be disregarded, for the present, as well as *many other* minute deviations, to be brought into view when we are better prepared to understand them.

These minute variations, for short periods of time, do not sensibly affect general results.

6

CHAP. III. half a year, we are sure the perigee is in this part of the orbit. If we continue observations round and round, and find where 180 degrees of longitude correspond with half a year, there will be the position of the longer axis; which is sometimes called the *line of the apsides*.

Difficulties, how avoided

We cannot determine the exact point of the apogee or perigee, by direct observations on the sun's apparent diameter; for about these points the variations are extremely slow and imperceptible.

If we take observations in respect to the sun's longitude, when the earth is at *b*, and watch for the opposite longitude, when the earth is about *a*, and find that the area *b D a* was described in little less than half a year, and the *area a C b*, in a little more than half a year, then we know that *b* is very near the apogee, and *a* very near the perigee.

If we take another point, *b'*, and its opposite, *a'*, and find converse results, then we know that the apogee is between the points *b'* and *b*, and we can proportion to it to great exactness.

Longitude of apogee and perigee.

(77.) The longitude of the apogee, for the year 1801, was 99° 31′ 9″, and, of course, the perigee was in longitude 279° 31′ 9″. These points move forward, in respect to the stars, about 12″ annually, and, in respect to the equinox, about 62″; more exactly 61″.905, and, of course, this is their annual increase of longitude.

In the year 1250, the perigee of the sun coincided with the winter solstice, and the apogee with the summer solstice; and at that time the sun was 178 days and about 17½ hours on the south side of the equator, and 186 days and about 12½ hours on the north side; being longer in the northern hemisphere than in the southern, by seven days and 19 hours: at present, the excess is seven days and near 17 hours.

The year unequally divided.

(78.) As the sun is a longer time in the northern than in the southern hemisphere, the first impression might be, that more solar heat is received in one hemisphere than in the other; but the amount is the same; for whatever is gained in time, is lost in distance; and what is lost in time, is gained by a decrease of distance. The amount of heat depends on the intensity multiplied by the time it is applied; and the product of the time and distance to the sun, is the same in either hemisphere; but the amount of heat received, for a single day, is different in the two hemispheres.

(79.) Conceive a line drawn through the sun, at right angles to the greater diameter of the orbit $D\,S\,C$ (see Fig. 18), the point C is 8° 21′ from the first point of Aries; and if we observe the time occupied by the sun in describing 180 degrees of longitude, from this point (or from any point very near this point), that time, taken from the whole year, will give the time of describing the other 180 degrees.

A method of obtaining the eccentricity of an orbit.

Without being very minute, we venture to state, that the time of describing the arc $D\,A\,C$ is 178 days 17½ hours; and the time of describing the arc $C\,B\,D$ is 186 days 12½ hours. But, as *areas* are in proportion to the times of their description; therefore,

$$\text{area } CSDA : \text{area } CBDS :: \overset{\text{d.}}{178}\ \overset{\text{h.}}{17\tfrac{1}{2}} : \overset{\text{d.}}{186}\ \overset{\text{h.}}{12\tfrac{1}{2}}.$$

By taking half of the greater axis of the ellipse equal *unity*, and the eccentricity an unknown quantity, e, the mathematician can soon obtain analytical expressions for the two *areas* in question; and then, from the proportion, he can find the value of the eccentricity e: but there is a better method — we only give an outside view of this, for the *light* it throws on the general principle.

(80.) Now let us conceive the orbit of the earth inclosed by a circle whose diameter is the greatest diameter of the ellipse, as represented by Fig. 19.

Preparation for finding the true variation in an ellipse.

For the sake of simplicity we will suppose the observer at rest at the point o (one focus of the ellipse), and the sun really to move round on the ellipse, describing equal *areas* in equal times round the point o.

Conceive, also, an imaginary sun to pass round the circle, describing equal angles, in equal times, round the center m. Now suppose the two suns to be together at the point B: — they depart, one on the ellipse, the other on the circle; and, on account of both describing equal areas, in equal times, round their respective centers of motion, they will be together

CHAP. III.

Fig. 19.

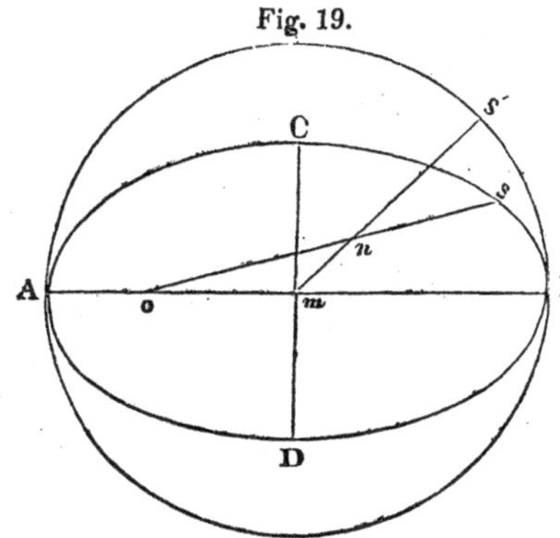

at the point *A*, and again at the point *B*, and so continue in each subsequent revolution.

The imaginary sun on the circle everywhere describes equal angles in equal times; and the true sun, on the ellipse, describes only equal *areas* in equal times; but the *angles* will be unequal. Conceive the two suns to depart, at the same time, from the point *B*, and, after a certain interval of time, one is at *s*, the other at *s'*. Then we must have

$$area\ oBs\ :\ area\ mBs'\ ::\ area\ ellipse\ :\ area\ circle.$$

Mean and true anomaly.

The angle Bms' is the angle the sun would make, or its increase in longitude from the apogee; provided the angular motion of the sun was uniform. The angle Bos is its true increase of longitude; the difference between these two angles is the angle mno.

The angle Bms' is always known *by the time;* and if to every degree of the angle Bms' we knew the corresponding angle mno, the two would give us the angle Bos; for,

$$Bms'-mno=mon, \text{ or } Bos.$$

The angle Bms' is called the *mean anomaly*, and the angle Bos is called the *true anomaly.*

The equation of the center.

The angle Bms' is greater than the angle Bos, all the way from the apogee to the perigee; but from the perigee to the apogee, the true sun, on the ellipse, is in advance of the imaginary sun on the circle.

The angle mno is called the *equation of the center;* that is, it is the angle to be applied to the angle about the center m, to make it equal to the *true anomaly.*

The angle mno depends on the eccentricity of the ellipse; and its amount is put in a table corresponding to every

CHAP. III.

degree of the *mean anomaly*; subtractive, from the apogee to the perigee, and additive from the perigee to the apogee.*

The greatest equation of the center gives the eccentricity of the orbit.

(81.) Again: conceive the two suns to set out from the same point, B; and as the angle $B\,m\,s'$ increases uniformly, it will increase and become greater and greater than the angle $B\,o\,s$, until the true sun attains its *mean* angular motion, and no longer. Then the angle $m\,n\,o$ attains its greatest value, and, at that time the side $mn{=}no$, and the point n is perpendicular over $o\,m$, and $o\,s$ is a *mean* proportional between $o\,B$ and $o\,A$. That is, when the sun, or any planet, attains the *greatest equation of the center*, the true sun is very near the extremity of the shorter axis of the ellipse: o, the greatest equation of the center, can be determined by observation; and, from the greatest equation, we have the most accurate method of computing the eccentricity of the ellipse, as we may see by the note below.†

† Let C (Fig. 20) be the place of the true sun, and G the place of the imaginary sun; the line $m\,F$ cuts off equal portions of the circle and the ellipse. Then we have to make the sector $m\,F\,G$ to the triangle $o\,m\,C$, as the circle is to the ellipse. Now let

Fig. 20.

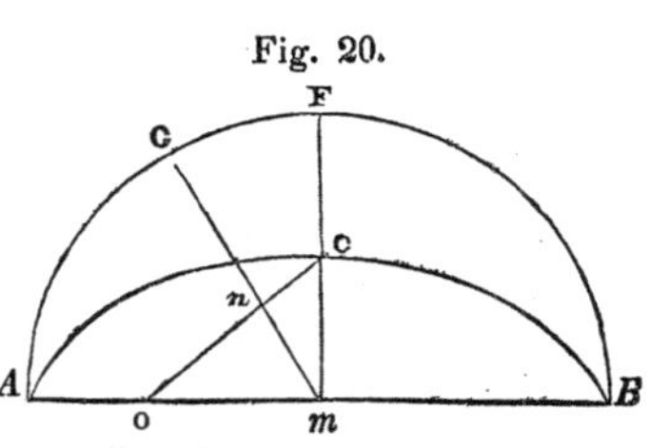

$$mB{=}a, \quad mC{=}b, \quad om{=}ea, \quad \pi{=}3.1416;$$

Then, the area of the circle is πa^2; the *area* of the ellipse is πab; that of the sector is $(GF)\frac{a}{2}$, and of the triangle $\frac{eab}{2}$.

$$\text{Hence,} \quad \frac{eab}{2} \ : \ GF\left(\frac{a}{2}\right) \ :: \ \pi ab \ : \ \pi a^2;$$

* By a mere mechanical contrivance, the modern astronomical tables are so arranged, that all corrections are rendered additive; so that the mechanical operator cannot make a mistake, as to signs, and he may continue to work without stopping to think. These arrangements have their advantages, *but they cover up and obscure principles.*

When once the eccentricity of any planetary ellipse becomes known, the equation of the center, corresponding to all degrees of the *mean anomaly*, can be computed and put into a table for future use; but this labor of constructing tables belongs exclusively to the mathematician.

Method of deducing the eccentricity from the greatest equation of the center.

Or, - - $eab : (GF)a :: b : a$;

Or, - - $ea : GF :: 1 : 1$.

Consequently, $GF=ea$, and $FG=om$; which shows that the angle $o\,Cm$ is nearly equal to $Fm\,G$, unless it is a very eccentric ellipse. Now we must compute the number of degrees in the arc FG. The whole circumference is $2\pi a$.

Therefore, $2\pi a : ea :: 360 :$ arc FG:

Hence, - - - arc $FG=\frac{180\,e}{\pi}=$angle $n\,m\,C$.

But the angle $o\,n\,m=n\,m\,C+n\,Cm=2\,nm\,C$, nearly;

Therefore, $\frac{360\,e}{\pi}=2\,n\,m\,C=o\,n\,m=$ greatest equation of center, nearly.

But the greatest equation of the center, for the solar orbit, is, by observation, $1^{\circ}\ 55'\ 30''$; and as the sun has not quite its greatest equation of the center, when at the point C, it will be more accurate to put

$$\frac{360\,e}{\pi}=1^{\circ}\ 55'\ 24''.$$

From this equation, it is true, we have only the approximate value of e; but it is a *very* approximate value, and sufficiently accurate.

Reducing both members to seconds, and we have,

$$3600\cdot360\,e=6924\,\pi, \quad \text{and} \quad e=0.0167842.$$

The greatest equation of the center is at present diminishing at the rate of $17''.17$ in *one hundred* years: this corresponds to a diminution of eccentricity by 0.00004166, which is determined by a solution of the following equation:

$$3600\frac{360\,e}{\pi}=17''.17.$$

CHAPTER IV.

THE CAUSES OF THE CHANGE OF SEASONS.

CHAP. IV.

(82.) THE annual revolution of the earth in its orbit, combined with the position of the earth's axis to the plane of its orbit, produces the change of the seasons.

The cause of the change of seasons.

If the axis were perpendicular to the plane of its orbit, there would be no change of seasons, and the sun would then be all the while in the celestial equator.

This will be understood by Fig. 21. Conceive the plane of the paper to be the plane of the earth's orbit, and conceive the several representations of the earth's axis, *NS*, to be inclined to the paper at an angle of 66° 32′.

Fig. 21.

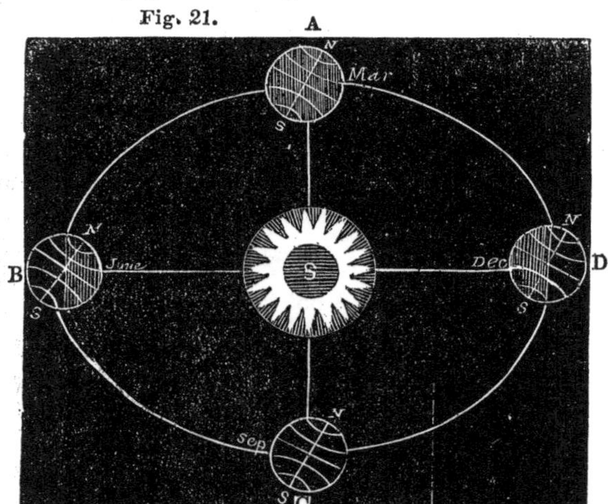

In all representations of *NS*, one half of it is supposed to be above the paper, the other half below it.

NS is always parallel to itself; that is, it is always in the same position* — always at the same inclination to the plane

* Except minute variations, which it would be improper to notice in this part of the work.

CHAP. IV. of its orbit — always directed to the same point in the heavens, in whatever part of the orbit it may be.

The plane of the equator, represented by Eq, is inclined to the plane of the orbit by an angle of 23° 28′.

Importance of inspecting the figure.

By inspecting the figure, the reader will gather a clearer view of the subject than by whole pages of description: he will perceive the reason why the sun must shine over the north pole, in one part of its orbit, and fall as far short of that point when in the opposite part of its orbit; and the number of degrees of this variation depends, of course, on the position of the axis to the plane of the orbit.

Position of the axis to cause no change of seasons.

Now conceive the line NS to stand perpendicular to the plane of the paper, and continue so; then Eq would lie on the paper, and the sun would at all times be in the plane of the equator, and there would be no change of seasons. If NS were more inclined from the perpendicular than it now is, then we should have a greater change of seasons.

By inspecting the figure, we perceive, also, that when it is summer in the northern hemisphere, it is winter in the southern; and conversely, when it is winter in the northern, it is summer in the southern.

When a line from the sun makes a right angle with the earth's axis, as it must do in two opposite points of its orbit, the sun will shine equally on both poles, and it is then in the plane of the equator; which gives equal day and night the world over.

Equal days and nights, for all places, happen on the 20th of March of each year, and on the 22d or 23d of September. At these times the sun crosses the celestial equator, and is said to be in the equinox.

The equinoctial and solstitial points.

The longitude of the sun, at the vernal equinox, is 0°; and at the autumnal equinox, its longitude is 180°.

The time of the greatest north declination is the 20th of June; the sun's longitude is then 90°, and is said to be at the *summer solstice.*

The time of the greatest south declination is the 22d of December; the sun's longitude, at that time, is 270°, and is said to be at the *winter solstice.*

By inspecting the figure, we perceive, that when the earth is at the summer solstice, the north pole, *P*, and a considerable portion of the earth's surface around, is within the enlightened half of the earth; and as the earth revolves on its axis *N S*, this portion constantly remains enlightened, giving a constant day—or a day of weeks and months duration, according as any particular point is nearer or more remote from the pole: the pole itself is enlightened full six months in the year, and the circle of more than 24 hours constant sunlight extends to 23° 28′ from the pole (not estimating the effects of refraction). On the other hand, the opposite, or south pole, *S*, is in a long season of darkness, from which it can be relieved only by the earth changing position in its orbit.

Long seasons of sunlight and darkness at and about the poles.

"Now, the temperature of any part of the earth's surface depends mainly, if not entirely, on its exposure to the sun's rays. Whenever the sun is above the horizon of any place, that place is receiving heat; when below, parting with it, by the process called radiation; and the whole quantities received and parted with in the year must balance each other at every station, or the equilibrium of temperature would not be supported. Whenever, then, the sun remains more than 12 hours above the horizon of any place, and less beneath, the general temperature of that place will be above the average; when the reverse, below. As the earth, then, moves from *A* to *B*, the days growing longer, and the nights shorter in the northern hemisphere, the temperature of every part of that hemisphere increases, and we pass from spring to summer, while at the same time the reverse obtains in the southern hemisphere. As the earth passes from *B* to *C*, the days and nights again approach to equality—the excess of temperature in the northern hemisphere, above the mean state, grows less, as well as its defect in the southern; and at the autumnal equinox, *C*, the mean state is once more attained. From thence to *D*, and, finally, round again to *A*, all the same phenomena, it is obvious, must again occur, but reversed; it being now winter in the northern, and summer in the southern hemisphere."

Temperature of the earth.

The inquiry is sometimes made why we do not have the warmest weather about the summer solstice, and the coldest weather about the time of the winter solstice.

Times of extreme temperature.

This would be the case if the sun immediately ceased to give extra warmth, on arriving at the summer solstice; but if it could radiate extra heat to warm the earth three weeks, before it came to the solstice, it would give the same extra heat three weeks after; and the northern portion of the earth must continue to increase in temperature as long as the sun continues to radiate more than its medium degree of heat over the surface, at any particular place. Conversely, the whole region of country continues to grow cold as long as the sun radiates less than its mean annual degree of heat over that region. The medium degree of heat, for the whole year, and for all places, of course, takes place when the sun is on the equator; the average temperature, at the time of the two equinoxes. The medium degree of heat, for our northern summer, considering only two seasons in the year, takes place when the sun's declination is about 12 degrees north; and the medium degree of heat, for winter, takes place when the sun's declination is about 12 degrees south; and if this be true, the heat of summer will begin to decrease about the 20th of August, and the cold of winter must essentially abate on or about the 16th of February, in all northern latitudes.

CHAPTER V.

EQUATION OF TIME.

(83.) WE now come to one of the most important subjects in astronomy — the equation of time.

Without a good knowledge of this subject, there will be constant confusion in the minds of the pupils; and such is the nature of the case, that it is difficult to understand even the *facts*, without investigating their causes.

Sidereal time perfect.

Sidereal time has no *equation;* it is uniform, and, of itself, perfect and complete.

The time, by a perfect clock, is theoretically perfect and complete, and is called *mean time*.

Solar time not uniform

The time, by the sun, *is not uniform;* and, to make it agree with the *perfect* clock, requires a correction — a quantity to make equality; and this quantity is called the equation of time.*

If the sun were stationary in the heavens, like a star, it would come to the meridian after exact and equal intervals of time; and, in that case, there would be no equation of time.

If the sun's motion, in right ascension, were uniform, then it would also come to the meridian after equal intervals of time, and there would still be no equation of time. But (speaking in relation to appearances) the sun is not stationary in the heavens, nor does it move uniformly; therefore it cannot come to the meridian at equal intervals of time, and, of course, the solar days must be *slightly unequal.*

Mean and apparent noon.

When the sun is on the meridian, it is then apparent noon, for that day: it is the *real solar noon,* or, as near as may be, half way between sunrise and sunset; but it may not be noon by the *perfect* clock, which runs hypothetically true and uniform throughout the whole year.

A fixed star comes to the meridian at the expiration of every 23 h. 56 m. 04.09 s. of mean solar time; and if the sun were stationary in the heavens, it would come to the meridian after every expiration of just that same interval. But the sun increases its right ascension every day, by its apparent eastward motion; and this increases the time of its coming to the meridian; and the *mean interval* between its successive transits over the meridian is just 24 hours; but the actual intervals are variable — some less, and some more than 24 hours.

On and about the 1st of April, the time from one meridian of the sun to another, as measured by a perfect clock, is 23 h. 59 m. 52.4 s.; less than 24 hours by about 8 seconds. Here, then, the sun and clock must be constantly separating. On

* In astronomy, the term equation is applied to all corrections to convert a mean to its true quantity.

CHAP. V. and about the 20th of December, the time from one meridian of the sun to another is 24 h. 0 m. 24.3 s., more than 24 seconds over 24 hours; and this, in a few days, increases to minutes — and thus we perceive the fact of equation of time.

Equation of time the result of two causes.

To detect the law of this variation, and find its amount, we must separate the cause into its two natural divisions.

1. *The unequal apparent motion of the sun along the ecliptic.*

2. *The variable inclination of this motion to the equator.*

If the sun's apparent motion along the ecliptic were uniform, still there would be an equation of time; for that motion, in some parts of the orbit, is oblique to the equator, and, in other parts, parallel with it; and its eastward motion, in right ascension, would be greatest when moving parallel with the equator.

From the first cause, separately considered, the sun and clock would agree two days in a year — the 1st of July and the 30th of December.

From the second cause, separately considered, the sun and clock agree *four* days in a year — the days when the sun crosses the equator, and the days he reaches the solstitial points.

When the results of these two causes are combined, the sun and clock will agree four days in the year; but it is on neither of those days marked out by the separate causes; and the intervals between the several periods, and the amount of the equation, appear to want regularity and symmetry.

Days in the year in which the sun and clock agree.

The four days in the year on which the sun and clock agree, that is, show noon at the same instant, are April 15th, June 16th, September 1st, and December 24th.

The greatest amount, arising from the first cause, is 7 m. 42 s., and the greatest amount, from the second cause, is 9 m. 53 s.; but as these maximum results never happen exactly at the same time, therefore the equation of time can never amount to 17 m. 35 s. In fact, the greatest amount is 16 m. 17 s., and takes place on the 3d of November; and, for a long time to come, the maximum value will take place on the same day of each year; but, in the course of ages, it will vary in its amount and in the time of the year in which the sun and clock agree, in consequence of the slow and gradual change in the position of the solar apogee. (See Art. 77.)

The equation of the sun's center, and the first part of the equation of time, have a common cause.

(84.) The elliptical form of the earth's orbit gives rise to the unequal motion of the earth in its orbit, and thence to the apparent unequal motion of the sun in the ecliptic; and this same *unequal* motion is what we have denominated the first cause of the equation of time. Indeed, this part of the equation of time is nothing more than the equation of the center (80), changed into time at the rate of four minutes to a degree.

The greatest equation for the sun's longitude (81, note), is by observation 1° 55′ 30″; and this, proportioned into time, gives 7 m. 42 s., for the maximum effect in the equation of time arising from the sun's unequal motion. When the sun departs from its perigee, its motion is greater than the mean rate, and, of course, comes to the meridian later than it otherwise would. In such cases, the sun is said to be slow — and it is slow all the way from its perigee to its apogee; and fast in the other half of its orbit.

For a more particular explanation of the second cause, we must call attention to Fig. 22.

Fig. 22.

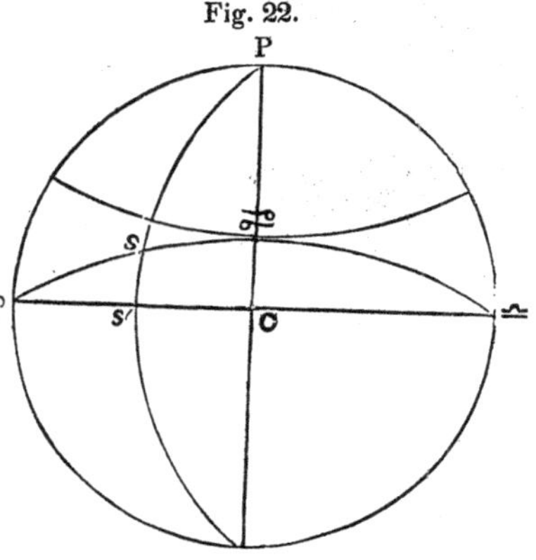

Let ♈ ♋ ♎ (Fig. 22) represent the ecliptic, and ♈ *C* ♎ the equator.

By the first correction, the apparent motion along the ecliptic is rendered uniform; and the sun is then supposed to pass over equal spaces in equal intervals of time along the arc ♈ *S* ♋. But equal spaces of arc, on the ecliptic, do not correspond with equal spaces on the equator. In short, the points on the ecliptic must be reduced to corresponding points on the equator. For instance, the number of degrees represented by ♈ *S* on

CHAP. V. the ecliptic, is greater than to the same meridian along the equator. The difference between ♈S and ♈S', turned into time, is the equation of time arising from the obliquity of the ecliptic corresponding to the point S.

At the points ♈, ♋, and ♎, and also at the southern tropic, the ecliptic and the equator correspond to the same meridian; but all *other equal distances*, on the ecliptic and equator, are included by different meridians.

How to compute the second part of the equation of time.

To compute the equation of time arising from this cause, we must solve the spherical triangle ♈$S S'$; ♈S is the sun's longitude, and the angle at ♈ is the obliquity of the ecliptic, and at S' is a right angle. Assume any longitude, as 32°, 35°, or 40°, or any other number of degrees, and compute the base. The difference between this base and the sun's longitude, converted into time, is the quantity sought corresponding to the assumed longitude; and by assuming every degree in the first quadrant, and putting the result in a table, we have the amount for every degree of the entire circle, for all the quadrants are symmetrical, and the same distance from either equinox will be the same amount.

What is meant by sun fast and slow of clock.

The perfect clock, or mean time, corresponds with the equator; and as uniform spaces along the equator, near the point ♈, will pass over more meridians than the same number of equal spaces on the ecliptic; therefore the sun, at S, will be *fast of clock*, or come to the meridian before it is noon by the clock — and this will be true all the way to the tropic, or to the 90th degree of longitude, where the sun and clock will agree. In the second quadrant, the sun will come to the meridian after the clock has marked noon. In the third quadrant the sun will again *be fast;* and, in the fourth quadrant, again slow of clock.

It will be observed, by inspecting the figure, that what the sun loses in eastward motion, by *oblique direction* near the equator, is made up, when near the tropics, by the diminished distances between the meridians.

For a more definite understanding of this matter, we give the following table.

Table showing the separate results of the two causes for the equation of time, corresponding to every fifth day of the second years after leap year; but is nearly correct for any year. CHAP. V.

	1st cause. Sun slow of Clock.	2d cause. Sun slow of Clock.		1st cause. Sun fast.	2d cause. Sun slow.
	m. s.	m. s.		m. s.	m. s.
January 5	0 41	5 8	July 1	0 0	3 32
10	1 22	6 35	7	0 40	5 8
15	2 2	7 48	12	1 19	6 35
20	2 41	8 45	17	1 57	7 48
25	3 19	9 26	22	2 35	8 45
29	3 56	9 49	28	3 12	9 26
Feb. 3	4 30	9 53	Aug. 2	3 47	9 49
8	5 2	9 40	7	4 21	9 53
13	5 32	9 9	12	4 52	9 40
18	5 39	8 23	17	5 22	9 9
23	6 24	7 22	22	5 50	8 23
28	6 45	6 9	28	6 14	7 22
March 5	7 3	4 46	Sept. 2	6 36	6 9
10	7 18	3 15	7	6 56	4 46
15	7 29	1 39	12	7 12	3 15
20	7 37	sun fast	17	7 24	1 39
25	7 42	1 39	23	7 34	sun fast
30	7 42	3 15	28	7 40	1 39
April 4	7 40	4 46	Oct. 3	7 42	3 15
9	7 34	6 9	8	7 40	4 46
14	7 24	7 22	13	7 34	6 9
19	7 12	8 23	18	7 24	7 22
24	6 56	9 9	23	7 12	8 23
30	6 36	9 40	28	6 56	9 9
May 5	6 14	9 53	Nov. 2	6 36	9 40
10	5 50	9 49	7	6 14	9 53
15	5 22	9 26	12	5 50	9 49
20	4 52	8 45	17	5 22	9 26
26	4 21	7 48	22	4 52	8 45
31	3 47	6 35	27	4 22	7 48
June 5	3 12	5 8	Dec. 2	3 47	6 35
10	2 35	3 32	7	3 12	5 8
16	1 57	1 48	12	2 35	3 32
21	1 19	sun slow	17	1 57	1 48
26	0 40	1 48	21	1 19	sun slow.
			26	0 40	1 48

By this table, the regular and symmetrical result of each cause is visible to the eye; but the actual value of the equation of time, for any particular day, is the combined results of these two causes. Thus, to find the equation of time for the 5th day of March, we look at the table and find that

Use of the preceding table.

CHAP. V.

The first cause gives sun slow, - - -	7m. 3 s.	
The second, " sun slow, - - -	4 46	
Their combined result (or algebraic sum) is	11 49	slow.

That is, the sun being *slow*, it does not come to the meridian until 11 m. 49 s. after the noon shown by a perfect clock; but whenever the sun is on the meridian, it is then noon, apparent time; and, to convert this into mean time, or to set the clock, we must *add* 11 m. 49 s.

Use of the equation of time.

By inspecting the table, we perceive, that on the 14th of April the two results nearly counteract each other; and consequently the sun and clock nearly agree, and indicate noon at the same instant. On the 2d of November the two results unite in making the sun *fast;* and the equation of time is then the sum of 6 36 and 9 40, or 16 m. 16 s.; the *maximum* result.

The sun at this time being *fast*, shows that it comes to the meridian 16 m. 16 s. before twelve o'clock, true mean time; or, when the sun is on the meridian, the clock ought to show 11 h. 43 m. 44 s.; and thus, generally, *when the sun is fast, we must subtract the equation of time from apparent time, to obtain mean time;* and conversely, when the sun is *slow*.

As no clock can be relied upon, to run to true mean time, or to any exact definite rate, therefore clocks must be frequently rectified by the *sun*. We can observe the apparent time, and then, by the application of the equation of time, we determine the true *mean time*.

A table for equation of time depending on the sun's longitude can be formed.

(85.) As the sun has a particular motion, corresponding to every particular point on the ecliptic, and, at the same time, the particular point on the ecliptic has a definite relation to the equator, therefore any point, as S (Fig. 22), on the ecliptic, has the two corrections for the equation of time; consequently a table can be formed for the equation of time, depending on the longitude of the sun; and such a table would be *perpetual*, if the longer axis of the solar orbit did not change its position in relation to the equinoxes. But as that change is *very slow*, a table of that kind will serve for many years, with a very trifling correction, and such a table is to be found in many astronomical works. CHAP. V.

It is very important that the *navigator*, *astronomer*, and *clock regulator*, should thoroughly understand the equation of time; and persons thus occupied pay great attention to it; but most people in common life are hardly aware of its existence. Utility of the equation of time.

CHAPTER VI.

THE APPARENT MOTIONS OF THE PLANETS.

(86.) WE have often reminded the reader of the great regularity of the fixed stars, and of their uniform positions in relation to each other; and by this very regularity and constancy of relative positions, we denominate them fixed; but there are certain other celestial bodies, that manifestly change their positions in space, and, among them, the sun and moon are most prominent. CHAP. VI.

In previous chapters, we have examined some facts concerning the sun and moon, which we briefly recapitulate, as follows: Recapitulation.

1. That the sun's distance from the earth is very great; but at present we cannot determine how great, for the want of one element — its horizontal parallax.

2. Its magnitude is much greater than that of the earth.

3. The distance between the sun and earth is slightly variable; but it is regular in its variations, both in distance and in apparent angular motion.

4. The moon is comparatively very near the earth; its distance is variable, and its mean distance and amount of variations are known. It is smaller than the earth, although, to the mere vision, it appears as large as the sun.

The apparent motions of both sun and moon are always in one direction; and the variations of their motions are never far above or below the mean.

But there are several other bodies that are not fixed stars; Other celestial bodies.

CHAP. VI. and although not as conspicuous as the sun and moon, have been known from time immemorial.

They appear to belong to one family; but, before the true system of the world was discovered, it was impossible to give any rational theory concerning their motions, so irregular and erratic did they appear; and this very irregularity of their apparent motions induced us to delay our investigations concerning them to the present chapter.

The planets.—Venus. In general terms, these bodies are called planets — and there are several of recent discovery — and some of very recent discovery; but as these are not conspicuous, nor well known, all our investigations of principles will refer to the larger planets, Venus, Mars, Jupiter, and Saturn. We now commence giving some *observed facts*, as extracted from the Cambridge astronomy.

The morning and evening star. (87.) "There are few who have not observed a beautiful star in the west, a little after sunset, and called, for this reason, the *evening-star*. This star is Venus. If we observe it for several days, we find that it does not remain constantly at the same distance from the sun. It departs to a certain distance, which is about 45°, or $\frac{1}{4}$th of the celestial hemisphere, after which it begins to return; and as we can ordinarily discern it with the naked eye only when the sun is below the horizon, it is visible only for a certain time immediately after sunset. By and by it sets with the sun, and then we are entirely prevented from seeing it by the sun's light. But after a few days, we perceive, in the morning, near the eastern horizon, a bright star which was not visible before. It is seen at first only a few minutes before sunrise, and is hence called the *morning star*. It departs from the sun from day to day, and precedes its rising more and more; but after departing to about 45°, it begins to return, and rises later each day; at length it rises with the sun, and we cease to distinguish it. In a few days the evening star again appears in the west, very near the sun; from which it departs in the same manner as before; again returns; disappears for a short time; and then the morning star presents itself.

These alternations, observed without interruption for more

than 2000 years, evidently indicate that the evening and morning star are one and the same body. They indicate, also, that this star has a proper motion, in virtue of which it oscillates about the sun, sometimes preceding and sometimes following it.

These are the phenomena exhibited to the naked eye; but the admirable invention of the telescope enables us to carry our observations much farther."

The phases of Venus.

(88.) On observing Venus with a telescope, the *irradiation* is, in a great measure, taken away, and we perceive that it has *phases, like the moon.* At evening, when approaching the sun, it presents a luminous crescent, the points of which are from the sun. The crescent diminishes as the planet draws nearer the sun; but after it has passed the sun, and appears on the other side, the crescent is turned in the other direction; the enlightened part always toward the sun, showing that it receives its light from that great luminary. The crescent now gradually increases to a semicircle, and finally to a full circle, as the planet again approaches the sun; *but, as the crescent increases, the apparent diameter of the planet diminishes;* and at every alternate approach of the planet to the sun, the phase of the planet is full, and the apparent diameter small; and at the other approaches to the sun, the crescent diminishes down to zero, and the apparent diameter increases to its maximum. When very near the sun, however, the planet is lost in the sunlight; but at some of these intervals, between disappearing in the evening, and reappearing in the morning, it appears to run over the sun's disc as a *round, black spot;* giving a fine opportunity to measure its greatest apparent diameter.* When Venus appears full, its apparent diameter is not more than 10″, and when a black spot on the sun, it is 59″.8, or very nearly 1′. Hence its greatest distance must be, to its least distance, as 59″.8 to 10, or nearly as 6 to 1.

The phases of Venus and its apparent diameter have corresponding changes.

* Astronomers do not measure the apparent diameters of the planets by the process described for the sun and moon, because they pass the meridian too quickly. Most of them will pass the meridian in a small fraction of a second. They use

CHAP. VI.

(89.) When we come to form *a theory* concerning the real motion of this planet, we must pay particular attention to the fact, that it is always in the same part of the heavens as the sun—never departing more than 47° on each side of it—called its greatest *elongation*. In consequence of being always in the neighborhood of the sun, it can never come to the meridian near midnight. Indeed, it always comes to the meridian *within* three hours 20 minutes of the sun, and, of course, in daylight. But this does not prevent meridian observations being taken upon it, through a good telescope;*

Venus always near the sun.

Greatest elongation.

a *micrometer*, which is two spider lines, always parallel, near the focus of a telescope, and so attached, by the *mechanism* of screws, as to open and close at pleasure.

To understand its grade of adjustment, bring the two lines together, so as to form one line. Then take any object, whose angular diameter is known at that time, as the diameter of the sun, and open the lines so as just to take in its disc, counting the turns, and parts of a turn given to the index screw to open to this object. From this we can compute the angle corresponding to *one turn*, or to any part of a turn, of the *index screw*.

Now if we wish to measure the apparent diameter of any planet, bring the lines together, and then open them, just to inclose the planet; and the number of turns, or the part of a turn, given to the screw, will determine the result.

This may not be the exact mechanism of every micrometer, but this is the principle of their construction.

* Perhaps we ought to have informed the reader before, "that the stars continue visible through telescopes, during the day, as well as the night; and that, in proportion to the power of the instrument, not only the largest and brightest of them, but even those of inferior luster, such as scarcely strike the eye, at night, as at all conspicuous, are readily found and followed even at noonday,—unless in that part of the sky which is very near the sun,—by those who possess the means of pointing a telescope accurately to the proper places. Indeed, from the bottoms of deep narrow pits, such as a well, or the shaft of a mine, such bright stars as pass the zenith may even be discerned by the naked eye; and we have ourselves heard it stated by a celebrated optician, that the

and, as to this particular planet, it is sometimes so bright as to be seen by the unassisted eye in the daytime. CHAP. VI.

Motion of Venus in respect to the stars.

(90.) Even without instruments and meridian observations, the attentive observer can determine that the motion of Venus, in relation to the stars, is very irregular — sometimes its motion is rapid — sometimes slow — sometimes *direct* — sometimes *stationary*, and sometimes *retrograde*;* but the direct motion prevails, and, as an attendant to the sun, and in its own irregular manner, as just described, it appears to traverse round and round among the stars.

Mercury similar in all appearances to Venus.

(91.) But Venus is not the only planet that exhibits the appearances we have just described. There is one other, and only one — *Mercury;* a very small planet, rarely visible to the naked eye, and not known to the very ancient astronomers. Whatever description we have given of Venus applies to Mercury, except in degree. Its variations of apparent diameter are not so great, and it never departs so far from the sun; and the interval of time, between its vibrations from one side to the other of the sun, is much less than that of Venus.

A conclusion.

(92.) *These appearances clearly indicate that the sun must be the center, or near the center, of these motions, and not the earth; and that Mercury must revolve in an orbit within that of Venus.*

So clear and so unavoidable were these inferences, that even the ancients (who were the most determined advocates for the immobility of the earth, and for considering it as the principal object in creation — the center of all motion, etc.) were compelled to admit them; but with this admission, they contended, that the sun moved round the earth, carrying these planets as attendants.

The apparent diame-

(93.) By taking observations on the other planets, the ancient astronomers found them variable in their apparent diam-

earliest circumstance which drew his attention to astronomy, was the regular appearance, at a certain hour, for several successive days, of a considerable star, through the shaft of a chimney."—*Herschel's Astronomy.*

* In astronomy, direct motion is eastward among the stars; *stationary* is no apparent motion, in respect to the stars; and retrograde is a westward motion.

CHAP. VI. ters of the planets are variable.

eters, and angular motions; *so much so, that it was impossible to reconcile appearances with the idea of a stationary point of observation;* unless the appearances were taken for realities, and that was against all true notions of philosophy.

The planet *Mars* is most remarkable for its variations; and the great distinction between this planet and Venus, is, that it does not always accompany the sun; but it sometimes, yea, at regular periods, is in the opposite part of the heavens from the sun — called Opposition — at which time it rises about sunset, and comes to the meridian about midnight.

The earth not the center of its motion.

The greatest apparent diameter of Mars takes place when the planet is in opposition to the sun, and it is then 17″.1; and its least apparent diameter takes place when in the neighborhood of the sun, and it is then but about 4″; *showing that the sun, and not the earth, is the center of its motion.*

Systematic irregularities

The general motion of all the planets, in respect to the stars, is *direct;* that is, eastward; but all the planets that attain opposition to the sun, while in opposition, and for some time before and after opposition, have a retrograde motion — and those planets which show the greatest change in apparent diameter, show also the greatest amount of retrograde motion — and all the observed irregularities are systematic in their irregularities, showing that they are governed, at least, by constant and invariable laws. If the earth is really stationary, we cannot account for this retrograde motion of the planets, unless that motion is real; and if real, why, and how can it change from direct to stationary, and from stationary to retrograde, and the reverse?

Retrograde motion of the planets accounted for.

But if we conceive the earth in motion, and going the same way with the planet, and moving more rapidly than the planet, then the planet will appear to run back; that is, retrograde.

And as this retrogradation takes place with every planet, when the earth and planet are both on the same side of the sun, and the planet in opposition to the sun; and as these circumstances take place in all positions from the sun, it is a sufficient explanation of these appearances; and conversely, then, these appearances show the motion of the earth.

(94.) When a planet appears stationary, it must be really

so, or be moving directly to or from the observer. And if it be moving to or from the observer, that circumstance will be indicated by the change in apparent diameter; and observations confirm this, and show that no planet is really stationary, although it may appear to be so.

CHAP VI.

Planets never stationary.

(95.) If we suppose the earth to be but one of a family of bodies, called planets — all circulating round the sun at different times — in the order of *Mercury, Venus, Earth, Mars* (omitting the small telescopic planets), Jupiter, Saturn, Herschel, or *Uranus*, we can then give a rational and simple account for every appearance observed, and without discussing the ancient objections to the true theory of the solar system, we shall adopt it at once, and thereby save time and labor, and introduce the reader into simplicity and truth.

The earth a planet.

(96.) The true solar system, as now known and acknowledged, is called the Copernican system, from its discoverer, Copernicus, a native of Prussia, who lived some time in the fifteenth century.

Copernicus and the Copernican system.

But this theory, simple and rational as it now appears, and capable of solving every difficulty, was not immediately adopted; for men had always regarded the earth as the chief object in God's creation; and consequently man, the lord of creation, a most important being. But when the earth was hurled from its imaginary, dignified position, to a more humble place, it was feared that the dignity and vain pride of man must fall with it; and it is probable that this was the root of the opposition to the theory.

Lost and revived.

So violent was the opposition to this theory, and so odious would any one have been who had dared to adopt it, that it appears to have been abandoned for more than one hundred years, and was revived by Galileo about the year 1620, who, to avoid persecution, presented his views under the garb of a dialogue between three fictitious persons, and the points left undecided.

Galileo and his dialogue

But the caution of Galileo was not sufficient, or his dialogue was too convincing, for it woke up the sacred guardians of truth, and he was forced to sign a paper denouncing the theory as heresy, on the pain of perpetual imprisonment.

CHAP. VI. But this is a digression. With the history of astronomy, as interesting as it may be, we design to have little to do, and to proceed only with the science itself.

CHAPTER VII.

FIRST APPROXIMATIONS TO THE RELATIVE DISTANCES OF THE PLANETS FROM THE SUN. HOW THE RESULTS ARE OBTAINED.

(97.) BEING convinced of the truth of the Copernican system, the next step seems to be, to find the periodical times of the revolutions of the planets, and at least their relative distances from the sun.

Distinction between inferior and superior planets.

Mercury and *Venus*, never coming in opposition to the sun, but revolving around that body in orbits that are within that of the earth, are called *inferior* planets.

Those that come in opposition, and thereby show that their orbits are outside of the earth, are called *superior* planets.

We shall show how to investigate and determine the position of one inferior planet; and the same principles will be sufficient to determine the position of any inferior planet.

It will be sufficient, also, to investigate and determine the orbit of one superior planet; and if that is understood, it may be considered as substantially determining the orbits of all the superior planets; and after that, it will be sufficient to state results.

For materials to operate with, we give the following table of the planetary irregularities (so called) drawn from observation:

Planets.	Greatest Apparent Diameters.	Least Apparent Diameters.	Angular Dist. from Sun at the instant of being stationary.	Mean arc of Retrogradation.
	″	″	° ′	° ′
Mercury.	11.3	5.0	18 00	13 30
Venus.	59.6	9.6	28 48	16 12
Earth.				
Mars.	17.1	3.6	136 48	16 12
Jupiter.	44.5	30.1	115 12	9 54
Saturn.	20.1	16.3	108 54	6 18
Uranus.	4.1	3.7	103 30	3 36

Planets.	Mean Duration of the Retrograde Motion.	Mean Duration of the Synodic Revolution, or interval between two successive oppositions.
Mercury.	23 days.	118 days.
Venus.	42 "	584 "
Earth.		
Mars.	73 "	780 "
Jupiter.	121 "	399 "
Saturn.	139 "	378 "
Uranus.	151 "	370 "

In the preceding table, the word *mean* is used at the head of several columns, because these elements are variable — sometimes more and sometimes less, than the numbers here given — which indicates that the planets do not revolve in circles round the sun, but most probably in ellipses, like the orbit of the earth.

Why the word MEAN should be used.

On the supposition, however, that the planets revolve in circles (which is not far from the truth), the greatest and least apparent diameters furnish us with sufficient data to compute the distances of the planets from the sun in *relation* to the distance of the earth, taken as *unity*.

(98.) In addition to the facts presented in the preceding table, we must not fail to note the important element of the *elongations* of *Mercury* and *Venus*. This term can be applied to no other planets.

The elongations of Mercury and Venus.

It is very variable in regard to Mercury — showing that the orbit of that planet is quite elliptical. The variation is much less in regard to *Venus*, showing that Venus moves round the sun more nearly in a circle.

This element variable and what it shows.

The least extreme elongation of Mercury is - 17° 37'.
The greatest " " " is - 28° 4'.
The mean (or the greatest elongation when both the earth and planet are at their mean distances from the sun) is - - - 22° 46'.
The least extreme elongation of Venus is - 44° 58'.
The greatest " " " is - 47° 30'.
The mean (or at mean distances), is - 46° 30'.

The least extremes must happen when the planet is in its perigee and the earth in its apogee, and the greatest when the earth is in perigee and the planet in apogee; but it is

CHAP. VII.

very seldom that these two circumstances take place at the same time.

How to find the comparative magnitudes of the orbits of Mercury, Venus, and the earth

Relying on these facts as established by observations, we can easily deduce the relative orbits of Mercury and Venus.

Fig. 23.

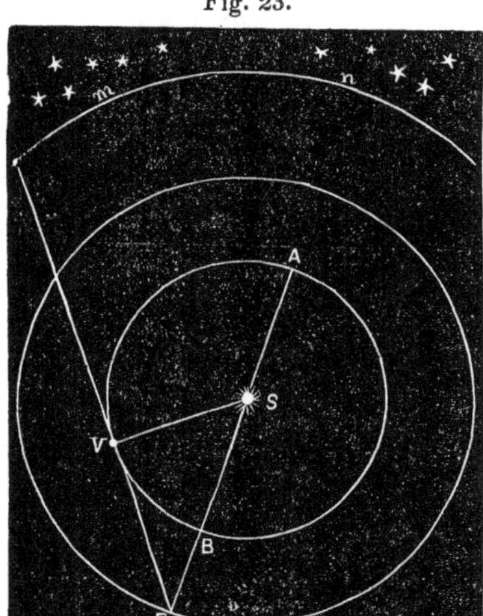

Let *S* (Fig. 23) represent the sun, *E* the earth, *V* Venus.

Conceive the planet to pass round the sun in the direction of *A V B*.

The earth moves also in the same direction, but not so rapidly as Venus.

Now it is clearly evident, from inspection, that when the planet is passing by the earth, as at *B*, it will appear to pass along in the heavens in the direction of *m* to *n*. But when the planet is passing along in its orbit, at *A*, and the earth about the position of *E*, the planet will appear to pass in the direction of *n* to *m*. When the planet is at *V*, as represented in the figure, its absolute motion is nearly toward the earth, and, of course, its appearance is nearly stationary.

What to understand by stationary.

It is *absolutely stationary* only at one point, and even then but for a moment; and that point is where its apparent motion changes from direct to retrograde, and from retrograde to direct; which takes place when the angle *S E V* is about 29 degrees on each side of the line *S E*.

When the line *E V* touches the circumference *A V B*, the angle *S E V*, or *angle of elongation*, is then greatest; and the triangle *S E V* is right angled at *V*; and if *S E* is made radius, *S V* will be the sine of the angle *S E V*.

But the line *S E* is assumed equal to *unity*, and then *S V* will be the natural sine of 46° 20′, and can be taken out of any table of natural sines; or it can be computed by logarithms, and the result is .72336.

For the planet Mercury, the mean of the same angle is 22° 46′; and the natural sine of that angle, or the mean radius of the planet's orbit, is .38698.

Thus we have found the relative mean distances of three planets from the sun, to stand as follows:

Mercury, - - - - - -	0.38698
Venus, - - - - - - -	0.72336
Earth, - - - - - -	1.00000

(99.) If the orbits were perfect circles, then the angle *S E V*, of greatest elongation, would always be the same; but it is an *observed fact* that it is not always the same; therefore the orbits are not circles; and when *S V* is least, and *S E* greatest, then the angle of elongation is *least*; and conversely, when *S V* is greatest and *S E* least, then the angle of elongation is the greatest possible; and by observing in what parts of the heavens the greatest and least elongations take place, we can approximate to the positions of the longer axis of the orbits.

The orbits of Mercury and Venus not circles.

(100.) By means of the apparent diameters, we can also find the approximate relations of their orbits. For instance, when the planet Venus is at *B*, and appears on the sun's disc, its apparent diameter is 59″.6; and when it is at *A*, or as near *A* as can be seen by a telescope, its apparent diameter is 9″.6. Now put

Computation of orbits from apparent diameters

$$SB=x; \quad \text{then} \quad EB=1-x, \quad \text{and} \quad AE=1+x.$$

By Art. 66, $1-x \ : \ 1+x \ :: \ 96 \ : \ 596$;

Hence, - - - - $x=0.72254$.

By a like computation, the mean distance of Mercury from the sun is 0.3864.

(101.) To determine the mean *relative* distances of the superior planets from the sun, we proceed as follows:

Let *S* (Fig. 24) represent the sun, *E* the earth, and *M* one of the superior planets, say *Mars*. It is easy to decide, from observation, when the planet is in opposition to the sun.

CHAP. VII.

Method of approximating to the orbits of the superior planets.

Fig. 24.

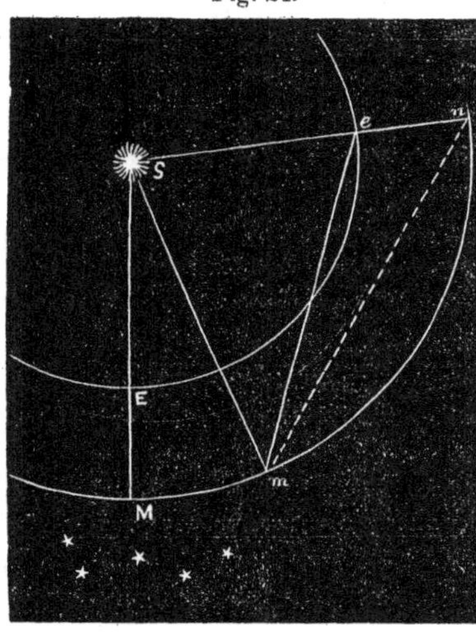

This gives the position of S, E, and M, in one right line, in respect to longitude. Now by knowing the true angular motion of the earth about the sun (73), and the mean angular motion of the planet,* we can determine the angle mSe, corresponding to any *definite future time;* for, by the motion of the earth round the sun, we can determine the angle ESe; and by the motion of the planet in the same time, we can determine the angle MSm; and the dif-

---

The relative distance of a planet from the sun determined by the variation in its apparent diameter

By means of apparent diameters, we can determine the values of the orbit. When the planet is in opposition to the sun, at E (Fig. 24), measure its apparent diameter; and, after a definite time, when the earth is at e, measure the apparent diameter again, and observe the angle Sem. Produce Se to n. Then, by the apparent diameters, we have the proportion of em and en (en is the same as EM, brought to this position); and in the triangle emn we have the proportion between the two sides and the included angle men. These are sufficient data to determine the angles enm and emn; and their difference is the angle Sme. Now we can determine the side Sm, of the triangle Sme, and the triangle Sem is completely known. Subtract the angle eSm from the whole angle eSM, and the angle MSm is left. That is, while the earth is describing the angle ESe, the planet describes the angle MSm. Put P for the periodical revo-

---

* Here we anticipate a little; for we have not shown how to determine the periodical time of revolution from observation: but this is shown in a future chapter, and in the above text note.

ference of these two angles is the angle $m\,S\,e$. By direct observation at e, we determine the angle $S\,e\,m$; and two angles, and the side $S\,e$, of the triangle $S\,m\,e$, are sufficient to determine the side Sm, the value sought. The triangle gives the following proportion:

CHAP. VII.

$$\sin.\ S\,m\,e \ :\ 1\ ::\ \sin.\ S\,e\,m\ :\ Sm=\frac{\sin.\ S\,e\,m}{\sin.\ S\,m\,e}.$$

This is a general solution, for any superior planet; but the result is only approximate; for, until we know the eccentricity of the orbit in question, and the part of the orbit in which the planet then is, we cannot accurately know the angle $M\,S\,m$.

Why the result is approximate.

---

lution of the planet; then, on the supposition of uniform motion, we have

$$\text{arc } MSm\ :\ \text{arc } ESe\ ::\ 365\tfrac{1}{4}\ :\ P$$

In this proportion the two arcs are known, and from thence P becomes known; *and thus, we perceive, that by the variations of the apparent diameter of a planet, we can determine its relative distance from the sun, and its periodical revolution.*

We give the following hypothetical example, for the purpose of further illustration.

A problem

The apparent diameter of Mars, when in opposition to the sun, was observed to be 17″.1. *One hundred and eleven days afterward, when the earth had passed over* 110° *of its orbit, the apparent diameter of Mars was again observed, and found to be* 7″.4, *and its angular position, in longitude, was* 87° *from the sun. From these data, it is required to find the relative approximate distance of the planet from the sun, and the approximate time of its revolution round the sun.*

Its solution. — Fig. 24.

From these data we have the angle $M\,S\,n=110°$, $S\,e\,m=87°$; therefore $n\,e\,m=93°$.

By the observed apparent diameter, we have $E\,M$ to $e\,m$ as 7″.4 to 17″.1; but $E\,M=e\,n$, therefore

$$en\ :\ em\ ::\ 74\ :\ 171.$$

In the triangle $n\,e\,m$ we can take $e\,n=74$, and $E\,m=171$, for the purpose merely of finding the angles. Then, by trigonometry, we have

CHAP. VII.

Results from variations in apparent diameters

(102.) By a perusal of the last text note, it will be seen, by those even who are not expert mathematicians, that it is not difficult to decide upon the relative distances of the planets from the sun, by observing their changes in apparent diameter, as seen from the earth. Such observations have been often made, and the following table shows the results; which are compared with the results deduced from Kepler's Third Law.*

Planets.	Deduced from apparent Diameters.	From Kepler's Law.	Difference or Error.
Mercury ...	0.386400	0.387098	—.000698
Venus	0.722540	0.723331	—.000791
Earth......	1.000000	1.000000	
Mars	1.533333	1.523692	+.009641
Jupiter.....	5.180777	5.202776	—.021999
Saturn	9.579000	9.538786	+.040214
Uranus.....	19.500000	19.182390	+.317610

Text note continued.

$171+74 : 171-74 :: \tan. \frac{87^\circ}{2} : \tan. \frac{1}{2}$, difference between the angle n and $n\,m\,e$.

That is, - $245 : 97 :: \tan. 43^\circ 30' : \tan \frac{1}{2} Sme$.

Whence, $Sme = 41^\circ 11'$. Now in the triangle Sme,

$\sin. 41^\circ 11' : 1 :: \sin. 87^\circ : Sm = 1.517$.

Secondly, as the angle $Sme = 41^\circ 11'$ and Sem 87°, therefore, - - $mSe = 51^\circ 49'$, and MSm $58^\circ 11'$.

But the times of revolution, between any two planets, must be inversely as the angles they describe in the same time; the *greater* the angle, the shorter the periodic time; and therefore if we put P to represent the periodical revolution of Mars, we shall have

$58\frac{2}{10} : 110 :: 365\frac{1}{4} : P$. Hence $P = 690\frac{2}{5}$ days.

The true time is 686.97964; showing an error of a little more than three days; but this is not a great error, considering the *remoteness* of the data, and the want of minuteness and unity in the *supposed observations.* Our object is only to teach principles; not, as yet, to establish minute results.

* A principle to be explained in Physical Astronomy.

The distances drawn from Kepler's law, are considered more accurate than conclusions drawn from most other considerations; and it is rather remarkable that these deductions from the apparent diameters agree as well as they do, owing to the difficulty of settling the exact apparent diameter, by observation. Take the apparent diameter of Uranus, for example, 3″.7 and 4″.1, and change either of them $\frac{1}{10}$ of a second, and it will make a great difference in the deduced result. CHAP. VII. Why the results from apparent diameters cannot be relied upon for accuracy.

CHAPTER VIII.

METHODS OF OBSERVING THE PERIODICAL REVOLUTIONS OF THE PLANETS, AND THEIR RELATIVE DISTANCES FROM THE SUN.

(103.) THE subject of this chapter will be to explain the principles of finding the periodical revolutions of the planets around the sun. If observers on the earth were at the center of motion, they could determine the times of revolution by simple observation. But as the earth is one of the planets, and all observers on its surface are carried with it, the observations here made must be subjected to mathematical corrections, to obtain true results; and this was an impossible problem to the ancients, as long as they contended for a stationary earth. CHAP. VIII. Why direct observations are not to the point.

If the observer could view the planets from the center of the sun, he would see them in their true places among the stars — and there are only two positions in which an observer on the earth will see a planet in the same place as though he viewed it from the center of the sun, and these positions are *conjunction* and *opposition*. Two important positions.

Thus, in Fig. 24, when the earth is at *E*, and a planet at *M*, the planet is in opposition to the sun; and it is seen projected among the stars at the same point, whether viewed from *S* or from *E*.

In Fig. 23, if the planet is at *B*, or *A*, it is said to be in conjunction with the sun; but a conjunction *cannot be ob-* Conjunctions cannot be observed.

CHAP. VIII.

served on account of the brilliancy of the sun, unless it be the two planets, Mercury and Venus, and then only when they pass directly before the face of the sun, and are projected on its surface as a *black spot*. *Such conjunctions are called transits.*

Revolution of inferior planets less, and of superior planets greater than a year.

(104.) All the planets move around the sun in the same direction, and not far from the same plane, and the rudest and most careless observations show that those planets nearest the sun, perform their revolutions in shorter periods than those more remote. From this, we decide at once that the mean angular motion of all the superior planets is less than the mean angular motion of the earth in its orbit; and the mean angular motion of the inferior planets, as seen from the sun, is greater than the mean motion of the earth.

Times of opposition can be observed

(105.) The time that any planet comes in opposition to the sun, can be very distinctly determined by observation. Its longitude is then 180 degrees from the longitude of the sun, and comes to the meridian nearly or exactly at midnight. If it is a little short of opposition at the time of one observation, and a little past at another, the observer can proportion to the exact time of opposition, and such time can be definitely recorded—and by such observation, we have the true position of the planet, as seen from the sun. Another opposition of the same kind and of the same planet, can be observed and recorded.

Fig. 25.

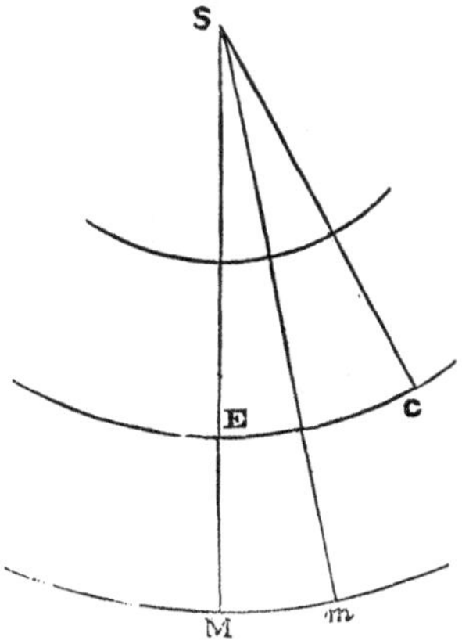

S
E
C
M
m

Synodical revolution.

The elapsed time between two such oppositions is called the synodical revolution of the planet.

Mean angular motion of the planets determined from their synodical revolutions.

We note the time that a planet is in opposition to the sun. Then *S*, *E* and *M* are in one plane as represented in Fig. 25. If the planet *M* should remain at rest while the earth *E* made its revolution, then the synodical revolution would be the same as the length of our year. But all the planets move in the same direction as the earth; and therefore the earth, after making a revolution, must pass onward and employ additional time to overtake the planet; and the more rapidly the planet moves, the longer time it will require. Hence, in case two planets have but a small difference in angular motion, their synodical period must be proportionately long. The planet Jupiter moves about 31° in its orbit in a year; and therefore, after one opposition, the earth is round to the same point in $365\frac{1}{4}$ days, and to gain the 31° requires about 32 days more; hence the synodical revolution of Jupiter must be about 397 days, by this *very rough* and imperfect computation. By inspecting the table on page 105, we perceive that the mean synodical revolution of Jupiter is 399 days, and this observed fact shows us that Jupiter passes over about 31° in a year, and of course its revolution must be a little less than 12 years; and by the same considerations, we can form a rough estimate of the periodical revolutions of all the planets.

General considerations.

(106.) The general principle being understood, we may now be more scientific. The mean motion of the earth in its orbit is very accurately known. Represent its daily motion by a. The angular motion of the planet (any superior planet that may be under consideration) is unknown; therefore, represent its daily motion by x. Let the angle $E\,SC$ represent a, and the angle $M\,S\,m$ represent x; then the angle $m\,SC$ or $(a—x)$ will represent the daily angular advance of the earth over the planet; and as many times as the angle $m\,SC$ is contained in 360°, will be the number of days in a synodical revolution. Therefore, $\dfrac{360}{a-x}=$ the observed time of a synodical revolution; and by taking the times from the table (page 105), we have the following equations:

Computation to determine the mean angular motion of the earth.

Mars.	Jupiter.	Saturn.	Uranus.
$\dfrac{360}{a-x}=780,$	$\dfrac{360}{a-x}=399,$	$\dfrac{360}{a-x}=378,$	$\dfrac{360}{a-x}=370.$*

* These equations correspond to the general equation $t=\dfrac{d}{a-x}$ in Robinson's Algebra, page 105, University edition.

CHAP. VIII.

The value of a is 59′ 8″, and then a solution of these several equations gives the mean angular motion, per day, of the several planets, as follows:

Mars.	Jupiter.	Saturn.	Uranus.
31′ 27″	4′ 59″.4	1′ 59″.5	45″.3

Times of revolution derived from the angular motion.

Dividing the whole circle 360° by the mean daily motion of each planet, will give their respective times of revolution, and the following are the results:

Mars.	Jupiter.	Saturn.	Uranus.
687 days.	4331 days.	10840 days.	28610 days.

(106.) For the inferior planets, *Mercury* and *Venus*, we have the same principle, only making x greater than a, and

For Mercury.	For Venus.
$\frac{360}{x-a}=118;$	$\frac{360}{x-a}=584.$
$x=4° 2' 11'';$	$x=1° 36' 7''.$

Mean angular motion of the inferior planets, and their revolution round the sun.

These diurnal angular motions correspond to 89 days for the revolution of Mercury, and 224.8 days for the revolution of Venus. All these results are, of course, understood as first approximations, and accuracy here is not attempted. We are only showing principles; and it will be noticed, that the times here taken in these considerations, are only to the nearest days, and not fractions of a day, as would be necessary for accurate results. By this method accuracy is never attempted, on account of the eccentricity of the orbits. No two synodical revolutions are exactly alike; and therefore it is very difficult to decide what the real mean values are.

(107.) To obtain accuracy, in astronomy, observations must be carried through a long series of years. The following is an example; and it will explain how accuracy can be attained in relation to any other planet.

On the 7th of November, 1631, M. Cassini observed Mercury passing over the sun; and from his observations then taken, deduced the time of conjunction to be at 7 h. 50 m., mean time, at Paris, and the true longitude of Mercury 44° 41′ 35″.

Observations carried through a long course

Comparing this occultation with that which took place in 1723, the true time of conjunction was November 9th, at 5 h. 29 m., P. M., and Mercury's longitude was 46° 47′ 20″.

The elapsed time was 92 years, 2 days, 9 h. 39 m. Twenty-two of these years were bissextile; therefore the elapsed time was (92×365) days, plus 24 d. 9 h. 39 m.

CHAP. VIII. of years, to secure accuracy.

In this interval, Mercury made 382 revolutions, and 2° 5′ 45″ over. That is, in 33604.402 days, Mercury described 137522.095826 degrees; and therefore, by *division*, we find that in one day it would describe 4°.0923, at a mean rate.

Thus, knowing the mean daily rate to great accuracy, the mean revolution, in time, must be expressed by the fraction $\frac{360}{4.0923}$; or, 87.9701 days, or 87 days 23 h. 15 m. 57 s.

Another method of observing the periodical revolutions of the planets.

(108.) The following is another method of observing the periodical times of the planets, *to which we call the student's special attention.*

The orbits of all the planets are a little inclined to the plane of the ecliptic.

The planes of all the planetary orbits pass through the center of the sun; the plane of the ecliptic is one of them, and therefore the plane of the ecliptic and the plane of any other planet must intersect each other by some line passing through the center of the sun. *The intersection of two planes is always a straight line.* (See Geometry.)

The reader must also recognize and acknowledge the following principle:

That a body cannot appear to be in the plane of an observer, unless it really is in that plane.

For example: an observer is always in the plane of his meridian, and no body can appear to be in that plane unless it really is in that plane; it cannot be projected in or out of that plane, by parallax or refraction.

Hence, when any one of the planets appears to be in the plane of the ecliptic, it actually is in that plane; and let the time be recorded when such a thing takes place.

What is meant by node.

The planet will immediately pass out of the plane, because the two planes do not coincide. Passing the plane of the ecliptic is called passing the *node.* Keep track of the planet until it comes into the same plane; that is, crosses the other *node:* in this interval of time the planet has described just

CHAP. VIII. 180°, as *seen from the sun* (unless the nodes themselves are in motion, which in fact they are; but such motion is not sensible for one or two revolutions of Venus or Mars).

Two nodes 180 degrees from each other, as seen from the sun.

Continue observations on the same planet. until it comes into the ecliptic the second time after the first observation, or to the same node again; and *the time elapsed, is the time of a revolution of that planet round the sun.* From such observations the periodical time of Venus became well known to astronomers, long before they had opportunities to decide it by comparing its transits across the sun's disc; and by thus knowing its periodical time and motion, they were enabled to calculate the times and circumstances of the transits which happened in 1761, and in 1769; save those resulting from parallax alone.

First idea of the perigee of the planets.

(109.) On comparing the time that a planet remains on each side of the ecliptic, we can form some idea of the position of its apogee and perigee. If it is observed to be on each side of the ecliptic the same length of time, then it is evident that the orbit of the planet is circular, or that its longer axis coincides with its nodes. If it is observed to be a shorter time north of the plane of the ecliptic than south of it, then it is evident that its perigee is north of the ecliptic; but nothing more definite can be drawn from this circumstance.

Final results.

(110.) Finally. By the combination of the different methods, explained in articles (98), (100), (101), (105), (107), and (108), and extending the observations through a long course of years, and from age to age, the times of revolution, the mean relative distances of the planets from the sun, were approximated to, step by step, until a great degree of exactness was attained, and the following were the results:

	Sidereal Revolution.	Mean distance from ☉
Mercury, - - -	87.969258	0.387098
Venus, - - -	224.700787	0.723332
Earth, - - -	365.256383	1.000000
Mars, - - -	686.979646	1.523692
Jupiter, - -	4332.584821	5.202776
Saturn, - - -	10759.219817	9.538786
Uranus, - -	30686.820830	19.182390

(111.) By inspecting the preceding table, we find that the greater the distance from the sun, the greater the time of revolution; but the *ratio* for the time is greater than the *ratio* corresponding to distance; yet we cannot doubt that some connection exists between these *ratios*. CHAP. VIII. Times of revolution and distances compared

For instance, let us compare the *Earth* with *Jupiter*. The *ratio* between their times of revolution, is *near* 12.

The *ratio* between their relative distances from the sun, as we perceive, is nearly 5.2.

The square of 12 is 144; the cube of 5.2 is near 141. But 12 is a little greater than the real ratio between the times of revolution, and 5.2 is not quite large enough for the ratio of distance; and by taking the correct *ratios*, they seem to bear the relation of *square* to *cube*.

Without a very rigid or close examination, we perceive that five revolutions of Jupiter are nearly equal to two revolutions of Saturn; that is, $\frac{5}{2}$ is nearly the ratio between their times of revolution.

By inspecting the column of distances, we perceive that the ratio of the distances of these two planets, is nearly $\frac{25}{52}$; and if we square the first ratio, and cube the second, we shall have nearly the same ratio.

Now let us compare two other planets, say *Venus* and *Mars*, more exactly. Result discovered.

Their ratio of revolution is	686.979 log. -	2.836948
	224.701 log. -	2.351601
Log. of the ratio,	- - -	0.485347
Multiply by	- - - -	2
Log. of the *square* of the ratio of time,		0.970694
Their ratio of distance is,	15.23692 log. -	1.182883
	7.23332 log. -	859323
Log. of the ratio,	- - -	0.323560
Multiply by	- - - - -	3
Log. of the *cube* of the ratio of distance,		0.970680

Thus we perceive that the squares of the times of revolution, are to each other as the cubes of the mean distances of

CHAP. VIII. Kepler's laws.

the planets from the sun,* and this is called *Kepler's third law*; and it was by such numerical comparisons that Kepler discovered the law.†

We may now recapitulate the three laws of the solar system, called Kepler's laws, as they were discovered by that philosopher.

1*st*. *The orbits of the planets are ellipses, of which the sun occupies one of the foci.*

2*d*. *The radius vector in each case describes areas about the focus, which are proportional to the times.*

3*d*. *The squares of the times of revolution are to each other as the cubes of the mean distances from the sun.*

---

* For a concise mathematical view of this subject, we give the following: Let d and D represent mean distances from the sun, and t and T the times of revolution. Then

$\frac{T}{t}=n$, $\frac{D}{d}=m$; n and m taken to represent the ratios. Square the 1st equation and cube the 2d. Then

$$\frac{T^2}{t^2}=n^2, \text{ and } \frac{D^3}{d^3}=m^3.$$

But by inspection we know that

$$n^2=m^3; \text{ therefore, } \frac{T^2}{t^2}=\frac{D^3}{d^3}, \text{ or, } t^2 : T^2 :: d^3 : D^3.$$

---

† It appears that Kepler did not compare ratios, as we have done; but took the more ponderous method of comparing the elements of the ratios (the numbers themselves); for, says the historian: — It was on the 8th of March, 1618, that it first came into Kepler's mind to compare the powers of the numbers which express their revolutions and distances; and by *chance* he compared the squares of the times with the cubes of the distances; but from too great anxiety and impatience, he made such *errors* in computation, that he rejected the hypothesis as false and useless; but on examining almost every other relation in vain, he returned to the same hypothesis, and on the 15th of May, of the same year, he renewed his calculation with complete success, and established this law, which has rendered his name immortal

CHAPTER IX.

TRANSITS OF VENUS AND MERCURY. — HOW SUN'S HORIZONTAL PARALLAX DEDUCED

CHAP. IX.

Attempts to find the sun's parallax.

(112.) WE have thus far been very patient in our investigations — groping along — finding the form of the planetary orbits, and their relative magnitudes; but, as yet, we know nothing of the distance to the sun; save the indefinite fact, that it must be very great, and its magnitude great; but how great we can never know, without the sun's parallax. Hence, to obtain this element, has always been an interesting problem to astronomers.

Difficulties of ancient astronomers.

The ancient astronomers had no instruments sufficiently refined to determine this parallax by direct observation, in the manner of finding that of the moon (Art. 60), and hence the ingenuity of men was called into exercise to find some artifice to obtain the desired result.

After Kepler's laws were established, and the relative distances of the planets made known, it was apparent that their *real* distance could be deduced, provided the distance between the earth and any planet could be made known.

Parallax of Mars.

(113.) The relative distances of the earth and Mars, from the sun (as determined by Kepler's law) are as 1 to 1.5237; and hence it follows that Mars, in its oppositions to the sun, is but about one half as far from the earth as the sun is; and therefore its parallax (Art. 60) must be about double that of the sun; and several partially successful attempts were made to obtain it by observation.

Maraldi obtains an approximation to the parallax of Mars.

On the 15th of August, 1719, Mars being very near its opposition to the sun, and very near a star of the 5th magnitude, its parallax became sensible; and Mr. Maraldi, an Italian astronomer, pronounced it to be 27″. The relative distance of Mars, at that time, was 1.37, as determined from its position and the eccentricity of its orbit.

But horizontal parallax is the angle under which the earth appears; and, at a greater distance, it will appear under a

9

CHAP. IX. less angle. The distance of Mars from the earth, at that time, was .37, and the distance of the sun was 1; therefore, 1 : .37 :: 27″ : 9″.99, or 10″, nearly, for the sun's horizontal parallax.

Observations by Wargentin and Lacaille. On the 6th of October, 1751, Mars was attentively observed by Wargentin and Lacaille (it being near its opposition to the sun), and they found its parallax to be 24″.6, from which they deduced the mean parallax of the sun, 10″.7. But at that time, if not at present, the parallax of Mars could not be *observed directly*, with sufficient accuracy to satisfy astronomers; for no observer could rely on an angular measure within 2″; for full that space was eclipsed by the micrometer wire.

Dr. Halley's suggestion. (114.) Not being satisfied with these results, Dr. Halley, an English astronomer, very happily conceived the idea of finding the sun's parallax by the comparisons of observations made from different parts of the earth, on a *transit of Venus* over the sun's disc. If the plane of the orbit of Venus coincided with the orbit of the earth, then Venus would come between the earth and sun, at every inferior conjunction, at intervals of 584.04 days. But the orbit of *Venus* is inclined to the orbit of the earth by an angle of 3° 23′ 28″; and, in the year 1800, the planet crossed the ecliptic from south to north, in longitude 74° 54′ 12″, and from north to south, in longitude 254° 54′ 12″: the first mentioned point is called the *ascending node*; the last, the *descending node*. The nodes retrograde 31′ 10″ in a century.

The nodes of Venus.

What times in the year transits may take place. (115.) The mean synodical revolution of 584 days corresponds with no aliquot part of a year; and therefore, in the course of time, these conjunctions will happen at different points along the ecliptic. The sun is in that part of the ecliptic near the *nodes* of Venus, June 5th and December 6th or 7th; and the two last transits happened in 1761 and in 1769; and from these periods we date our knowledge of the solar parallax.

Revolutions compared. (116.) The periodical revolution of the earth is 365.256383 days, and that of Venus is 224.700787; and as numbers they *are nearly* in proportion of 13 to 8.

From this it follows, that eight revolutions of the earth

require nearly the same time as 13 revolutions of Venus; and, of course, whenever a conjunction takes place, eight years afterward another conjunction will take place very near the same point in the ecliptic.*

Comparative motions of Venus and the earth.

* The ratio of the times of these revolutions is directly compared, as terms of a fraction, thus, $\frac{224.700787}{365.256381}$; and it is manifest that 365.256383 days, multiplied by the number 224700787, will give the same product as 224.700787 days multiplied by the number 365256383; that is, after an elapse of 224700787 years, the conjunction will take place at the same point in the heavens; and all intermediate conjunctions will be but approximations to the same point: and to obtain these approximate intervals, we reduce the above fraction to its approximating fractions, by the principle of continued fractions. (See Robinson's Arithmetic.)

The approximating fractions are

$$\frac{1}{1}, \quad \frac{1}{2}, \quad \frac{2}{3}, \quad \frac{3}{5}, \quad \frac{8}{13}, \quad \frac{235}{382}.$$

To say nothing of the first two terms, these fractions show that two revolutions of the earth are near, in length of time, to three revolutions of Venus; three revolutions of the earth a nearer value to five revolutions of Venus; and eight revolutions of the earth a still nearer value to 13 revolutions of Venus; and 235 revolutions of the earth a very near value to 382 revolutions of Venus.

The period of eight years, under favorable circumstances, will bring a second transit at the same node; but if not in eight years, it will be 235 years, or 235+8=243 years.

For a transit at the other node, we must take a period of 235—8 years, divided by 2, or 113 years; and sometimes the period will be eight years less than this, or 105 years. The first transit known to have been observed was in 1639, December 4th; to this add 235 years, and we have the time of the next transit, at the same node, 1874, December 8th; and eight years after that will be another, 1882, December 6th. The first transit observed at the ascending node, was

CHAP. IX.

Periods of conjunctions at the same time of the year.

If the proportion had been exactly as 13 to 8, then the conjunctions would always take place exactly at the same point; but, as it is, the points of conjunction in the heavens are east and west of a given point, and approximate nearer and nearer to that point as the periods are greater and greater.

Only two transits can happen at intervals of 8 years.

To be more practical, however, the intervals between conjunctions are such, combined with a *slight motion of the nodes*, that the geocentric latitude of Venus, at inferior conjunctions near the ascending node, changes about 19′ 30″ to the north, in the period of about eight years. At the descending node, it changes about the same quantity to the southward, in the same period; and as the disc of the sun is but little over 32′, it is impossible that a third transit should happen 16 years after the first; hence only two transits can happen, at the same node, separated by the short interval of eight years.

Periods between the transits of Venus.

(117.) If at any transit we suppose Venus to pass directly over the center of the sun, as seen from the center of the earth — that is, pass conjunction and node at the same time — at the end of another period of about eight years, Venus would be 19′ 30″ north or south of the sun's center; but as the semidiameter of the sun is but about 16′, no transit could happen in such a case; and there would be but one transit at that node until after the expiration of a long period of 235 or 243 years.

After passing the period of eight years, we take a lapse of 105 or 113 years, or thereabouts, to look for a transit at the other node.

Transits can be computed.

Dr. Halley shows how to find the sun's parallax.

(118.) Knowing the relative distances of Venus, and the earth, from the sun — the positions and eccentricities of both orbits—also their angular motions and periodical revolutions—every circumstance attending a transit, as seen from the earth's center, can be calculated; and Dr. Halley, in 1677, read a paper before the London Astronomical Society, in

---

Text note continued.

in 1761, June 5th; eight years after, 1769, June 3d, there was another; and the next that will occur, at that node, will be in 2004, June 7th, 235 years after 1769.

which he explained the manner of deducing the parallax of the sun, from observations taken on a transit of Venus or Mercury across the sun's disc, compared with computations made for the earth's center, or by comparing observations made on the earth at great distances from each other.

Why the transits of Venus are better adapted to give the solar parallax than those of Mercury.

The transits of Venus are much better, for this purpose, than those of Mercury; as Venus is larger, and nearer the earth, and its parallax at such times much greater than that of Mercury; and so important did it appear, to the learned world, to have correct observations on the last transit of Venus, in 1769, at remote stations, that the British, French, and Russian governments were induced to send out expeditions to various parts of the globe, to observe it. "The famous expedition of Captain Cook, to Otaheite, was one of them."

The result

(119.) The mean result of all the observations made on that memorable occasion, gave the sun's parallax, on the day of the transit (3d of June), 8''.5776. The horizontal parallax, at mean distance, may be taken at 8''.6; which places the sun, at its mean distance, no less than 23984 times the length of the earth's semidiameter, or about 95 millions of miles.

The importance of this problem.

This problem of the sun's horizontal parallax, as deduced from observations on a transit of Venus, we regard as the most important, for a student to understand, of any in astronomy; for without it, the dimensions of the solar system, and the magnitudes of the heavenly bodies, must be taken wholly on trust; and we have often protested against mere facts being taken for knowledge.

A general explanation.

(120.) We shall now attempt to explain this whole matter on general principles, avoiding all the little minutiæ which render the subject intricate and tedious; for our only object is to give a clear idea of the nature and philosophy of the problem.

Let S (Fig. 26) represent the sun, and mn and PQ small portions of the orbits of Venus and the earth.

As these two bodies move the same way, and nearly in the same plane, we may suppose the earth stationary, and Venus

CHAP. IX.

The case simplified.

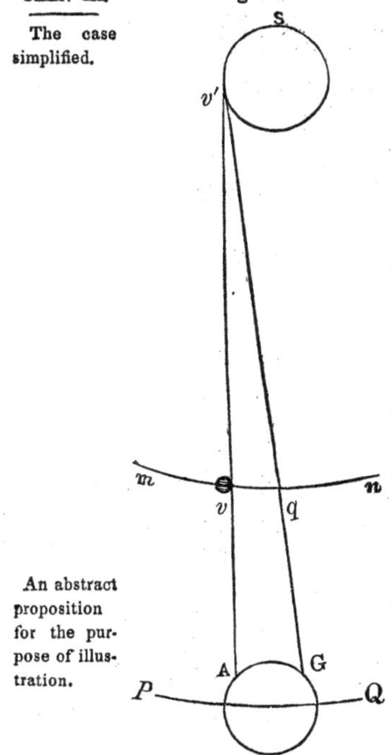

Fig. 26.

to move with an angular velocity, equal to the difference of the two.

When the planet arrives at v, an observer at A would see the planet projected on the sun, making a dent at v'.

But an observer at G would not see the same thing until after the planet had passed over the small arc $v\,q$, with a velocity equal to the difference between the angular motion of the two bodies; and as this will require quite an interval of absolute time, it can be detected; and it measures the angle $A\,v'\,G$; an angle under which a definite portion of the earth appears as seen from the sun.

An abstract proposition for the purpose of illustration.

(121.) To have a more definite idea of the practicability of this method, let us suppose the parallactic angle, $A\,v'\,G$, equal to 10″, and inquire how long Venus would be in passing the *relative* arc $v\,q$.

Venus, at its mean rate, passes	-	1° 36′ 8″ in a day.
The earth, " "		59′ 8″ "

The relative, or excess motion of Venus for a mean solar day is then 37′.

Now, as 37′ is to 24h. so is 10″ to a fourth term; or, as 2220″ : 1440m. :: 10″ : 6 m. 29 s.

Now if observation gave more than 6 minutes and 29 seconds, we shall conclude that the parallactic angle was more than 10″; if less, less. But this is an abstract proposition. When treating of an actual case in place of the mean motion, we must take the actual angular motions of the earth and Venus at that time, and we must know the actual position of the observers A and G in respect to each other, and the position of each in relation to a line joining the center of the earth and the center of the sun; and then by comparing the local time of observation made at A, with the time at *G*, and referring both to one and the same meridian, we shall have the interval of time occupied by the planet in passing from *v to q*, from which we deduce the parallactic angle *A v' G*, and from thence the horizontal parallax.

A combination of many observations

The same observations can be made when the planet passes off the sun, and a great many stations can be compared with *A*, as well as the station *G*. In this way, the mean result of a great many stations was found in 1761, and in 1769, and the mean of all cannot materially differ from the truth.

Another method of deducing the problem.

(122.) There is another method of considering this whole subject, which is in some respects more simple and preferable to the one just explained. It is for the observers at every station to keep the track of the transit all the way across the sun's disc, and take every precaution to measure the length of chord upon the disc, which can be done by carefully noting the times of external and internal contacts, and the beginning and end of the transit, and at short intervals carefully measuring the distance of the planet to the nearest edge of the sun by a micrometer.

Situation of different observers.

If the parallax is sensible, it is evident that two observers, situated in different hemispheres, will not obtain the same chord. For example, an observer in the northern hemisphere, as in Sweden or Norway, will see Venus traversing a more southern chord than an observer in the southern hemisphere.

Now if each observer gives us the length of the chord as observed by himself, and, knowing the angular diameter of the sun, we can compute the distance of each chord from the sun's center, and of course we then have the angular breadth of the zone on the sun's disc between them. But as this zone is formed by straight lines passing through the same point, the center of Venus, its *absolute breadth* will depend on its distance from the point *v*; that is, the two triangles *A B v* and *a b v* (Fig. 27) will be proportional, and we have

The result.

$$A\,v : a\,v :: A\,B : a\,b.$$

But the first three of these terms are known; therefore the fourth, *a b*, is known also; and if any definite angular space

CHAP. IX.

Fig. 27.

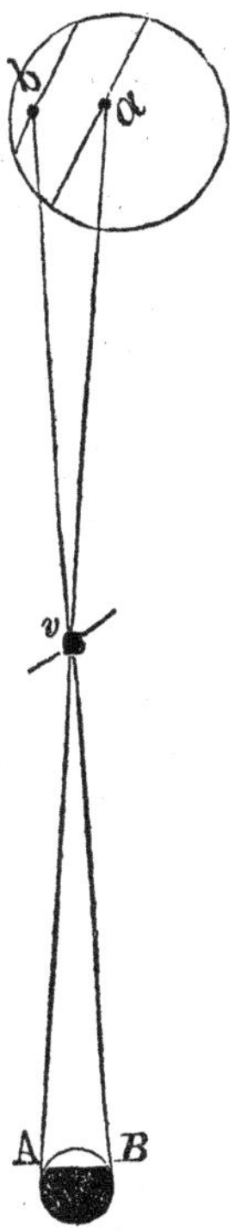

on the sun becomes known, the whole semidiameter becomes known, and from thence the horizontal parallax is immediately deduced.*

Under what circumstances this method should not be used.

(123.) The accuracy of this method should be questioned when Venus passes near the sun's center, for the two chords are never more than 30″ asunder, and hence they will not perceptibly differ in length when passing near the sun's center, and Venus will be upon the sun nearly the same length of time to all observers.

(124.) The apparent diameter of Mercury and Venus can be very accurately measured when passing the sun's disc. In 1769 the diameter of Venus was observed to be 59″.

Transits of Mercury not important.

(125.) The same general principles apply to the transits of Mercury and Venus; but those of Mercury are not important, on account of the smaller parallax and smaller size of that planet; but owing to the more rapid revolution of Mercury, its transits occur more frequently. The frequent appearance of this planet on the face of the sun, gives to astronomers fine opportunities to determine the position of its node and the inclination of its orbit.

Revolutions of Mercury and the earth compared.

In 1779, M. Delambre, from observations on the transit of May 7, placed the ascending node, as seen from the sun, in longitude 45° 57′ 3″. From the transit of the 8th of May, 1845, as observed at Cincinnati, it must have been in longitude 46° 31′ 10″; this gives it a progressive motion of about 1° 10′ in a century. The inclination of the orbit is 7° 0′ 13″. The periodical time of revolution is 87.96925 days; that of the earth is 365.25638 days, and by making a fraction of these numbers, and reducing as in the last text note, we find

* That is, as the real diameter of the sun, is to the real diameter of the earth, so is the sun's angular semidiameter to its horizontal parallax. (See 66).

that 6, 7, 13, 33, 46, 79, and 520 years, or revolutions of the earth *nearly* correspond to complete revolutions of Mercury. Hence we may look for a transit in 6, 7, 13, 33, 46, &c., years, or at the expiration of any combination of these years after any transit has been observed to take place; and by examining the following table, the years will be found to follow each other by some combination of these numbers.

CHAP. IX.

Intervals between transits.

The following is a list of all the transits of Mercury that have occurred, or will occur, between the years 1800 and 1900:

At the ascending node.		At the descending node.	
1802, - - -	Nov. 8.	1799, - - -	May 7.
1822, - - -	Nov. 4.	1832, - - -	May 5.
1835, - - -	Nov. 7.	1845, - - -	May 8.
1848, - - -	Nov. 9.	1878, - - -	May 6.
1861, - - -	Nov. 11.	1891, - - -	May 9.
1868, - - -	Nov. 4.		
1881, - - -	Nov. 7.		
1894, - - -	Nov. 10.		

CHAPTER X.

THE HORIZONTAL PARALLAXES OF THE PLANETS COMPUTED, AND FROM THENCE THEIR REAL DIAMETERS AND MAGNITUDES.

CHAP. X.

Real magnitudes and distances can now be determined

(126.) HAVING found the real distance to the sun, and the sun's horizontal parallax, we have now sufficient data to find the real distance, diameter, and magnitude, of every planet in the solar system.

In Art. 60 we have explained, or rather defined, the horizontal parallax of any body to be the angle under which the semidiameter of the earth appears, as seen from that body; and if the earth were as large as the body, the apparent diameter of the body, and its horizontal parallax, would have the same value. And, in general, the diameter of the earth is to the diameter of any other planetary body, as the horizontal parallax of that body is to its apparent semidiameter.

The mean horizontal parallax of the sun, as determined in

CHAP. X. the last chapter, is 8″.6; the semidiameter of the sun, at the corresponding mean distance, is 16′ 1″, or 961″. Now let d represent the real diameter of the earth, and D that of the sun, then we shall have the following proportion:

Real diameter of the sun determined.

$$d : D :: 8''.6 : 961''.0.$$

But d is 7912 miles; and the ratio of the last two terms is 111.66; therefore $D=(111.66)(7912)=883454$ miles.

Real distance between the earth and sun determined.

(127.) The sun's horizontal parallax is the angle at the base of a right angled triangle; and the side opposite to it is the radius of the earth (which, for the sake of convenience, we now call unity). Let x represent the radius of the earth's orbit; then, by trigonometry,

$$\sin. 8''.6 : 1 :: \sin. 90^\circ : x;$$

$$\text{Therefore, } x=\frac{\sin. 90^\circ}{\sin. 8''.6}=\log. 10.00000-\log. 5.620073.*$$

That is, the log. of $x=4.379927$, or $x=23984$; which is the distance between the earth and sun, when the semidiameter of the earth is taken for the *unit* of measure; but, for general reference, and to aid the memory, we may say the distance is 24000 times the earth's semidiameter.

(128.) Now let us *change the unit* from the semidiameter of the earth to an English mile; and then the distance between the earth and sun is

Distance in round numbers.

$$(3956)(23984)=94880706;$$

and, in round numbers, we say 95 millions of miles.

By Kepler's third law, we know the relative distances of

---

* Students generally would be unable to find the sine of 8″.6, or the sine of any other very small arc; for the directions given in common works of trigonometry are too gross, and, indeed, inaccurate, to meet the demands of astronomy.

On the principle that the sines of small arcs vary as the arcs themselves, we can find the sine of any small arc as follows:

Sine of 1′, taken from the tables, is - - - - -	6.463726
Divide by 60, that is, subtract the log. of 60, - -	1.778151
The sine of 1″, therefore, is - - - - - -	4.685575
Multiply by the number 8.6; that is, add log. -	0.934498
The sine of 8″.6, therefore, must be, - - - -	5.620073

In the same manner, find the sine of any other *small arc*.

all the planets from the sun; and now, having found the real distance of the earth, we may have the distance in miles, by multiplying the distance of the earth by the ratio corresponding to any other planet. Thus, for the distance of Venus, we multiply 94880706 by .72333; and the result is 68629960 miles, for the distance of Venus: and proceed, in the same manner, for the distance of any other planet.

CHAP. X.

How to find the distance of any planet from the sun in miles.

(129.) By observations taken on the transit of Venus, in 1769, it was concluded that the horizontal parallax of that planet was 30″.4; and its semidiameter, at the same time, was 29″.2. Hence (Art. 127), 304 : 292 : : 7912 : to a fourth term; which gives 7599 miles for the diameter of Venus.

To find the diameter of Venus.

(130.) We cannot observe the horizontal parallax of Jupiter, Saturn, or any other very remote planet: if known at all, it becomes known by computation; but the parallax can be known, when the *real* distance is known; and, by Kepler's third law, and the solar parallax, we do know all the planetary distances; and can, of course, compute any particular horizontal parallax.

Parallax of the planets cannot be observed.

For the horizontal parallax of Jupiter, when at a distance from the earth equal to its mean distance from the sun, we proceed as follows:

The parallax, or the semidiameter of the earth, when seen at the distance of the sun, is 8″.6. When seen from a greater distance, the angle would be *proportionally less.*

Put h equal to the horizontal parallax of Jupiter; then we have, - $5.202776 : 1 :: 8''.6 : h$; or $h = \frac{8''.6}{5.202776}$.

From this, we perceive, *that if we divide the sun's horizontal parallax by the ratio of a planet's distance from the sun, the quotient will be the horizontal parallax of the planet, when at a distance from the earth equal to its mean distance from the sun.*

How to compute the parallax of the planet.

(131.) To find the diameter of a planet, in relation to the diameter of the earth, we have a similar proportion as in Art. 126; and to find the diameter of Jupiter, we proceed as follows:

How to find the real diameters of the planets.

The greatest apparent diameter of Jupiter, as seen from

CHAP. X. the earth, is 44″.5; the least is 30″.1; therefore the mean, as seen from the sun, cannot be far from 37″.3, and the semidiameter 18″.65; La Place says it is 18″.35; and this value we shall use. Now, as in Art. 126, let $d=7912$, $D=$ the unknown diameter of Jupiter; $\frac{8''.6}{5.202776}$ is its horizontal parallax, and 18″.35 its corresponding semidiameter; then, as in Art. 126, - $7912. : D :: \frac{8.6}{5.202776} : 18.35$;

Therefore $D=\frac{7912\times 18.35\times 5.202776}{8.6}=7912\times 11.11=$ 87900 miles.

In the same manner, we may find the diameter of any other planet.

Jupiter not spherical. We have just seen that the diameter of Jupiter is 11.11 times the diameter of the earth; but this is the equatorial diameter of the planet. Its polar diameter is less, in the proportion of 167 to 177, as determined by the mean of many micrometrical measurements; which proportion gives 82930 miles, for the polar diameter of Jupiter. These extremes give the mean diameter of Jupiter, to the mean diameter of the earth, as 10.8 to 1.

How to find the magnitude of the planets. (132.) But the magnitudes of similar bodies are to one another as the cubes of their like dimensions; therefore the magnitude of Jupiter is to that of the earth, as $(10.8)^3$ to 1, and from thence we learn that Jupiter is 1260 times greater than the earth.

In the same manner we may find the magnitude of any other planet, and it is thus that their magnitudes have often been determined, and the results may be seen in a concise form in Table IV, which gives a summary view of the solar system.

The masses and attractions of the different planets will be investigated in physical astronomy, after we become acquainted with the theory of universal gravity.

CHAPTER XI.

A GENERAL DESCRIPTION OF THE PLANETS.

(133.) We conclude this section of astronomy by a brief description of the solar system, which we have purposely delayed lest we might interrupt the course of reasoning respecting the planetary motions. The reader is referred to Table IV, for a concise and comparative view of all the facts that can be numerically expressed; and aside from these facts, little can be said by way of explanation or description. Chap. XI.

The fact, that the sun or a planet revolves on an axis, must be determined by observing the motion of spots on the visible disc; and if no spots are visible, the fact of revolution cannot be ascertained.* But when spots are visible, their motion and apparent paths will not only point out the time of revolution, but the position of the axis. Facts revealed by spots on the sun or planets.

THE SUN.

(134.) The sun is the central body in the system, of immense magnitude, comparatively stationary, the dispenser of light and heat, and apparently the repository of that force which governs the motion of all other bodies in the system. The sun the repository of force.

"Spots on the sun seem first to have been observed in the year 1611, since which time they have constantly attracted attention, and have been the subject of investigation among astronomers. These spots change their appearance as the sun revolves on its axis, and become greater or less, to an observer on the earth, as they are turned to, or from him; they also change in respect to real magnitude and number; one spot, seen by Dr. Herschel, was estimated to be more than six times the size of our earth, being 50000 miles in diameter. Sometimes forty or fifty spots may be seen at the same time, and sometimes only one. They are often so large as to be seen with the naked eye; this was the case in 1816.

"In two instances, these spots have been seen to burst into several parts, and the parts to fly in several directions, like a piece of ice thrown upon the ground.

---

* Mercury is an exception to this principle.

CHAP. XI.

"In respect to the nature and design of these spots, almost every astronomer has formed a different theory. Some have supposed them to be solid opaque masses of scoriæ, floating in the liquid fire of the sun; others as satellites, revolving round him, and hiding his light from us; others as immense masses, which have fallen on his disc, and which are dark colored, because they have not yet become sufficiently heated.

"Dr. Herschel, from many observations with his great telescope, concludes, that the shining matter of the sun consists of a mass of phosphoric clouds, and that the spots on his surface are owing to disturbances in the equilibrium of this luminous matter, by which openings are made through it. There are, however, objections to this theory, as indeed there are to all the others, and at present it can only be said, that no satisfactory explanation of the cause of these spots has been given."

Singular means of discovering rotation.

(135.) *Mercury*. This planet is the nearest to the sun, and has been the subject of considerable remark in the preceding pages It is rarely visible, owing to its small size and proximity to the sun, and it never appears larger to the naked eye than a star of the fifth magnitude.

Mercury is too near the sun to admit of any observations on the spots on its surface; but its period of rotation has been determined by the variations in its *horns*—the same *ragged* corner comes round at regular intervals of time—24h. 5m.

Times when Mercury may be seen.

The best time to see Mercury, in the *evening*, is in the spring of the year, when the planet is at its greatest elongation *east* of the sun. It will then be visible to the naked eye about fifteen minutes, and will set about an hour and fifty minutes after the sun. When the planet is *west* of the sun, and at its greatest distance, it may be seen in the morning, most advantageously in August and September. The symbol for the greatest elongation of Mercury, as written in the common almanacs, is ☿ Gr. Elon.

High mountains on Venus.

(136.) *Venus*. This planet is second in order from the sun, and in relation to its position and motion, has been sufficiently described. The period of its rotation on its axis is 23h. 21m. The position of the axis is always the same, and is not at right angles to the plane of its orbit, which gives it a change of seasons. The tangent position of the sun's light across this

CHAP. XI.

Telescopic views of Venus.

planet shows a very rough surface; indeed, high mountains. By the radiating and glimmering nature of the light of this planet, we infer that it must have a deep and dense atmosphere.

The earth a planet.

(137.) *The Earth* is the next planet in the system; but it would be only formality to give any description of it in this place. As a planet, it seems to be highly favored above its neighboring planets, by being furnished with an attendant, the *moon*; and insignificant as this latter body is, compared to the whole solar system, it is the most important and interesting to the inhabitants of our earth. The two bodies, the *earth* and the *moon*, as seen from the sun, are very small: the former subtending an angle of about 17″ in diameter, the latter about 4″, and their distance asunder never greater than between seven and eight minutes of a degree.

The earth's attendant.

Contrary to the general impression, the moon's motion in absolute space is always concave toward the sun.*

Mars; his physical appearance, &c.

(138.) *Mars*—the first superior planet—is of a red color, and very variable in its apparent magnitude. About every

---

The moon's motion concave toward the sun.

* This may be shown thus—the moon is inside the earth's orbit from the last quarter to the first quarter, on an average 14 days and 18 hours. During this time the earth moves in its orbit 14° 30′. Let $L\,n\,F$ be a portion of the earth's orbit equal to 14° 30′, L the position of the earth at the *First Quarter* of the moon, and F its position at the *Last Quarter*. Draw the chord $L\,F$, and compute $m\,n$ the versed sine of the arc 7° 15′.

Fig. 28.

The mean radius of the earth's orbit is 397 times the radius of the lunar orbit. A radius of 397 and an angle 7° 15′ gives a versed sine of 3.49; but on this scale the distance from the earth to the moon is *unity*, or less than one third of $n\,m$; hence, the moon's path must be between the chord $L\,F$ and the arc $L\,n\,F$—that is, *always concave toward the sun.*

CHAP. XI. other year, when it comes to the meridian near midnight, it is then most conspicuous; and the next year it is scarcely noticed by the common observer.

Telescopic View of Mars.

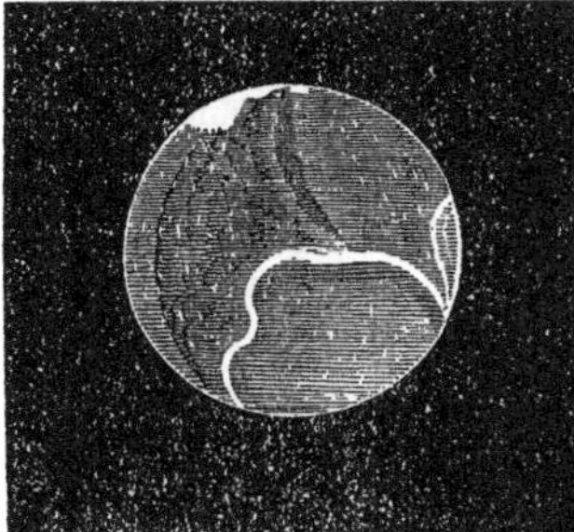

"The physical appearance of Mars is somewhat remarkable. His polar regions, when seen through a telescope, have a brilliancy so much greater than the rest of his disc, that there can be little doubt that, as with the earth so with this planet, accumulations of ice or snow take place during the winters of those regions. In 1781 the south polar spot was extremely bright; for a year it had not been exposed to the solar rays. The color of the planet most probably arises from a dense atmosphere which surrounds him, of the existence of which there is other proof depending on the appearance of stars as they approach him; they grow dim and are sometimes wholly extinguished as their rays pass through that medium."

Apparent imperfection in the system.

(139.) The next planet, as known to ancient astronomers, is Jupiter; but its distance is so great beyond the orbit of Mars, that the void space between the two had often been considered as an *imperfection*, and it was a general impression among astronomers that a *planet ought to occupy* that vacant space.

Bode's law.

Professor Bode, of Berlin, on comparing the relative distances of the planets from the sun, discovered the following remarkable fact—that if we take the following series of numbers:

0, 3, 6, 12, 24, 48, 96, 192, &c.,

and then add the number 4 to each, and we have,

4, 7, 10, 16, 28, 52, 100, 196, &c.,

The reason why it should not be called a law.

and this last series of numbers very nearly, *though not exactly*, corresponds to the relative distances of the planets from the sun, with the exception of the number 28. This is sometimes called Bode's law; but remarkable as it certainly is, it should not be dignified by the term *law*, until some better account of it can be given than its mere existence; for, at present, all that can be said of it is, "here is an astonishing coincidence." But, mere accident as it may be, it suggested the possibility of some small, undiscovered planet revolving in this region, and we can easily imagine the astonishment of astronomers, on finding *four* in place of *one*, revolving in orbits tolerably well corresponding to this law, or rather coincidence. Had they even found but *one*, it would seem to indicate something more than mere coincidence; but finding *four*, proves the series to be simply accidental — *unless* the *four* or more planets there discovered were originally *one* planet; and then came the inquiry, is not this the case? Thus originated the idea that these *new* and *newly* discovered small planets are but fragments of a larger one, which formerly circulated in that interval, and was blown to pieces by some internal explosion — and *we shall examine this hypothesis in a text note, under physical astronomy.*

CHAP. XI.

A bold hypothesis.

The names of these planets, in the order of the times of their discovery, are, *Ceres, Pallas, Juno, Vesta.* The order of their distances from the sun, is *Vesta, Juno, Ceres, Pallas.*

Planets.	Names of Discoverers.	Residence of Discoverers.	Date of Discovery.
Ceres ...	M. Piazzi,	Palermo, Sicily,	1st Jan., 1801.
Pallas...	Dr. Olbers,	Bremen, Germany,	28th Mar., 1802.
Juno ...	M. Harding,	Lilienthal, near Bremen,	1st Sept. 1804.
Vesta...	Dr. Olbers,	Bremen,	29th Mar., 1807.

If a planet has really burst, it is but reasonable to suppose that it separated into many fragments; and, agreeably to this view of the subject, astronomers have been constantly on the alert for new planets, in the same regions of space; and every discovery of the kind greatly increases the probability of the theory. The following very recent discoveries are said to have been made, but the elements of the orbits are not regarded as sufficiently accurate to demand a place in the table.

Recent discoveries favorable to this hypothesis.

On the 8th of December, 1845, Mr. Hencke, of Dreisen, claims to have discovered a planet which he calls *Astrea;* and the same observer also claims another, discovered in 1847, called *Hebe.* His success induced others to a like examination, and a Mr. Hind, of London, within the past year,

New planets discovered in 1845 and 1846.

CHAP. XI. 1848, claims a seventh and eighth asteroid, named *Iris* and *Flora*.

Thus we have eight miniature worlds, supposed to have once composed a planet; and if the four last named are veritable discoveries, we shall soon have the elements of their orbits in an unquestionable shape.

The elements of the orbits of the four known asteroids, as given for the epoch 1820, are not as accurate as the following, which were deduced from the Nautical Almanac for 1846 and 1847; which have been corrected from more modern, extended, and accurate observations. (Epoch Jan., 1847.)

On account of the small magnitude of these new planets, and their recent discovery, nothing is known of them save the following tabular facts, and these are only approximation to the truth.

Planets.	Sidereal Revolutions.	Mean Distance from the Sun.	Eccentricity of Orbits.
	Days.		
Vesta.......	1324. 289	2. 36120	0. 08913
Juno	1594. 721	2. 66514	0. 25385
Ceres.......	1683. 064	2. 76910	0. 07844
Pallas.......	1685. 162	2. 77125	0. 24050

Planets.	Longitude of Ascending Node.	Inclination of Orbits.	Longitude of Perihelion.
	° ′ ″	° ′ ″	° ′ ″
Vesta.......	103 20 47	7 8 29	251 4 34
Juno	170 53 0	13 2 53	54 18 32
Ceres.......	80 47 56	10 37 17	147 25 41
Pallas.......	172 42 38	34 37 42	121 20 13

Object of Fig. 29. (140.) With the two elements, the longitude of the ascending nodes, and the inclination of the orbits to the ecliptic, we are enabled to give a general projection of these orbits around the celestial sphere, in relation to the ecliptic, as represented on page 37; and our object is to show that there are two points in the heavens, nearly opposite to each other, near to which all these planets pass. One of these points is about the longitude of 185 degrees, and the latitude of 15 degrees north; and the other is the opposite point on the celestial sphere. If these planets are but fragments of one original planet, which burst or exploded by its internal fires, from that

Where the original planet must have exploded, if the hypothesis of an original planet is true

moment they must have started from the same point, *and the orbits of all have one common distance from the sun;* and for ages after such a catastrophe, these fragments must have had nearly a *common node;* and the fact that they do not, *at present*, pass through a common point, nor have a common node, does not prove that they were not originally in one body; for, owing to mutual disturbances, and the disturbances of other planets, the nodes must change positions; and the longer axis of the orbits, especially the very eccentric ones, must change positions; and now (after we know not how many ages), it is not inconsistent with the theory of an explosion, that we find the orbits as they are.

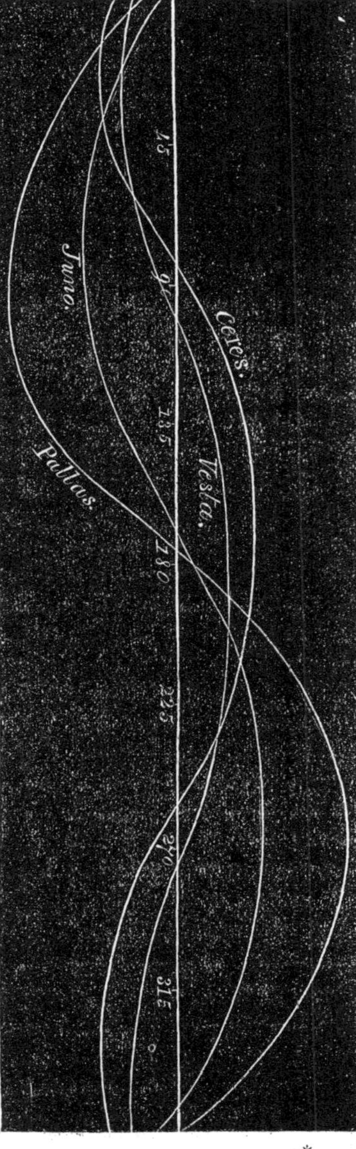

Fig. 29.

The hypothesis that these planets were originally one, and must, therefore, have two common points in the heavens near which they must all pass, led to the discovery of Juno and

 Vesta, by carefully observing these two portions of the heavens.

The apparent diameters of these planets are too small to be accurately measured; and therefore we have only a very rough or conjectural knowledge of their real diameters.

All of these planets are invisible to the naked eye, except Vesta, which sometimes can be seen as a star of the 5th or 6th magnitude.

(141.) *Jupiter.* We now come to the most magnificent planet in the system — the well-known Jupiter — which is nearly 1300 times the magnitude of the earth.

Jupiter's belts. The disc of Jupiter is always observed to be crossed, in an eastern and western direction, by dark bands, as represented in Fig. 30.

Fig. 30. — Telescopic View of Jupiter.

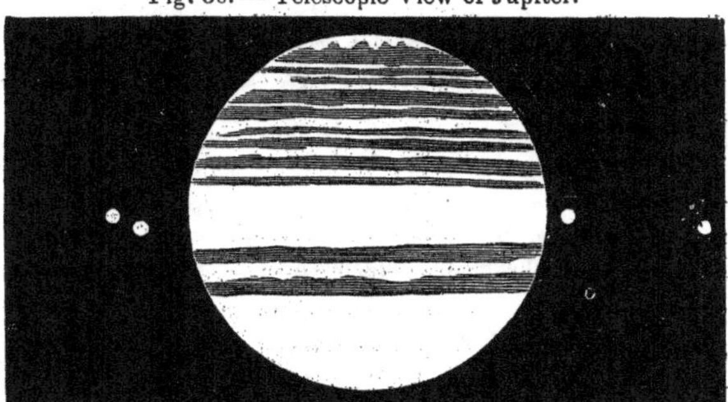

"These belts are, however, by no means alike at all times; they vary in breadth and in situation on the disc (though never in their general direction). They have even been seen broken up, and distributed over the whole face of the planet: but this phenomenon is extremely rare. Branches running out from them, and subdivisions, as represented in the figure, as well as evident dark spots, like strings of clouds, are by no means uncommon; and from these, attentively watched, it is concluded that this planet revolves in the surprisingly short period of 9 h. 55 m. 50 s. (sid. time), on an axis perpendicular to the direction of the belts. Now, it is very remarkable, and forms a most satisfactory comment on the reasoning by which the spheroidal figure of the earth has been deduced from its diurnal rotation, that the outline of Jupiter's disc is evidently not circular, but elliptic, being considerably flattened in the direction of its axis of rotation.

Diurnal revolution.

"The parallelism of the belts to the equator of Jupiter, their occasional variations, and the appearances of spots seen upon them, render it extremely probable that they subsist in the atmosphere of the planet, forming tracts of comparatively clear sky, determined by currents analogous to our tradewinds, but of a much more steady and decided character, as might indeed be expected from the immense velocity of its rotation. That it is the comparatively darker body of the planet which appears in the belts, is evident from this,—that they do not come up in all their strength to the edge of the disc, but fade away gradually before they reach it. Atmosphere of Jupiter.

(142.) "When Jupiter is viewed with a telescope, even of moderate power, it is seen accompanied by four small stars, nearly in a straight line parallel to the ecliptic. These always accompany the planet, and are called its *Satellites*. They are continually changing their positions with respect to one another, and to the planet, being sometimes all to the right, and sometimes all to the left; but more frequently some on each side. The greatest distances to which they recede from the planet, on each side, are different for the different satellites, and they are thus distinguished: that being called the ***First*** satellite, which recedes to the least distance; that the *Second*, which recedes to the next greater distance, and so on. The satellites of Jupiter were discovered by Galileo, in 1610. Jupiter's satellites.

"Sometimes a satellite is observed to pass between the sun and Jupiter, and to cast a shadow which describes a chord across the disc. This produces an eclipse of the sun, to Jupiter, analogous to those which the moon produces on the earth. It follows that Jupiter and its satellites are opake bodies, which shine by reflecting the sun's light.

"Careful and repeated observations show that the motions of the satellites are from west to east, in orbits nearly circular, and making small angles with the plane of Jupiter's orbit. Observations on the eclipses of the satellites make known their synodic revolutions, from which their sidereal revolutions are easily deduced. From measurements of the greatest apparent distances of the satellites from the planet, their real distances are determined.

"A comparison of the mean distances of the satellites, with their sidereal revolutions, proves that Kepler's third law, with respect to the planets, applies also to the satellites of Jupiter. The squares of their sidereal revolutions are as the cubes of their mean distances from the planet.

"The planets Saturn and Uranus are also attended by satellites, and the same law has place with them."

(143.) By the eclipses of Jupiter's satellites, the progressive nature of light was discovered; which we illustrate in the following manner: Progressive nature of light.

CHAP. XI.

Fig. 31.

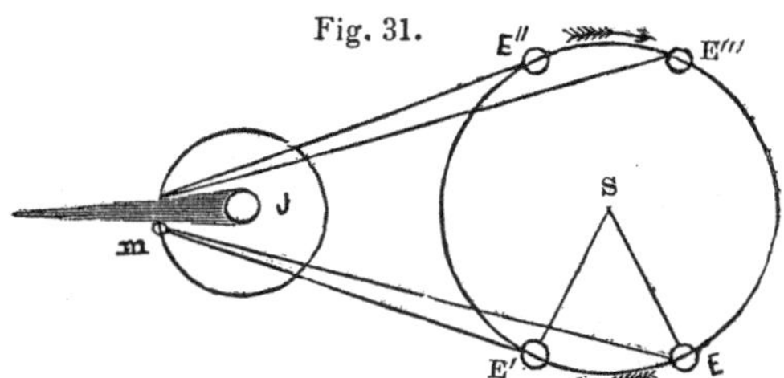

Let S (Fig. 31) represent the sun, J Jupiter, E earth, and m Jupiter's first satellite. By careful and accurate observations astronomers have decided that the mean revolution of this satellite round its primary, is performed in 42 h. 28 m. and 35 s.; that is, the mean time from one eclipse to another.

Velocity of light, how determined.

But when the earth is at E, and moving in a direction toward, or nearly toward, the planet as represented in the figure, the mean time between two consecutive eclipses is shortened about 15 seconds; and we can explain this on no other hypothesis than that the earth has advanced and met the successive progression of light. When the earth is in position as respects the sun and Jupiter, as represented in our figure at E'', and moving from Jupiter, then the interval between two consecutive eclipses of Jupiter's first satellite is prolonged or increased about 15 seconds.

But during the interval of one revolution of Jupiter's first satellite, the earth moves in its orbit about 2880000 miles; this, divided by 15, gives 192000 miles for the motion of light in one second of time; and this velocity will carry light from the sun to the earth in about eight and one-fourth minutes.

Longitude found by the eclipses of Jupiter's satellites.

(144.) As an eclipse of one of Jupiter's satellites may be seen from all places where the planet is there visible, two observers viewing it will have a signal for the same moment, at their respective places; and their difference in local time will give their difference in longitude. For example, if one observer saw one of these eclipses at 10 h. in the evening, and another at 8 h. 30 m., the difference of longitude between the observers would be 1 h. 30 m. in time, or 22° 30′ of arc.

The absolute time that the eclipse takes place, is the same to all observers; and he who has the latest local time is the most eastward.

These eclipses cannot be observed at sea, by reason of the motion of the vessel.

(145.) *Saturn.* The next planet in order of remoteness from the sun, is Saturn, the most wonderful object in the solar system. Though less than Jupiter, it is about 79000 miles in diameter, and 1000 times greater than our earth.

Saturn — his rings.

"This stupendous globe, besides being attended by no less than seven satellites, or moons, is surrounded with two broad, flat, extremely thin rings, concentric with the planet and with each other; both lying in one plane, and separated by a very narrow interval from each other throughout their whole circumference, as they are from the planet by a much wider. The dimensions of this extraordinary appendage are as follows:

Exterior diameter of exterior ring,................	=	176418.
Interior ditto,................................	=	155272.
Exterior diameter of interior ring,................	=	151690.
Interior ditto,................................	=	117339.
Equatorial diameter of the body,................	=	79160.
Interval between the planet and interior ring,.......	=	19090.
Interval of the rings	=	1791.
Thickness of the rings not exceeding,............	=	100.

Dimensions of the rings.

Fig. 32. — Telescopic View of Saturn.

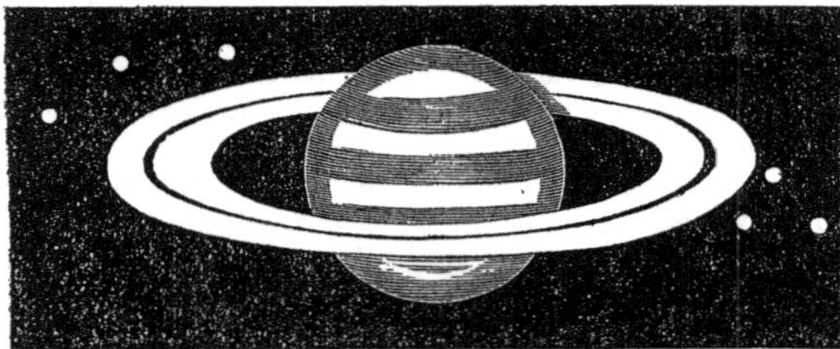

The rings are opake.

"The figure represents Saturn surrounded by its rings, and having its body striped with dark belts, somewhat similar, but broader and less strongly marked than those of Jupiter, and owing, doubtless, to a similar cause. That the ring is a solid opake substance, is shown by its throwing its shadow on the body of the planet, on the side nearest the sun, and on the other side receiving that of the body, as shown in the figure. From the parallelism of the belts with the plane of the ring, it may be conjectured that the axis of rotation of the planet is perpendicular to that plane; and this conjecture is confirmed by the occasional appearance of extensive dusky spots on its surface, which when watched, like the spots on Mars or Jupiter, indicate a rotation in 10 h. 29 m. 17 s. about an axis so situated.

"It will naturally be asked how so stupendous an arch, if composed of solid and ponderous materials, can be sustained without collapsing

CHAP. XI.

The stability of the rings.

and falling in upon the planet? The answer to this is to be found in a swift rotation of the ring in its own plane, which observation has detected, owing to some portions of the ring being a little less bright than others, and assigned its period at 10 h. 29 m. 17 s., which, from what we know of its dimensions, and of the force of gravity in the Saturnian system, is very nearly the periodic time of a satellite revolving at the same distance as the middle of its breadth. It is the centrifugal force, then, arising from this rotation, which sustains it; and, although no observation nice enough to exhibit a difference of periods between the outer and inner rings have hitherto been made, it is more than probable that such a difference does subsist as to place each independently of the other in a similar state of equilibrium.

The rings revolve around the planet like satellites.

"Although the rings are, as we have said, very nearly concentric with the body of Saturn, yet recent micrometrical measurements, of extreme delicacy, have demonstrated that the coincidence is not mathematically exact, but that the center of gravity of the rings oscillates round that of the body, describing a very minute orbit, probably under laws of much complexity. Trifling as this remark may appear, it is of the utmost importance to the stability of the system of the rings. Supposing them mathematically perfect in their circular form, and exactly concentric with the planet, it is demonstrable that they would form (in spite of their centrifugal force) a system in a state of *unstable equilibrium*, which the slightest external power would subvert — not by causing a rupture in the substance of the rings — but by precipitating them, *unbroken*, on the surface of the planet. For the attraction of such a ring or rings on a point or sphere eccentrically situate within them, is not the same in all directions, but tends to draw the point or sphere toward the nearest part of the ring, or away from the center. Hence, supposing the body to become, from any cause, ever so little eccentric to the ring, the tendency of their mutual gravity is, not to correct, but to increase this eccentricity, and to bring the nearest parts of them together."

Uranus alias Herschel.

(146.) *Uranus.* The next planet, beyond Saturn, was discovered by Sir W. F. Herschel, in 1781, and, for a time, was called Herschel, in honor of its discoverer; but, according to custom, the name of a heathen deity has been substituted, and the planet is now called Uranus — *the father of Saturn.*

This planet rarely visible to the naked eye.

This planet is rarely to be seen, without a telescope. In a clear night, and in the absence of the moon, when in a favorable position above the horizon, it may be seen as a star of about the 6th magnitude. Its real diameter is about 35000 miles, and about 80 times the magnitude of the earth.

The existence of this planet was suggested by some of the perturbations of Saturn; which could not be accounted for by the action of the then known planets; but it does not appear that any computations were made, as a guide to the place where the unknown disturbing body ought to exist; and, as far as we know, the discovery by Herschel was mere accident.

CHAP. XI.

Facts led to the discovery of Neptune.

But not so with the planet Neptune, discovered in the latter part of September, 1846, by a French astronomer, *Leverrier*; and also a Mr. Adams, of Cambridge, England, who has put in his claim as the discoverer. The truth is, that the attention of the astronomers of Europe had been called to some extraordinary perturbations of *Uranus*; which could not be accounted for without supposing an attracting body to be situated in space, beyond the orbit of Uranus; and so distinct and clear were these irregularities, that both geometers, Leverrier and Adams, fixed on the same region of the heavens, for the then position of their *hypothetical* planet; and by diligent search, the planet was actually discovered about the same time, in both France and England.

At present, we can know very little of this planet; and according to the best authority I can gather, its longitude, January 1, 1847, was 327° 24′. Mean distance from the sun, 30.2 (the earth's distance being unity); period of revolution 166 years. Eccentricity of orbit 0.0084; mass, $\frac{1}{23000}$.

According to Bode's law, the distance of the next planet from the sun, beyond Uranus, must be 38.8; and if Neptune really is at 30.2, it shows Bode's law to be only a remarkable coincidence; for there can be *no exceptions* to positive physical laws.

How to obtain a correct conception of the solar system

"We shall close this chapter with an illustration calculated to convey to the minds of our readers a general impression of the relative magnitudes and distances of the parts of our system. Choose any well-leveled field or bowling green. On it place a globe, two feet in diameter; this will represent the sun; Mercury will be represented by a grain of mustard seed, on the circumference of a circle 164 feet in diameter, for its orbit; Venus a pea, on a circle 284 feet in diameter; the earth

CHAP. XI. also a pea, on a circle of 430 feet; Mars a rather large pin's head, on a circle of 654 feet; Juno, Ceres, Vesta, and Pallas, grains of sand, in orbits of from 1000 to 1200 feet; Jupiter a moderate-sized orange, in a circle nearly half a mile across; Saturn a small orange, on a circle of four-fifths of a mile; and Uranus a full-sized cherry, or small plum, upon the circumference of a circle more than a mile and a half in diameter. As to getting correct notions on this subject by drawing circles on paper, or, still worse, from those very childish toys called orreries, it is out of the question. To imitate the motions of the planets in the above-mentioned orbits, Mercury must describe its own diameter in 41 seconds; Venus, in 4 m. 14 s.; the earth, in 7 minutes; Mars, in 4 m. 48 s.; Jupiter, in 2 h. 56 m.; Saturn, in 3 h. 13 m.; and Uranus, in 2 h. 16 m."—*Herschel's Astronomy.*

View of the planetary motions.

CHAPTER XII.

ON COMETS.

CHAP. XII.

Comets formerly inspired terror.

(147.) BESIDES the planets, and their satellites, there are great numbers of other bodies, which gradually come into view, increasing in brightness and velocity, until they attain a maximum, and then as gradually diminish, pass off, and are lost in the distance.

Knowledge banishes dread.

"*These bodies are comets.* From their singular and unusual appearance, they were for a long time objects of terror to mankind, and were regarded as harbingers of some great calamity.

"The luminous train which accompanied them was particularly alarming, and the more so in proportion to its length. It is but little more than half a century since these superstitious fears were dissipated by a sound philosophy; and comets, being now better understood, excite only the curiosity of astronomers and of mankind in general. These discoveries which give fortitude to the human mind are not among the least useful.

"It was formerly doubted whether comets belonged to the class of heavenly bodies, or were only meteors engendered fortuitously in the air by the inflammation of certain vapors. Before the invention of the telescope, there were no means of observing the progressive increase and diminution of their light. They were seen but for a short time, and their appearance and disappearance took place suddenly. Their light and vapory tails, through which the stars were visible, and their whiteness often intense, seemed to give them a strong resemblance to those transient fires, which we call shooting stars. Apparently, they differed from these only in duration. They might be only composed of a more compact substance capable of retarding for a longer time their dissolution. But these opinions are no longer maintained; more accurate observations have led to a different theory.

Parallax of comets.

"All the comets hitherto observed have a small parallax,* which places them far beyond the orbit of the moon; they are not, therefore, formed in our atmosphere. Moreover, their apparent motion among the stars is subject to regular laws, which enable us to predict their whole course from a small number of observations. This regularity and constancy evidently indicate durable bodies; and it is natural to conclude that comets are as permanent as the planets, but subject to a different kind of movement.

Comets are apparently mere masses of vapor.

"When we observe these bodies with a telescope, they resemble a mass of vapor, at the center of which is commonly seen a nucleus more or less distinctly terminated. Some, however, have appeared to consist of merely a light vapor, without a sensible nucleus, since the stars are visible through it. During their revolution, they experience progressive variations in their brightness, which appear to depend upon their distance from the sun, either because the sun inflames them by its heat, or simply on account of a stronger illumination. When their brightness is greatest, we may conclude from this very circumstance that they are near their perihelion. Their light is at first very feeble, but becomes gradually more vivid, until it sometimes surpasses that of the brightest planets; after which it declines by the same degrees until it becomes imperceptible. We are hence led to the conclusion that comets, coming from the remote regions of the heavens, approach, in many instances, much nearer the sun than the planets, and then recede to much greater distances.

Orbits of comets.

"Since comets are bodies which seem to belong to our planetary system, it is natural to suppose that they move about the sun like planets, but in orbits extremely elongated. These orbits must, therefore, still be ellipses, having their foci at the center of the sun, but having their major axes almost infinite, especially with respect to us, who observe only a small portion of the orbit, namely, that in which the comet becomes visible as it approaches the sun. Accordingly the orbits of comets must take the form of a *parabola*, for we thus designate the curve into which the ellipse passes, when indefinitely elongated.

"If we introduce this modification into the laws of Kepler, which

* The parallaxes of comets are known to be small, by two observers, at distant stations on the earth, comparing their observations taken on the same comet at near the same time. At the times the observations are made, neither observer can know how great the parallax is. It is only *afterward*, when comparisons are made, that judgment, in this particular, can be formed; and it is not common that any more definite conclusion can be drawn, than *that the parallax is small*, and, of course, the body distant.

CHAP. XII.

relate to the elliptical motion, we obtain those of the parabolic motion of comets.

Comets describe equal areas in equal times.

"Hence it follows that the areas described by the same comet, in its parabolic orbit, are proportional to the times. The areas described by different comets in the same time, are proportional to the square roots of their perihelion distances.

"Lastly, if we suppose a planet moving in a circular orbit, whose radius is equal to the perihelion distance of a comet, the areas described by these two bodies in the same time, will be to each other as 1 to $\sqrt{2}$. Thus are the motions of comets and planets connected.

"By means of these laws we can determine the area described by a comet in a given time after passing the perihelion, and fix its position in the parabola. It only remains then to bring this theory to the test of observation. Now we have a rigorous method of verifying it, by causing a parabola to pass through several observed places of a comet, and then ascertaining whether all the others are contained in it

Three observations sufficient to find the orbit of a comet.

"For this purpose three observations are requisite. If we observe the right ascension and declination of a comet at three different times, and thence deduce its geocentric longitude and latitude, we shall have the direction of three visual rays drawn at these times from the earth to the comet, and in the prolongation of which it must necessarily be found. The corresponding places of the sun are also known; it remains then to construct a parabola, having its focus at the center of the sun, and cutting the visual rays in points, the intervals of which correspond to the number of days between the observations.

Fig. 33.

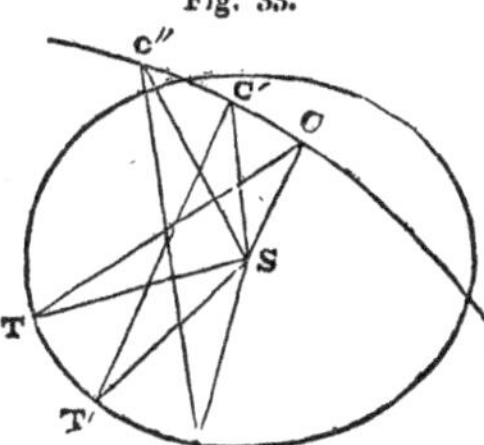

"Or if we suppose the earth in motion and the sun at rest, let T, T', T'', represent three successive positions of the earth, and TC, $T'C'$, $T''C''$, three visual rays drawn to the comet. The question is to find a parabola $CC'C''$, having its focus in S at the center of the sun, and cutting the three visual rays conformably to the conditions required.

The orbit of a comet found by three observations.

"These conditions are more than sufficient to determine completely the elements of the parabolic motion, that is, the perihelion distance of the comet, the position of the perihelion, the instant of passing this point, the inclination of the orbit to the ecliptic, and the position of its nodes. These five elements being known, we can assign the position of the comet for any time whatever, and compare it with the results of observation. But the calculation of the elements is very difficult, and can be performed only by a very delicate analysis, which cannot here be made known.

"About 120 comets have been calculated upon the theory of the parabolic motion, and the observed places are found to answer to such a supposition. We can have no doubt, therefore, that this is conformable to the law of nature. We have thus obtained precise knowledge of the motions of these bodies, and are enabled to follow them in space. This discovery has given additional confirmation to the laws of Kepler, and led to several other important results.

Inclinations of their orbits.

"Comets do not all move from west to east like the planets. Some have a direct, and some a retrograde motion.

"Their orbits are not comprehended within a narrow zone of the heavens, like those of the principal planets. They vary through all degrees of inclination. There are some whose plane is nearly coincident with that of the ecliptic, and others have their planes perpendicular to it.

"It is farther to be observed that the tails of comets begin to appear, as the bodies approach near the sun; their length increases with this proximity, and they do not acquire their greatest extent, until after passing the perihelion. The direction is generally opposite to the sun, forming a curve slightly concave, the sun on the concave side.

"The portion of the comet nearest to the sun must move more rapidly than its remoter parts, and this will account for the lengthening of the tail.

Some comets have no tails.

"The tail is, however, by no means an invariable appendage of comets. Many of the brightest have been observed to have short and feeble tails, and not a few have been entirely without them. Those of 1585 and 1763 offered no vestige of a tail; and Cassini describes the comet of 1682 as being as round and as bright as Jupiter. On the other hand, instances are not wanting of comets furnished with many tails, or streams of diverging light. That of 1744 had no less than six, spread out like an immense fan, extending to a distance of nearly 30 degrees in length.

"The smaller comets, such as are visible only in telescopes, or with difficulty by the naked eye, and which are by far the most numerous, offer very frequently no appearance of a tail, and appear only as round or somewhat oval vaporous masses, more dense toward the center; where, however, they appear to have no distinct nucleus, or anything which seems entitled to be considered as a solid body.

Others have several tails.

"The tail of the comet of 1456 was 60 degrees long. That of 1618, 100 degrees, so that its tail had not all risen when its head reached the middle of the heavens. The comet of 1680 was so great, that though its head set soon after the sun, its tail, 70 degrees long, continued visible all night. The comet of 1689 had a tail 68 degrees long. That of 1769 had a tail more than 90 degrees in length. That of 1811 had a tail 23 degrees long. The recent comet of 1843 had a tail 60 degrees in length."

The following figure gives a telescopic view of the comet of 1811.

CHAP. XII.

Elements of comets how determined.

"When we have determined the elements of a comet's orbit, we compare them with those of comets before observed, and see whether there is an agreement with respect to any of them. If there is a perfect identity as to the elements, we should have no hesitation in concluding that they belonged to different appearances of the same comet. But this condition is not rigorously necessary; for the elements of the orbit may, like those of other heavenly bodies, have undergone changes from the perturbations of the planets or their mutual attractions. Consequently, we have only to see whether the actual elements are nearly the same with those of any comet before observed, and then, by the doctrine of chances, we can judge what reliance is to be placed upon this resemblance."

Comet of 1811.

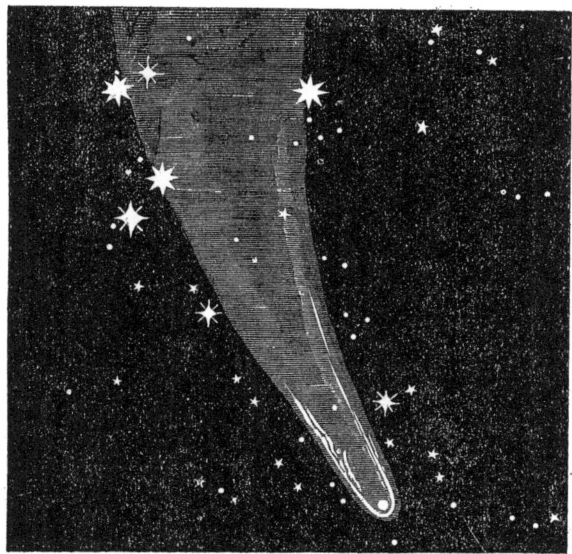

Dr. Halley's prediction verified.

"Dr. Halley remarked that the comets observed in 1531, 1607, 1682, had nearly the same elements; and he hence concluded that they belonged to the same comet, which, in 151 years, made two revolutions, its period being about 76 years. It actually appeared in 1759, agreeably to the prediction of this great astronomer; and again in 1832, by the computation of several eminent astronomers. According to Kepler's third law, if we take for unity half the major axis of the earth's orbit, the mean distance of this comet must be equal to the cube root of the square of 76, that is, to 17.95. The major axis of its orbit must, therefore, be 35.9; and as its observed perihelion distance is found to be 0.58, it follows that its aphelion distance is equal to 35.32. It

Particulars of comets.

departs, therefore, from the sun to thirty-five times the distance of the earth, and afterward approaches nearly twice as near the sun as the earth is, thus describing an ellipse extremely elongated.

"The intervals of its return to its perihelion are not constantly the same. That between 1531 and 1607 was three months longer than that between 1607 and 1682 ; and this last was 18 months shorter than the one between 1682 and 1759. It appears, therefore, that the motions of comets are subject to perturbations, like those of the planets, and to a much more sensible degree.

"Elements of the Orbits of the three Comets, which have appeared according to prediction, taken from the work of Professor Littrow.

	Halley.	Encke.	Biela.
Longitude of the ascending node, -	54°	335°	249°
Inclination of the orbit to the ecliptic,	162°	13°	13°
Longitude of the perihelion, - -	303°	157°	108°
Greatest semidiameter, that of the earth being called 1, - - - -	18	2.2	3.6
Least semidiameter, - - -	4.6	1.2	2.4
Time of revolution in years, -	76	3.29	6.74
Time of the perihelion passage, -	Nov. 16. 1835	May 4. 1832	Nov. 27 1832

"The comets of Encke and Biela move according to the order of the signs of the zodiac, or have their motions *direct;* the motion of that of Halley is *retrograde.*

Jupiter, and his satellites, a great stumbling-block to the comets.

"Comets, in passing among and near the planets, are materially drawn aside from their courses, and in some cases have their orbits entirely changed. This is remarkably the case with Jupiter, which seems, by some strange fatality, to be constantly in their way, and to serve as a perpetual stumbling-block to them. In the case of the remarkable comet of 1770, which was found by Lexell to revolve in a moderate ellipse in the period of about five years, and whose return was predicted by him accordingly, the prediction was disappointed by the comet actually getting entangled among the satellites of Jupiter, and being completely thrown out of its orbit by the attraction of that planet, and forced into a much larger ellipse. By this extraordinary renconter, *the motions of the satellites suffered not the least perceptible derangement*— a sufficient proof of the smallness of the comet's mass."

The comet of 1456, represented as having a tail of 60° in length, is now found to be Halley's comet, which has made several returns — in 1531, 1607, 1682, 1759, and recently, in 1835. In 1607 the tail was said to have been over 30° in length ; but in 1835 the tail did not exceed 12° Does it lose substance, or does the matter composing the tail condense ? or, have we received only exaggerated and distorted accounts from the earlier times, such as fear, superstition, and awe, always put forth ? We ask these questions, but cannot answer them.

CHAP. XII.

The following cut represents the appearance of the comet of 1819.

Fears entertained, by some, that comets may ultimately come into collision with our earth.

"Professor Kendall, in his Uranography, speaking of the fears occasioned by comets, says: "Another source of apprehension, with regard to comets, arises from the possibility of their striking our earth. It is quite probable that even in the historical period the earth has been enveloped in the tail of a comet. It is not likely that the effect would be sensible at the time. The actual shock of the head of a comet against the earth is extremely improbable. It is not likely to happen once in a million of years.

"If such a shock should occur, the consequences might perhaps be very trivial. It is quite possible that many of the comets are not heavier than a single mountain on the surface of the earth. It is well known that the size of mountains on the earth is illustrated by comparing them to particles of dust on a common globe."

CHAPTER XIII.

ON THE PECULIARITIES OF THE FIXED STARS.

CHAP. XIII.

FOR the facts as contained in the subject matter of this chapter, we must depend wholly on authority; for that reason we give only a compilation, made in as brief a manner as the nature of the subject will admit.

In the first part of this work it was soon discovered that the fixed stars were more remote than the sun or planets; and now, having determined their distances, we may make further inquiries as to the distances to the stars, which will give some index by which to judge of their magnitudes, nature, and peculiarities.

Base from which to measure to the stars.

"It would be idle to inquire whether the fixed stars have a sensible parallax, when observed from different parts of the earth. We have already had abundant evidence that their distance is almost infinite. It is only by taking the longest base accessible to us, that we can hope to arrive at any satisfactory result.

"Accordingly, we employ the major axis of the earth's orbit, which is nearly 200 millions of miles in extent. By observing a star from the two extremities of this axis, at intervals of six months, and applying a correction for all the small inequalities, the effect of which we have calculated, we shall know whether the longitude and latitude are the same or not at these two epochs.

Annual parallax.

"It is obvious, indeed, that the star must appear more elevated above the plane of the ecliptic when the earth is in the part of its orbit which is nearest to the star, and more depressed when the contrary takes place. The visual rays drawn from the earth to the star, in these two positions, differ from the straight line drawn from the star to the center of the earth's orbit; and the angle which either of them forms with this straight line, is called the *annual parallax*.

The effect of a sensible parallax.

"As the earth does not pass suddenly from one point of its orbit to the opposite, but proceeds gradually, if we observe the positions of a star at the intermediate epochs, we ought, if the annual parallax is sensible, to see its effects developed in the same gradual manner. For example, if the star is placed at the pole of the ecliptic, the visual rays drawn from it to the earth, will form a conical surface, having its apex at the star, and for its base, the earth's orbit. This conical surface being produced beyond the star, will form another opposite to the first, and the intersection of this last with the celestial sphere, will constitute a small ellipse, in which the star will always appear diametrically opposite to the earth, and in the prolongation of the visual rays drawn to the apex of the cones.

The annual parallax must be less than one second.

"But notwithstanding all the pains that have been taken to multiply observations, and all the care that has been used to render them perfectly exact, we have been able to discover nothing which indicates, with certainty, even the existence of an annual parallax, to say nothing of its magnitude. Yet the precision of modern observations is such, that if this parallax were only 1″, it is altogether probable that it would not have escaped the multiplied efforts of observers, and especially those of Dr. Bradley, who made many observations to discover it, and who, in this undertaking, fell unexpectedly upon the phenomena of aberration * and nutation. These admirable discoveries have themselves served to show, by the perfect agreement which is thus found to take

* Subject to be explained hereafter.

CHAP. XIII. place among observations, that it is hardly to be supposed that the annual parallax can amount to 1″. The numerous observations of the pole star, recently employed in measuring an arc of the meridian through France, have been attended with a similar result, as to the amount of the annual parallax. From all this we may conclude, that as yet there are strong reasons for believing that the annual parallax is less than 1″, at least with respect to the stars hitherto observed.

"Thus the semidiameter of the earth's orbit, seen from the nearest star, would not appear to subtend an angle of 1″; and to an observer placed at this distance, our sun, with the whole planetary system, would occupy a space scarcely exceeding the thickness of a spider's thread.

Conclusion to be drawn from these facts.

"If these results do not make known the distance of the stars from the earth, they at least teach us the limit beyond which the stars must necessarily be situated. If we conceive a right-angled triangle, having for its base half the major axis of the earth's orbit, and for its vertex an angle of 1″, the distance of this vertex from the earth, or the length of the visual ray, will be expressed by 212207, the radius of the earth's orbit being unity; and as this radius contains 23987 times the semidiameter of the earth, it follows that if the annual parallax of a star were only 1″, its distance from the earth would be equal to 5090209309 radii of the earth, or 20086868036404 miles; that is, more than 20 billions. But if the annual parallax is less than 1″, the stars are beyond the limit which we have assigned.

Changes in individual stars.

"It is evident that the stars undergo considerable changes, since these changes are sensible even at the distance at which we are placed. There are some which gradually lose their light, as the star δ of Ursa Major. Others, as β of Cetus, become more brilliant. Finally, there are some which have been observed to assume suddenly a new splendor, and then gradually fade away. Such was the new star which appeared in 1572,

A new star.

in the constellation Cassiopeia. It became all at once so brilliant that it surpassed the brightest stars, and even Venus and Jupiter when nearest the earth. It could be seen at midday. Gradually this great brilliancy began to diminish, and the star disappeared in sixteen months from the time it was first seen, without having changed its place in the heavens. Its color, during this time, suffered great variations. At first it was of a dazzling white, like Venus; then of a reddish yellow, like Mars and Aldebaran; and lastly, of a leaden white, like Saturn. An-

Another new star.

other star which appeared suddenly in 1604, in the constellation Serpentarius, presented similar variations, and disappeared after several months. These phenomena seem to indicate vast flames which burst forth suddenly in these great bodies. Who knows that our sun may not be subject to similar changes, by which great revolutions have perhaps taken place in the state of our globe, and are yet to take place.

Periodical changes.

"Some stars, without entirely disappearing, exhibit variations not less remarkable. Their light increases and decreases alternately in regular periods. They are called for this reason *variable stars*. Such is the star Algol, in the head of Medusa, which has a period of about three days ; δ of Cepheus, which has one of five days ; β of Lyra, six ; μ of Antinous, seven ; o of Cetus, 334 ; and many others.

Attempts to explain periodical changes.

"Several attempts have been made to explain these periodical variations. It is supposed that the stars which are subject to them, are, like all the other stars, self-luminous bodies, or true suns, turning on their axes, and having their surfaces partly covered with dark spots, which may be supposed to present themselves to us at certain times only, in consequence of their rotation. Other astronomers have attempted to account for the facts under consideration, by supposing these stars to have a form extremely oblate, by which a great difference would take place in the light emitted by them under different aspects. Lastly, it has been supposed that the effect in question is owing to large opake bodies, revolving about these stars, and occasionally intercepting a part of their light. Time and the multiplication of observations may perhaps decide which of these hypotheses is the true one.

Order in these observations.

"One of the best methods of observing these phenomena is to compare the stars together, designating them by letters or numbers, and disposing them in the order of their brilliancy. If we find, by observation, that this order changes, it is a proof that one of the stars thus compared, has likewise changed ; and a few trials of this kind will enable us to ascertain which it is that has undergone a variation. In this manner, we can only compare each star with those which are in the neighborhood, and visible at the same time. But by afterward comparing these with others, we can, by a series of intermediate terms, connect together the most distant extremes. This method, which is now practiced, is far preferable to that of the ancient astronomers, who classed the stars after a very vague comparison, according to what they called the *order of their magnitudes*, but which was, in reality, nothing but that of their brightness, estimated in a very imperfect manner.

Suggestion of Dr. Halley.

"By comparing the places of some of the fixed stars, as determined from ancient and modern observations, Dr. Halley discovered that they had a proper motion, which could not arise from parallax, precession, or aberration. This remarkable circumstance was afterward noticed by Cassini and Le Monnier, and was completely confirmed by Tobias Mayer, who compared the places of 80 stars, as determined by Roemer, with his own observations, and found that the greater part of them had a proper motion. He suggested that the change of place might arise from a progressive motion of the sun toward one quarter of the heavens ; but as the result of his observation did not accord with his theory, he remarks that many centuries must elapse before the true cause of this motion could be explained.

"The probability of a progressive motion of the sun was suggested upon theoretical principles by the late Dr. Wilson of Glasgow ; and Lalande deduced a similar opinion from the rotatory motion of the sun, by supposing, that the same mechanical force which gives it a motion

CHAP. XIII.

round its axis, would also displace its center, and give it a motion of translation in absolute space

Consequences of such a theory.

"If the sun has a motion in absolute space, directed toward any quarter of the heavens, it is obvious that the stars in that quarter must appear to recede from each other, while those in the opposite region would seem gradually to approach, in the same manner as when walking through a forest, the trees toward which we advance are constantly separating, while the distance of those which we leave behind is gradually contracting. The proper motion of the stars, therefore, in opposite regions, as ascertained by a comparison of ancient with modern observations, ought to correspond with this hypothesis; and Sir W. Herschel found, that the greater part of them are nearly in the direction which would result from a motion of the sun toward the constellation Hercules, or rather to a part of the heavens whose right ascension is 250° 52′ 30″, and whose north polar distance is 40° 22′. Klugel found the right ascension of this point to be 260°, and Prevost made it 230°, with 65° of north polar distance. Sir W. Herschel supposes that the motion of the sun, and the solar system, is not slower than that of the earth in its orbit, and that it is performed round some distant center. The attractive force capable of producing such an effect, he does not suppose to be lodged in one large body, but in the center of gravity of a cluster of stars, or the common center of gravity of several clusters."

The following figures, taken from Norton's Astronomy, represent the telescopic appearance of some of the double stars.

Double and multiple stars.

"There are stars which, when viewed by the naked eye, and even by the help of a telescope of moderate power, have the appearance of only a single star; but, being seen through a good telescope, they are found to be double, and in some cases a very marked difference is perceptible, both as to their brilliancy and the color of their light. These Sir W. Herschel supposed to be so near each other, as to obey reciprocally the power of each other's attraction, revolving about their common center of gravity, in certain determinate periods.

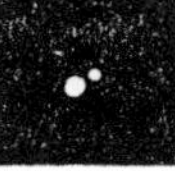
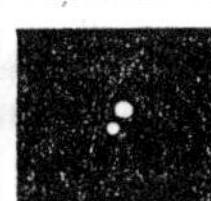
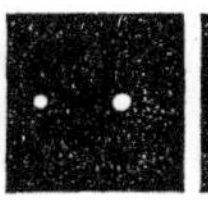
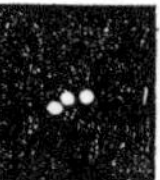
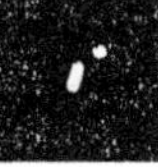

Castor, γ Leonis, Rigel, Pole Star, π Monoc, ϱ Cancri.

Revolutions of the multiple stars.

"The two stars, for example, which form the double star Castor, have varied in their angular situation more than 45° since they were observed by Dr. Bradley, in 1759, and appear to perform a retrograde revolution in 342 years, in a plane perpendicular to the direction of the sun. Sir W. Herschel found them in intermediate angular positions, at intermediate times, but never could perceive any change in their distance. The retrograde revolution of γ in Leo, another double star, is supposed to be in a plane considerably inclined to the line in which we view it, and to be completed in 1200 years. The stars ϵ of Bootes, perform a direct revolution in 1681 years, in a plane oblique to the sun. The stars ξ of Serpens, perform a retrograde revolution in about 375 years; and those of γ in Virgo in 708 years, without any change of their distance. In 1802, the large star ζ of Hercules, eclipsed the smaller one, though they were separate in 1782. Other stars are supposed to be united in triple, quadruple, and still more complicated systems.

Description of nebulæ.

"With respect to the determination of the real magnitude of the stars, and their respective distances, we have as yet made but little progress. Researches of this kind must be left to future astronomers. It appears, however, that the stars are not uniformly distributed through the heavens, but collected into groups, each containing many millions of stars. We can form some idea of them from those small whitish spots called Nebulæ, which appear in the heavens as represented in the accompanying illustration. By means of the telescope, we distinguish in these collections an almost infinite number of small stars, so near each other, that their rays are ordinarily blended by irradiation, and thus present to the eye only a faint uniform sheet of light. That large, white, luminous track, which traverses the heavens from one pole to the other, under the name of the Milky Way, is probably nothing but a nebula of this kind, which appears larger than the others, because it is nearer to us. With the aid of the telescope we discover in this zone of light such a prodigious number of stars that the imagination is bewildered in attempting to represent them. Yet from the angular distances of these stars, it is certain that the space which separates those which seem nearest to each other, is at least a hundred thousand times as great as the radius of the earth's orbit. This will give us some idea of the immense extent of the group. To what distance then must we withdraw, in order that this whole collection may appear as small as the other nebulæ which we perceive, some of which cannot, by the assistance of the best telescopes, be made to present anything but a bright speck, or a simple mass of light, of the nature of which we are able to form some idea only by analogy? When we attempt, in imagination, to fathom this abyss, it is in vain to think of prescribing any limits to

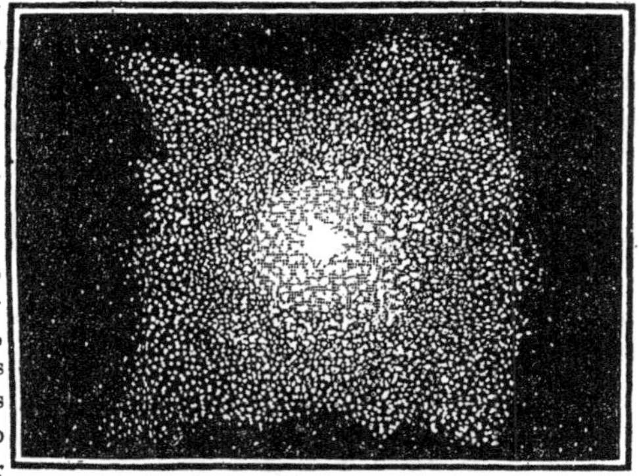

The Milky Way a nebula.

CHAP. XIII. the universe, and the mind reverts involuntarily to the insignificant portion of it which we are destined to occupy."

Observations on table II. Before we close this chapter, we think it important to call the attention of the reader to table II , in which will be seen, at a glance (in the columns marked annual variation), the general effect of the precession of the equinoxes; and although we have called particular attention to the fact elsewhere, we here notice that all the stars, from the 6th to the 18th hour of right ascension, have a progressive motion to the southward (—), and all the stars from the 18th to the 6th hour of right ascension have a progressive motion to the northward (+), and the greatest variations are at 0 h. and 12 h. But these motions are not, in reality, the motions of the stars; they result from motions of the earth. Whenever the annual motion of any star does not correspond with this common displacement of the equinox, we say the star has a proper motion; and by such discrepancy it has been decided, that those stars marked with an asterisk, in the catalogue, have proper motions; and the star 61 Cygni, near the close of the table, has the greatest proper motion.

The parallax of 61 Cygni discovered. From this circumstance, and from the fact of its being a double star, it was selected by Bessel as a fit subject for the investigation of stellar parallax; and it is now contended, and in a measure granted, that the annual parallax of this star is 0''.35, which makes its distance more than 592.000 times the radius of the earth's orbit; a distance that light could not traverse in less than nine and one-fourth years.

SECTION III.

PHYSICAL ASTRONOMY.

CHAPTER I.

GENERAL LAWS OF MOTION — THE THEORY OF GRAVITY.

CHAP. I.

What should be expected in this work.

(148.) IN a work like this, designed for elementary instruction, it cannot be expected that a full investigation of physical astronomy shall be entered into; for that subject alone would require volumes; and to fully appreciate and comprehend it, requires the matured philosopher combined with the accomplished mathematician.

We shall give, however, a sufficient amount to impart a good general idea of the subject — if one or two points are taken on trust.

Elementary principles.

For elementary principles we must turn a moment to natural philosophy, and consider the laws of *inertia*, *motion*, and *force*. Motion is a change of place in relation to other bodies which we conceive to be at rest; and the extent of change in *the time taken for unity* is called *velocity*, and the *essential cause* of motion we denominate *force*.

Velocity the measure of force.

A *double force* will give a *double velocity* to bodies moving freely in void space, or in an unresisting medium — a *triple force*, a *triple velocity*, &c. This is taken as an axiom — and hence, when we consider mere material points in motion, the relative velocities measure the relative amounts of force.

There are *three elements* to motion, which the philosopher never loses sight of; or we may say that he never thinks of motion without the *three distinct elements* of *time*, *velocity*, and *distance*, coming into his mind.

Algebraically, we put t, v, and d, to represent the three elements, and then we have this important and general equation,

$$tv = d \qquad (1)$$

CHAP. I.

Expression for force.

From this we derive $v=\frac{d}{t}$ (2) and $t=\frac{d}{v}$ (3)

(149.) As forces are in proportion to velocities (when momentum is not in question), therefore, if we put f and F to represent two forces corresponding to the distances d and D, which are described in the times t and T, then by making use of equation (2), in place of the velocities, we have

$$f : F :: \frac{d}{t} : \frac{D}{T} \qquad (4)*$$

The law of inertia.

(150.) A body at rest, has no power to put itself in motion; and having no *self power*, no internal force or *will*, in any shape, it cannot increase or diminish the motion it may have, or change the direction it may be moving. This is the law of *inertia*. It cannot of itself change its state; and if it is changed it must be acted upon by some external force; and this accords with universal experience; and this law is the most natural and simple of any we can imagine, but it is only in the motion of the heavenly bodies that it is fully exemplified.

Some central force must act on the motions of the earth, moon, and planets.

The earth, moon, and planets move in curves — not in right lines. The directions of their motions *are changed.* Something external from them must, therefore, change them; for the law of *inertia* would continue a motion once obtained in a straight line. Now this force must exist within the orbit of every curve; we therefore naturally refer it to the body round which others circulate. The earth and planets go round the sun, and if we could suppose a force residing in the sun to extend throughout the system sufficient to draw bodies to it, this would at once account not only for the planets deviating from a right line, but would account for a constant deviation of all bodies to that point, and the preservation of the system.

The moon's motion considered.

The moon goes round the earth, constantly deviating from the tangent of its orbit, and the law of inertia is constantly

* We number the proportions the same as equations, for a proportion is but an equation in another form.

urging it to rise from the center; the two on an average balancing each other, retains the moon in an orbit about the earth.

Now what and where is this force? Is it around the earth, or within the earth? Is it electrical or magnetic? or is it that same force (call it what we may) that makes a body fall toward the earth's center when unsupported on a resting base?

Contemplations of Sir Isaac Newton.

A trifling incident, the fall of an apple from a tree, seems to have led the mind of Newton to the contemplation of this *force* which compels and causes bodies to fall, and he at once conceived this force to extend to the moon and to cause it to deviate from the tangent of its orbit.

The next consideration was, whether if this were the force, it was the same at the distance of the moon, as on the surface of the earth; or if it extended with a diminished amount, what was the *law* of diminution?

Incipient steps to the theory of gravity.

Newton now resorted to computation, and for a test he conceived the force in question to extend to the moon, undiminished by the distance; and corresponding thereto he decided that the moon must then make a revolution in its orbit in 10 h. 55 m. But the actual time is 27 d. 7 h. 43 m., which shows that if the force is the same which pervades a falling body on the surface of the earth, it must be greatly diminished.

Important computations.

Now by making a reverse computation, taking the actual time of revolution, and finding how far the moon did really fall from the tangent of its orbit in one second of time, it was found to be about $\frac{1}{3600}$ part of 16 $\frac{1}{12}$ feet — the distance a body falls the first second of time.

But the distance to the moon is about 60 times the radius of the earth, and the inverse square of this is $\frac{1}{3600}$, which corresponds to the actual fall of the moon in one second.

A principle in philosophy

(151.) It is a well-established fact in philosophy, and geometrically demonstrated, that any force or influence existing at a point, must diminish as it spreads over a larger space, and in proportion to the increase of space. But space increases as the square of linear distance, as we see by Fig. 28.

CHAP. I.

A double distance spreads the influence over four times the space, whatever that influence may be; a triple distance, nine times the space, etc., the space increasing as the square of

Fig. 28.

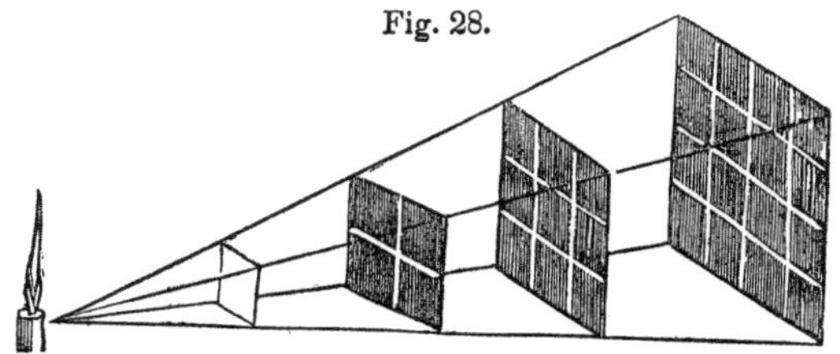

the distance. Therefore, any influence spreading in all directions from its central point must be *enfeebled* as the square of the distance.

The theory of universal gravity.

From observations and considerations like these, Newton established the all-important and now universally admitted theory of gravity.

This theory may be summarily stated in the following words:

Every body of matter in the universe attracts every other body, in direct proportion to its mass, and in the inverse proportion to the square of the distance.

This theory well established.

Some attempts have been made, from time to time, to call the truth of this theory in question, and substitute in its place the influence of light, caloric, and electricity; but any thing like a close application shows how feebly all such substitutes stand the test.

The theory of gravity so exactly accounts for all the physical phenomena of the solar system, that it is *impossible* it should be false; and although we cannot determine its nature or its essence, it is as unreasonable to doubt its existence, as to doubt the existence of animate beings, because we know nothing of the principle of life.

Attraction of an irregular body.

(152.) According to the theory of gravity, every particle composing a body has its influence, and a very irregular body may be divided in imagination into many smaller bodies, and the center of gravity of each taken as the point of attraction, and all the forces resolved into one will be the attraction of the whole body.

In a sphere composed of homogeneous particles, the aggregate attraction of all of them will be the same as if all were compressed at the center; but this will be true of no other body. The earth is not a perfect sphere, and two lines of attraction from distant points on its surface may not, yea, will not, cross each other at the earth's center of gravity. (See Fig. 10.) Attraction of a sphere.

(153.) A particle anywhere inside of a spherical shell of equal thickness and density, is attracted every way alike, and of course would show no indication of being attracted at all. Hence a body below the surface of the earth, as in a deep pit or well, will be less attracted than on the surface, as it will be attracted only by the diminished sphere below it. At the center of the earth a body would be attracted by the earth every way alike, and there would be no unbalanced force, and of course no perceptible or sensible attraction.* Attraction inside of a spherical shell. Attraction at the center of a sphere.

(154.) The attractive power on the surface of any perfect and homogeneous sphere may be expressed by *the mass of the sphere divided by the square of the radius.* Expression for the attraction on the surface of a sphere.

Consider the earth a sphere (as it is very nearly), and put E to represent its mass, and r its mean radius, then

$$\frac{E}{r^2} = g = 16\tfrac{1}{12} \text{ feet.}$$

This attractive force, algebraically expressed by $\frac{E}{r^2}$ we call g, and it is sufficient to cause bodies to fall $16\frac{1}{12}$ feet during the first second of time. If the earth had contained more matter, bodies would have fallen more than $16\frac{1}{12}$ feet the first second; if less, a less distance.

With the same matter, but more compact, so that r^2 would be less with E the same, $\frac{E}{r^2}$ would be greater, and the attractive power at the surface greater, and bodies would then fall more than $16\frac{1}{12}$ feet the first second of their fall. The definite attraction of the earth.

Now we say this $16\frac{1}{12}$ feet is the measure of the earth's attraction at its surface, and it is made the unit and standard measure, directly or indirectly, for all *astronomical forces.*

---

* See Robinson's Natural Philosophy, page 16.

CHAP. I.

For this reason, we call the undivided attention to this force, the known — the noted — the *all-important* $16\frac{1}{12}$ *feet.*

To find the attraction of a sphere at any distance.

(155.) By the theory of gravity, we can readily obtain an analytical expression for the attraction of a sphere at any distance from the center, after knowing the attraction at the surface. For example. Find the value of the attraction of the earth, at the distance of D from its center; r being the radius of the earth, and g the gravity at the surface; put x to represent the attraction sought. Then by the theory,

$$g : x :: \frac{1}{r^2} : \frac{1}{D^2}; \quad \text{Or, } x = g\left(\frac{r^2}{D^2}\right) \qquad (5)$$

As g and r are constant quantities, the variations to x will correspond entirely to the variations of D^2. We shall often refer to this equation.

An expression for the mutual attraction of two bodies.

(156.) As every particle of matter in the universe attracts every other particle, therefore the moon attracts the earth as well as the earth attracts the moon; and the extent by which they will *draw together*, depends on their *mutual* attraction. If m represents the mass of the moon, and R the radius of the lunar orbit; then,

The earth will attract the moon by the force $\frac{E}{R^2}$.

The moon will attract the earth by the force $\frac{m}{R^2}$.

The two bodies will draw together by the force $\frac{E+m}{R^2}$.

If we substitute the value of g, as found in (154), in equation (5), and making $R = D$, then we have the expression $\frac{E}{r^2}$.

The spirit of these expressions will be more apparent when we make some practical applications of them, as we intend soon to do.

CHAPTER II.

KEPLER'S LAWS — DEMONSTRATION OF THE SECOND AND THIRD — HOW A PLANETARY BODY WILL FIND ITS ORBIT.

CHAP. II.

Examinations of Kepler's laws.

(157.) In this chapter we design to make some examination of Kepler's laws, recapitulating them in order.

The orbits of the planets are ellipses, having the sun at one of their foci.

This law is but a concise statement of an *observed fact*, which never could have been drawn from any other source than observation; but the second law, namely,

That the radius vector of any planet (conceived to be in motion) *sweeps over equal areas in equal times* is susceptible of a rigid mathematical demonstration, under the following general theorem.

A general theorem.

Any body, being in motion, and constantly urged toward any fixed point, not in a line with its motion, must describe equal areas in equal times round that point.

Its demonstration.

Fig. 29.

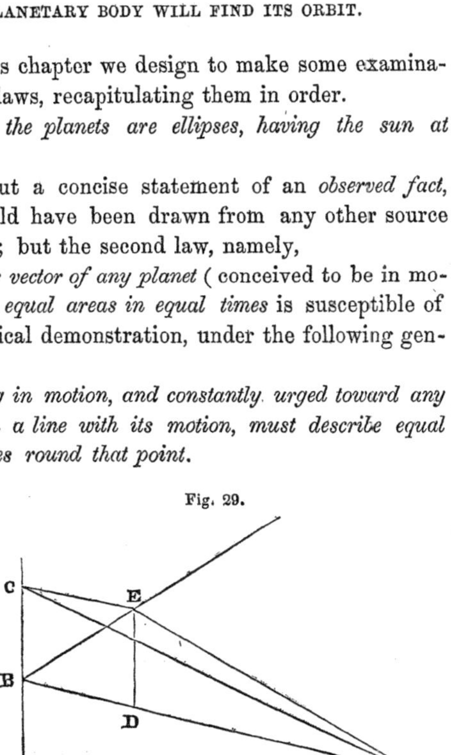

Let a moving body be at *A*, having a velocity which would carry it to *B*, say in one second of time. *By the law of inertia*, it would move from *B* to *C*, an equal distance, in the next second of time. But during this second interval of time, let us suppose it must obey an impulse or force from the point *S*, sufficient to carry it to *D*. It must then, by the composition of forces explained in natural philosophy, describe the diagonal *B E*, of the parallelogram *B D E C*.

Now in the first interval of time, we supposed the moving body described the triangle $S\ A\ B$. The second interval, it would have described the triangle $S\ B\ C$, if undisturbed by any force at S, but by such a force it describes the triangle $S\ B\ E$; but the triangle $S\ B\ E$ is equal to the triangle $S\ B\ C$, because they have the *same base* $S\ B$, and lie between the parallels $S\ B$ and $E\ C$. Also the triangle $S\ B\ C$ is equal to the triangle $S\ A\ B$, because they terminate in the same point S, and have equal bases $A\ B$ and $B\ C$. Therefore the triangle $S\ A\ B$ is equal to the triangle $S\ B\ E$, because they are both equal to the triangle $S\ B\ C$; that is, the moving body describes equal *areas* in equal times about the point S, and this is entirely independent of the nature of the force at S; it may be directly or inversely as the distance, or as the square of the distance.

The converse of the theorem.

The converse of this theorem is, that when a body describes equal areas in equal times round any point, the body is constantly urged *toward that point*; and therefore as the planets are observed to describe equal areas in equal times round the sun, their tendency is toward the sun, and *not toward any other point* within the orbits.

Kepler's third law proves that the sun's attraction is inversely as the square of the distance.

(158.) The third law of Kepler is most important of all, namely — *The squares of the times of revolution are to each other as the cubes of the distances from the sun.* By this law it is proved, that it is the same force which urges all the planets to the same point, and that its *intensity* is inversely as the square of the distance from that point (the center of the sun), confirming the Newtonian theory of gravity.

Fig. 30.

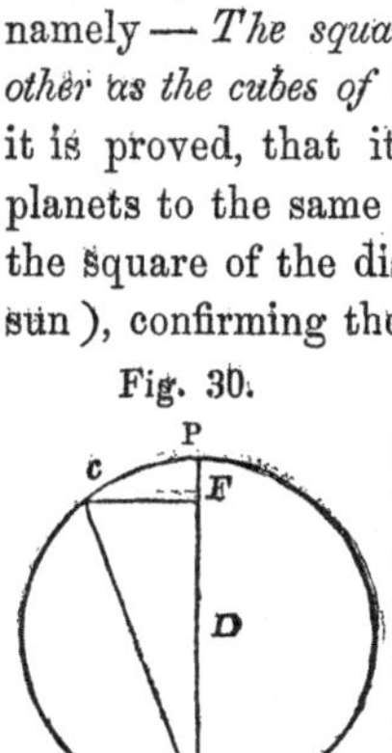

To show this, let us suppose that the planets revolve round the sun in circular orbits (which is not far from the truth), and let P (Fig. 30) represent the position of a planet; F the distance which the planet is drawn from a tangent during *unity of time;* in the same time that it describes the indefinite small arc c; and the number of times that c is contained in the whole circumference, so many units of time, then, must be in one revolution.

If D is the diameter of the orbit and t the time of revolution, then will

$$t=\frac{\pi D}{c}. \quad . \quad . \quad . \qquad (1)$$

An important truth demonstrated.

So for any other planet. If f is the force urging it toward the sun, a its corresponding arc, T its time of revolution, and R the radius of its orbit; then, reasoning as before,

$$T=\frac{2\pi R}{a}. \quad . \quad . \quad . \qquad (2)$$

By comparing (1) and (2) we have

$$t : T :: \frac{D}{c} : \frac{2R}{a}.$$

By squaring, $\quad t^2 : T^2 :: \frac{D^2}{c^2} : \frac{4R^2}{a^2}.$

By Kepler's law, $t^2 : T^2 :: r^3 : R^3$.

By comparing the two last proportions, and observing that $2r$ may be put for D, and reducing, we have

$$\frac{1}{c^2} : \frac{1}{a^2} :: r : R.$$

But by the well-known property of the circle, we have

$$F : c :: c : 2r; \text{ or, } c^2=2rF.$$

In like manner, ... $a^2=2Rf.$

Substituting these values in the last proportion, and reducing, we have

$$\frac{1}{rF} : \frac{1}{Rf} :: r : R;$$

Or, . . $Rf : rF :: r : R.$

Hence, . $R^2f=r^2F;$ or, $F : f :: R^2 : r^2.$

Or, ... $F : f :: \frac{1}{r^2} : \frac{1}{R^2}.$

That is, the attractive force of the sun is reciprocally proportional to the square of the distance.

The theory of gravity

(159.) If we commence with the hypothesis, that bodies tend toward a central point with a force inversely propor-

CHAP. II.

and laws of motion result in Kepler's third law.

tional to the squares of their distances, and then compute the corresponding times of revolution, we shall find that *the squares of the times must be as the cubes of the distances.* Hence Kepler's third law is but the natural mathematical relation which must exist between times and distances among bodies moving freely, in circular orbits, animated by one central force which varies as the inverse square of the distance.

An inquiry.

(160.) Having shown that Kepler's third law is but a mathematical theorem when the planets move in circles and their masses inappreciable in comparison to that of the sun's, we now inquire whether the law is true, or only approximately true, when the orbits are ellipses, and their masses considerable.

How answered.

On one of these points of inquiry, the reader must take our assertion; for its demonstration requires the use of the *integral calculus*, a method that we designed not to employ in this work. Kepler's third law supposes all the force to be in the central body, and the planets only moving points. But we have seen in Art. (120) that the attracting force on any planet is the mass of both sun and planet divided by the square of their mutual distance; and therefore when the mass of the planet is appreciable, the force is increased, and the time of revolution a little shortened. But the fact that Kepler's law corresponds so well with other observations proves that the masses of all the planets are inappreciable compared to the mass of the sun.

Masses of the planets very small compared to the sun.

Kepler's third law mathematically true in elliptic orbits.

(161.) As to the other point, we state distinctly that the planets (considered as bodies without masses) revolving in ellipses of ever so great eccentricity, *the squares of the times of revolution are to each other as the cubes of half the greater axes of the orbits.*

We shall not attempt a demonstration of this truth; but hope the following explanation will give the reader a clear view of the subject.

Bodies revolving in ellipses round one of the foci, may be considered to have a rising and a falling motion; something like the motion of a pendulum. The motion of a pendulum depends on the *force of gravity*, the length of the pendulum, and the distance the pendulum was first drawn aside. The motion of a planet depends on the *force* of gravity, its mean distance from the sun, and the original impulse first given to it. Most persons, who have not investigated this subject, imagine that each planet must originally have had precisely the impulse it did have to maintain itself in its orbit; and so it must, to maintain itself in just that definite orbit in which it moves. *But had the original impulse been different, either as to amount or direction, or as to both, then by the action of gravity and inertia, the planet would have found a corresponding orbit.*

A common error of opinion.

(162.) The force of gravity, from the action of any attracting body, is always as the *mass of the body divided by the square of its distance.* Algebraically, if M is the mass of the body, r its distance, and F the force at that distance, then (see 118) we have - - - $\frac{M}{r^2}=F$. (See Fig. 28.)

Examination of the planetary motions in elliptic orbits

Now if the planet has such a velocity, c, as to correspond with the proportion $F : c :: c : 2r$,

Or, - - - - $c=\sqrt{2rF}=\sqrt{\frac{2M}{r}}$, and that velocity at right angles to r (Fig. 28), then the planet's orbit would be a circle, with the radius r. If the velocity had been *less in amount* than this expression, and still at right angles to r, then the planet would fall within the circle, and the action of gravity would increase the motion of the planet; and the motion would increase *faster* than the increased action of gravity: there would be a point, then, where the motion would be sufficient to maintain the planet in a circle, at its *then distance;* but the *direction* of the motion will not permit the planet to run into the circle, and it must fall within it.

ABOVE is further from the sun. BELOW is nearer to it.

The motion continues to increase until its position becomes at right angles to the radius vector; the motion is then as much *more than sufficient* to maintain the planet in a circle, as it was insufficient in the first instance; it therefore *rises*, by the law of inertia, and returns to the original point P, where it will have the same velocity as before; and thus the planet *vibrates* between two extreme distances.

CHAP. II

Gravity and original velocity determine the eccentricity and mean distances of the orbits.

If the velocity, on starting from the point P, were very much less than sufficient to maintain a circle, at that distance, then the orbit it would take would be very eccentric, and its mean distance much less than r. If the original velocity at P were greater than to maintain it in a circle, it would pass outside of this circle, and the point P would be the perihelion point of the orbit.

Thus, we perceive, that the eccentricity of orbits and mean distances from the sun depend on the amount and direction of the original impulse, or velocity which the planet has in some way obtained; *and it is not necessary that the planet should have any definite impulse, either in amount or direction, to move in an orbit, if the direction is not directly to or from the sun.*

A hypothetical case.

(163.) For a more definite explanation of this subject, let us conceive a planet launched out into space with a velocity sufficient to maintain it in a circle at the distance it then happened to be, but the direction of such velocity *not at right angles* to the sun, then the orbit will be elliptical, and the degree of eccentricity will depend on the direction of the motion; but the longer axis of the orbit will be equal to the diameter of the circle, to which its velocity corresponds; and the time of its revolution will be the same, whether the orbit is circular or more or less elliptical.

Fig. 31.

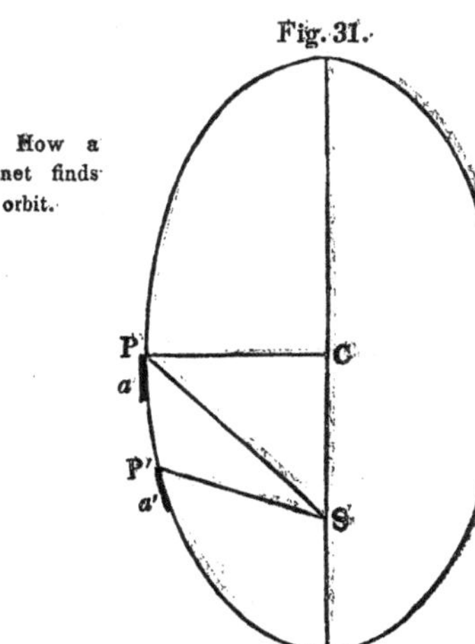

How a planet finds its orbit.

Let P (Fig. 31) be the position of a planet, S the sun; and let the velocity, a, be just sufficient to maintain the planet in a circle, if it were at right angles to SP.

Now to find the orbit that this planet would describe, draw the line PC at right angles to a, and from S let fall a perpendicular on PC; SC will be the eccentricity of the orbit, and PC will be the half of its conjugate axis; and with these lines the whole orbit is known.

(164.) Now let us suppose that a planet is rather *carelessly* launched into space, with a velocity neither at right angles to the sun, nor of sufficient amount to maintain it in a circle, at that distance from the sun.

Planets will find their orbits, whatever be the direction and force of their original motion.

Fig. 32.

Let P (Fig. 32) represent the position of the planet, a the amount and direction of its *haphazard* velocity during the first unit of time. The direction of the motion being within a right angle to $S P$, the action of gravity increases the velocity of the planet, on the same principle that a falling body increases in velocity; and the planet goes on in a curve, describing equal areas in equal times round the point S; and it will find a point, p, where its increased velocity will be just equal to the velocity in a circle whose radius is the diminished distance $S p$. From the point p, and at right angles to a, draw $p\ C$, &c., forming the right angled triangle $p\ C\ S$. $S\ C$ is the eccentricity, $S\ a$ the mean distance, and $p\ C$ half the conjugate axis of the orbit.

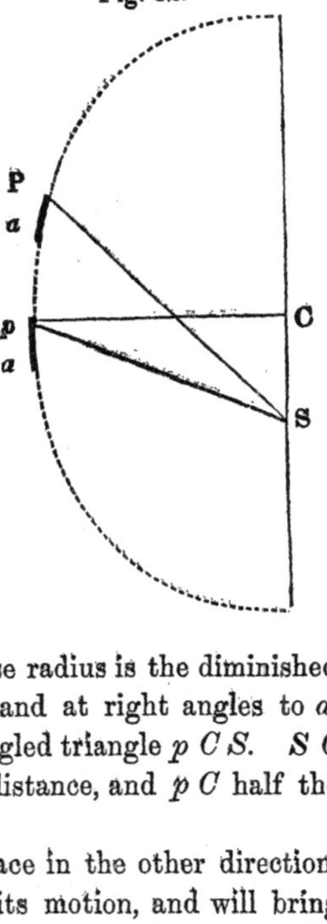

If the planet is launched into space in the other direction, the action of gravity will diminish its motion, and will bring it at right angles to the line joining the sun; it is then at its apogee, with a motion too feeble to maintain a circle at that distance; and it will, of course, approach nearer and nearer to the sun by the same laws of motion and force that it receded from the sun; hence the curve on each side of the apogee will be symmetrical; and the same reasoning will apply to the curve on each side of the perigee; and, in short, we shall have an ellipse.

The orbits will be symmetrical on each side of apogee and perigee.

To sum up the whole matter, it is found by a strict examination of the laws of *gravity*, *motion*, and *inertia*, that whatever

An important conclusion.

CHAP. II. may be the primary force and direction given to a planetary body (if not directly to or from the sun), *the planet will find a corresponding orbit, of a greater or less eccentricity, and of a greater or less mean distance; and whatever be the eccentricity of the orbit, the real velocity, at the extremity of the shorter axis, will be just sufficient to maintain the planet in a circular orbit, at that mean distance from the sun.**

Theory of Dr. Olbers concerning the asteroids

* Let S be the sun, and P the position of a planet as represented in the annexed figure, and we may now suppose it to burst into fragments, the figure representing three fragments only; the velocity and direction of one represented by a; of another by b, and of a third by c, &c.

Fig. 33.

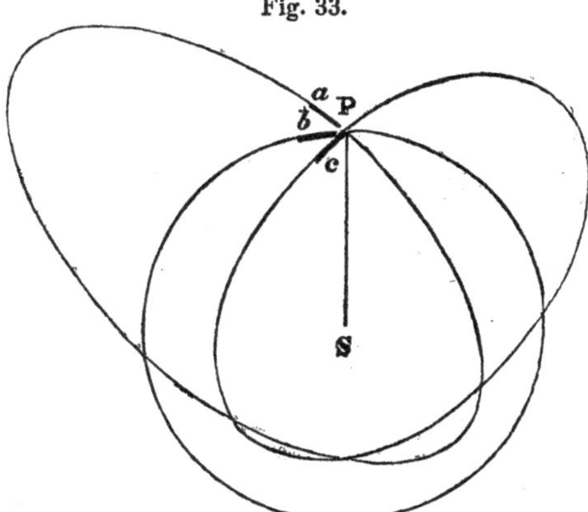

As action is just equal to reaction, under all circumstances, therefore the bursting of a planet can give the whole mass no additional velocity; a small mass may be blown off at a great velocity, but there will be an equal reaction on other masses, in the opposite direction.

On the bursting of a planet, the fragments would take orbits corresponding to their velocities and positions.

The whole might simply burst into about equal parts, and then they would but separate, and all the parts move along in the same general direction, and with the same *aggregate* velocity as the original planet. The bursting of a rocket is a very minute, but a very faithful representation of such an explosion.

Kepler's third law rigorously true in relation to ellipses, as well as to circles.

(165.) To see whether Kepler's third law applies to ellipses, we represent half the greater axis of any ellipse by A, and half the shorter axis by B, and then $(3.1416)AB$ is the *area* of the ellipse. Also, let a represent the velocity or distance

---

If the velocities of the several fragments were equal, the times of their revolutions would be equal; but the eccentricities of the several orbits would depend on the angles of a, b, c, &c., with SP. If a is at right angles to SP, and just sufficient to maintain the planet in a circle at that distance, then its orbit would have no eccentricity. If still at right angles, but not sufficient to maintain a circle at that distance, then SP would be the greatest radius of the orbit. Hence, we perceive, there is an abundance of room to have a multitude of orbits passing through the same point, during the first one or two revolutions; and the times of such revolutions may be equal, or very unequal. In short, *there is no physical impossibility* to be urged against the theory of Dr. Olbers, that the asteroids are but fragments of a planet.

The objection is (if an objection it can be called) that these planets have not, in fact, a common node, nor have an approximation to one; nor have they an approximation to a common radius vector, as SP. But the objection vanishes when we consider that the elements of the different orbits must be variable; and time, a *sufficient length of time*, would separate the nodes and change the positions of the orbits so as to hide the common origin, as is now the case.

But if it be true that these planets once had a common origin in one large planet, it is possible to find the variable nature of the elements of their orbits to such a degree of exactness as to trace them back to that origin — define the place where, and the time when, the separation must have occurred.

If, however, a planet should burst at one time, and afterward one or more of the fragments burst, there could be no tracing to a common origin; hence it is possible that the asteroids in question may have a common origin, and it be wholly beyond the power of man to show it.

CHAP. II. that the planet will move in a unit of time, when at the extremity of its shorter axis; then $\frac{1}{2}aB$ will express the *area* described in that unit of time.

But as equal areas are described in equal times, as often as this area is contained in the whole ellipse will be the number of such units in a revolution. Put $t=$ that number, or the time of revolution; then

$$t=\frac{(3.1416)AB}{\frac{1}{2}aB}=\frac{2(3.1416)A}{a}.$$

Let A' and B' be the semiaxes of any other ellipse; a' the velocity at the extremity of B', and t' the time of revolution;

then will - - $t'=\frac{2.(3.1416)A'}{a'}$

By comparing these equations, and rejecting common factors, we have - $t : t' :: \frac{A}{a} : \frac{A'}{a'}$.

But by Art. 162, $a=\sqrt{\frac{2M}{A}}$, and $a'=\sqrt{\frac{2M}{A'}}$

M mass of sun); and putting the values of a and a', in the above proportion, we have

$$t : t' :: A\frac{\sqrt{A}}{\sqrt{2M}} : \frac{A'\sqrt{A'}}{\sqrt{2M}};$$

Or, - - $t : t' :: A\sqrt{A} : A'\sqrt{A'}$.

By squaring $t^2 : t'^2 :: A^3 : A'^3$; which is Kepler's third law.

Eccentricities of the planetary orbits change by their mutual attractions.

(166.) We have seen, in articles 163 and 164, that the eccentricity of an orbit depends on the direction of the motion to the radius vector, when the planet is at mean distance. If that direction is at right angles to the radius vector at that time, then the eccentricity is nothing. If its direction is very acute, then the eccentricity is very great, &c.

Now suppose another planet to be situated at B (Fig. 32); its attraction on the planet, passing along in the orbit $p\,a$, is to give the velocity, a, a direction more at right angles to

Sp, and thus to diminish the eccentricity of the orbit. If the disturbing body, *B*, were anywhere near the line *C S*, its tendency would be to increase the eccentricity; and thus, in general, *A disturbing body near a line of the shorter axis of an orbit, has a tendency to diminish the eccentricity of the orbit of the disturbed body; and, anywhere near a line of the greater axis, has a tendency to increase the eccentricity.* Hence the eccentricities of the planets change in consequence of their mutual attractions; but their *mean distances never change.*

The mean distances never vary.

(167.) As the time of revolution is always the same for the same mean distance, whatever be the eccentricity of the orbit, therefore if we conceive a planet to turn into an infinitely eccentric orbit, and fall directly to the sun, the time of such fall would be *half a revolution*, in an orbit of half its present mean distance, as we perceive, by inspecting Fig. 34.

The principles and the computation of the time required for the planets to fall to the sun.

Fig. 34.

Hence, by Kepler's third law, we can compute the time that would be required for any planet to fall to the sun. Let x represent the time a planet would revolve in this new and infinitely eccentric orbit; then, by Kepler's law,

$$t^2 : x^2 :: 2^3 : 1^3, \text{ or, } x^2 = \frac{t^2}{8}.$$

Therefore half of the revolution, or simply the time of the fall, must be expressed by $\frac{t}{2\sqrt{8}}$, or, $\frac{t}{4\sqrt{2}}$; that is, to find the time in which any planet would fall to the sun, if simply abandoned to its gravity, or the time in which any secondary planet would fall to its primary, *divide its time of revolution by four times the square root of two.*

By applying this rule, we find that

	Days.	h.	m.
Mercury would fall to the sun in..............	15	13	13
Venus, ...	39	17	19
Earth, ...	64	13	39
Mars, ...	121	10	36
Jupiter,......................................	765	21	36
Saturn, ...	1901	23	24
Uranus, ...	5424	16	52

The moon would fall to the earth in 4 d. 19 h. 54 m. 36 s.

CHAPTER III.

MASSES OF THE PLANETS — DENSITIES — PRESSURE ON THEIR SURFACES.

CHAP. III.

Masses measured by attraction.

(168.) If the earth contained more matter, it would attract with greater force; and if the sun has a greater power of attraction than the earth, it is because it contains more *matter* than the earth; and therefore, if we can find the relative degree of attraction between two bodies, we have their relative masses of matter.

If the earth and sun have the same amount of matter, they will attract equally at equal distances. Let M be the mass of the sun, and E the mass of the earth, then (at the same unit of distance), *the attraction of the sun is, to the attraction of the earth, as M to E.*

But attraction is inversely as the square of the distance. Hence the attraction of the sun at D distance, is $\frac{M}{D^2}$; and the attraction of the earth at R distance is $\frac{E}{R^2}$.

Gravity of the sun is measured by the deviation of the earth from a tangent of its orbit.

The earth is made to deviate from a tangent of its orbit by the attraction of the sun; and the moon is made to deviate from a tangent of its orbit by the attraction of the earth, and the amount of these deviations will give the respective amounts of solar and terrestrial gravity.

If we take any small period of time, as a minute or a second, and compute the versed sine of the arc which the earth describes in its orbit during that time, such a quantity will express the sun's attraction; and if we compute the versed sine of the arc which the moon describes in the same time, that quantity will express the attraction of the earth.

How to compute the comparative masses of the sun and earth

In Figure 30, Art. 158, F represents the versed sine of an arc; and if we take D to represent the mean distance between the earth and sun, and consider the orbit a circle (as we may without error, 164), the whole circumference is

πD ($\pi = 6.2832$). Divide the whole circumference by the number of minutes in a revolution; say T, and the quotient will represent the arc a (Fig. 30). When T is very small, and of course a very small, the chord and arc *practically* coincide; and by the well known property of the circle, we have

CHAP. III.

$$2D : a :: a : F; \quad \text{Or,} \quad F = \frac{a^2}{2D} \quad . \qquad (1)$$

$$\text{But } a = \frac{\pi D}{T}; \text{ hence, } a^2 = \frac{\pi^2 D^2}{T^2}, \quad \text{and } \frac{a^2}{2D} = \frac{D\pi^2}{2T^2};$$

That is, $F = \frac{\pi^2 D}{2T^2}$; which is an expression for the sun's attraction at the distance of the earth. But $\frac{M}{D^2}$ is also an expression for the sun's attraction at the same distance; therefore,

$$\frac{M}{D^2} = \frac{\pi^2 D}{2T^2}; \quad \text{Or,} \quad M = \frac{\pi^2 D^3}{2T^2}.$$

In the same manner, if R represents the radius of the lunar orbit; t the number of minutes in the revolution of the moon; the mass of the central attracting body (in this case the earth) must be expressed by

$$E = \frac{\pi^2 R^3}{2t^2}.$$

Therefore,
$$E : M :: \frac{R^3}{t^2} : \frac{D^3}{T^2}.$$

This proportion gives a relation between the masses of the earth and sun *expressed in known quantities*.

If we assume unity for the mass of the earth, we shall have for the mass of the sun,

$$M = \frac{t^2 D^3}{T^2 R^3}. \quad . \quad . \quad . \quad (A)$$

(169.) This is a very general equation, for D may represent the radius of the earth's orbit, or the orbit of Jupiter or Saturn, and T will be the corresponding time of revolution. Also R may represent the radius of the lunar orbit, or the

The general application of this equation.

CHAP. III. orbit of one of Jupiter's or Saturn's moons, and then t will be its corresponding time of revolution.

The results of the equation will not be perfectly accurate, and why?

This equation, however, is not one of strict accuracy, as the distance a planet falls from the tangent of its orbit, in a definite moment of time, is not, accurately $\frac{M}{D^2}$, but $\frac{M+E}{D^2}$ (see 156), E being the mass of the planet. The force which retains a moon in its orbit is not only the attracting mass of the central body, but that of the moon also. But the planets being very small in relation to the sun, and in general the masses of satellites being very small in respect to their primaries, the errors in using this equation will in general be very small. The error will be greatest in obtaining the mass of the earth, as in that case the equation involves the periodic time of the moon; which period is different from what it would be were the moon governed by the attraction of the earth alone; but the mass of the moon is no inconsiderable part of the entire mass of both earth and moon; and also the attraction of the sun on the combined mass of the earth and moon, prolongs the moon's periodical time by about its 179th part.

Corrections for equation (A).

With these corrections the equation will give the mass of the sun to a great degree of accuracy; but we can determine the mass of the sun by the following method:

A more accurate equation.

From Art. 155, we learn that the attraction of the earth at the distance to the sun, is $g\left(\frac{r^2}{D^2}\right)$.

By Art. 168, we have just seen that the attraction of the sun on the earth, is $\frac{\pi^2 D}{2T^2}$; therefore,

$$E : M :: g\frac{r^2}{D^2} : \frac{\pi^2 D}{2T^2}.$$

Taking the mass of the earth as unity, we have

$$M=\frac{\pi^2 D^3}{2gr^2 T^2}, \qquad (B)$$

Equation (B) is more accurate than equation (A),

because (B) does not involve the periodical revolution of the moon, which requires correction to free it from the effects of the sun's attraction. To obtain a numerical expression for the mass of the sun, M, the numerator and denominator of the right hand member of equation (B) must be rendered homogeneous; and as g, the force of gravity of the earth, is expressed in feet (corresponding to T in seconds), therefore r the mean radius of the earth, and D the distance to the sun, must be expressed in feet. But from the sun's horizontal parallax, we have the ratio between r and D (see 127), which gives $D = 23984\,r$.

Chap. III.

How to obtain the numerical result.

This reduces the fraction to $\frac{\pi^2(23984)^3 r}{2gT^2}$. But to express the whole in numbers, we must give each symbol its value; that is, $\pi = 6.2832$; $r = (3956)(5280)$; $g = 16.1$; $T = 31558150$, the number of seconds in a sidereal year.

Therefore, $$M = \frac{(6.2832)^2(23984)^3(3956)(5280)}{(32.2)(31558150)^2}.$$

It would be too tedious to carry this out, arithmetically, without the aid of logarithms, and accordingly we give the logarithmetical solution, thus,

An example showing the great utility of logarithms

6 .2832 log. 0.798178×2	1 .596356
23 .984 log. 4.380000×3 . .	13 .140000
3956 log.	3 .597256
5280 log.	3 .722632
Logarithm of the numerator, ...	22 .056244

32.2 log.	1 .507856
31558150 log. 7.499114×2	14 .998228
Logarithm of the denominator, . .	16 .506084
Therefore M = 354945, whose log. is	5 .550160

The mass of the sun determined.

That is, the mass or force of attraction in the sun is 354945 times the mass or attraction of the earth. La Place

12

CHAP. III. says it is 354936 times; but the difference is of no consequence.

Equation (A) gives 350750; but equation (B), as we have before remarked, is far more accurate, and the result here given, agrees, within a few units, with the best authorities.

Equation (B) is not general; it will only apply to the relative masses of sun and moon, because we do not know the element g, the attraction, on the surface of any other planet, except the earth. That is, we do not know it as a primary fact; we can deduce it after we shall have determined the mass of a planet.

Equation (A) is general, and although not accurate, when applied to the earth and sun, is sufficiently so when applied to finding the masses of Jupiter, Saturn, or Uranus; because these planets are so remote from the sun, that the revolutions of their satellites are not *troubled* by the sun's attraction.

To find the masses of Jupiter, Saturn, and Uranus.

(170.) To find the mass of Jupiter (or which is the same thing, the mass of the sun when Jupiter is taken as unity), we conceive the earth to be a moon revolving about the sun, and compare it with one of Jupiter's satellites revolving round that body. To apply equation (A), let the radius of the earth equal *unity*, then the radius of Jupiter must be 11.11 (Art. 131); and as observation shows the radius of Jupiter's 4th satellite is 26.9983 times its equatorial radius, therefore the distance from the center of Jupiter to the orbit of its 4th satellite, must be the following product (11.11) (26.9983), which corresponds to R in the equation. $D=23984$; $T=365.256$; $t=16.6888$.

Therefore, by applying equation (A), ($M=\frac{t^2D^3}{T^2R^3}$); we have

$$M=\frac{(16.6888)^2(23984)^3}{(365.256)^2(11.11)^3(26.9983)^3}.$$

By logarithms	16.6888 log.	1.222410×2 .	2.444820
	23984 log.	4.380000×3 .	13.140000
Logarithm of the numerator,		. .	15.584820

365.256, log.	2 .562600×2	.	5 .125200
11.11, log.	1 .045714×3	.	3 .137142
26.9983, log.	1 .431320×3	.	4 .293960
Logarithm of the denominator,	.	.	12 .556302
Therefore $M = 1068$*, log. .	.	.	3 .028518

This result shows that the mass of the sun is 1068 times the mass of Jupiter; but we previously found the mass of the sun to be 354945 times the mass of the earth, and if unity is taken for the mass of the earth, and J for the mass of Jupiter, we shall have

$$1068\,J = 354945;$$

because each member of this equation is equal to the mass of the sun.

By dividing both members of this equation by 1068, we find the mass of Jupiter to be 332 times that of the earth; but in Art. 132, we found the *bulk* of Jupiter to be 1260 times the bulk of the earth; therefore the density of Jupiter is much less than the density of the earth. The mass of Jupiter compared to that of the earth.

In the same manner we may find the masses of Saturn and Uranus — the former is 105.6 times, and the latter 18.2 times the mass of the earth. The masses of Saturn and Uranus.

The principles embraced in equation (A) apply only to those planets that have satellites; for it is by the rapid or slow motion of such satellites that we determine the amount of the attractive force of the planet.

In short, the masses of those planets which have satellites, are known to great accuracy; but the results attached to others in table IV, must be regarded as near approximations. What results may be considered accurate.

The slight variations which the earth's motion experiences by the attractions of Venus and Mars, are sufficiently sensible to make known the masses of these planets; and M. Burckhardt gives $\frac{1}{405871}$ for Venus, and $\frac{1}{2546320}$ for Mars (the mass of the sun being unity). Mercury he put down at The masses of Venus, Mars, and Mercury.

---

* This is a correct result according to these data; but more modern observations, in relation to the micrometic measure of Jupiter, and the distance of his satellites, give results a little different, as expressed in table IV.

CHAP. III

$\frac{1}{2025810}$; but this result is little more than hypothetical, as it is drawn from its volume, on the supposition that the densities of the planets are reciprocal to their mean distances from the sun; which is nearly true for Venus, the earth, and Mars.

By means of gravity and the lunar parallax, we may find the diameter of the earth.

(171.) It may be astonishing, but it is nevertheless true, that by means of equations (A) and (B) *we can find the diameter of the earth* to a greater degree of exactness than by any one actual measurement.

We have several times observed that equation (A) is not accurate when used to find the masses of the earth and sun, because it contained the time of the revolution of the moon; which revolution is *accelerated* by the gravity of the moon, and *retarded* by the action of the sun.

Therefore, to make equation (A) accurately express the mass of the sun, the element t^2 requires two corrections, which will be determined by subsequent investigation. The first is an increase of $\frac{1}{75}$th part; the second is a diminution of $\frac{1}{358}$th part, and both corrections will be made if we take $\frac{76{\cdot}358}{75{\cdot}359}t^2$ in place of t^2.

A common axiom.

Then having two correct expressions for the mass of the sun, those two expressions must equal each other; that is,

$$\frac{76{\cdot}358\, t^2 D^3}{75{\cdot}359\, T^2 R^3} = \frac{\pi^2 D^3}{2 g r^2 T^2}.$$

By suppressing common factors, we have

$$\frac{76{\cdot}358\, t^2}{75{\cdot}359\, R^3} = \frac{\pi^2}{2 g r^2}.$$

In this equation r represents the mean radius of the earth, and we will suppose it unknown; the equation will then make it known.

The relation between R, the mean radius of the lunar orbit, and r, the mean radius of the earth, is given by means of the moon's horizontal parallax.

Equatorial horizontal parallax and

The moon's *equatorial* horizontal parallax, as we have seen, (65) is 57′ 3″; but the horizontal parallax for the mean radius, is 56′ 57″; this makes $R = (60.36)\,r$, whatever the numerical value of r may be. Put this value of R in the preceding equation, and suppress the common factor r^2, we then have

CHAP. III. Mean horizontal parallax.

$$\frac{76{\cdot}358\,t^2}{75{\cdot}359\,(60.36)^3 r} = \frac{\pi^2}{2g}.$$

Therefore,

$$r = \frac{2g\,76{\cdot}358\,t^2}{75{\cdot}359\,(60.36)^3 \pi^2}.$$

As g is expressed *in feet*, and corresponds to t in seconds, the numerical value of r will be in *feet*, which divided by 5280, the number of feet in a mile, will give the number of miles in the mean radius or mean semidiameter of the earth; and by applying the preceding equation, giving g, t, and π, their proper values; and by the help of logarithms, we readily find $r = 3953$ miles; only three miles from the most approved result; and we do not hesitate to say, that this result is more to be relied upon than any other.

Confidence in the result.

MASS OF THE MOON.

(172.) Approximations to the mass of the moon have been determined, from time to time, by careful observations on the tides; but it is in vain to look for mathematical results from this source; for it is impossible to decide whether any particular tide has been accelerated or retarded, augmented or diminished, by the winds and weather; and if not affected at the place of observation, it might have been at remote distances; but notwithstanding this objection, the mass of the moon can be pretty accurately determined by means of the tides, owing to the great number and variety of observations that can be brought into the account; and we shall give an exposition of this deduction hereafter; but at present we shall confine our attention to the following simple and elegant method of obtaining the same result.

The mass of the moon originally determined from observations on the tides.

If the moon had no mass; that is, if it were a mere material point, and was not disturbed by the attraction of the sun, then the distance that the moon would fall from a tangent of its orbit, in one second of time, would be just equal

CHAP. III.

to $\frac{gr^2}{R^2}$. (Art. 155.) In this expression g, r, and R, represent the same quantities as in the last article. The distance that the moon actually falls from a tangent of its orbit, in one second of time, is equal to the versed sine of the arc it describes in that time, and the analytical expression for it is found thus:

Let πR represent the circumference of the lunar orbit, and if t is put for the number of seconds in a mean revolution, then $\frac{\pi R}{t}$ represents the arc corresponding to the moon's motion in one second (Fig. 30), and as this so nearly coincides with a chord, we have

$$2R \; : \; \frac{\pi R}{t} \; :: \; \frac{\pi R}{t} \; : \; \frac{\pi^2 R}{2 t^2}.$$

An expression for the distance the moon falls in one second of time.

Hence, we perceive, that $\frac{\pi^2 R}{2 t^2}$ is the distance that the moon would fall from the tangent of its orbit in one second of time, if it were undisturbed by the action of the sun; but we can free it from such action by multiplying it by $\frac{359}{358}$, as we shall show in a subsequent chapter. That is, the attraction of both the earth and moon, at the distance of the lunar orbit, is $\frac{359 \pi^2 R}{358 \cdot 2 t^2}$.

But the attraction of the earth alone, at the same distance, is $\frac{g r^2}{R^2}$; and comparing these quantities with the more general expressions in Art. 156, we have

$$\frac{E}{R^2} \; : \; \frac{E+m}{R^2} \; :: \; \frac{g r^2}{R^2} \; : \; \frac{359 \pi^2 R}{358 \cdot 2 t^2}.$$

By suppressing the common denominator, in the first couplet, and calling E, the mass of the earth, unity, the proportion reduces to

$$1 \; : \; 1+m \; :: \; g r^2 \; : \; \frac{359 \pi^2 R^3}{358 \cdot 2 t^2}.$$

As in the last article, $R=(60.36)r$, and this value put for R^3, and reduced, gives

$$1 \ : \ 1+m \ :: \ g \ : \ \frac{359\pi^2(60.36)^3 r}{388{\cdot}2t^2};$$

The result.

Therefore, - - $1+m=\dfrac{359\pi^2(60.36)^3 r}{358{\cdot}2t^2 g}$.

This fraction, as well as the one in the last article, can be reduced arithmetically; but the operation would be too tedious; they are both readily reduced by logarithms, by which we found $1+m=1.01301$; hence $m=.01301$, which is a little greater than $\frac{1}{78}$th. Laplace says $\frac{1}{75}$th of the earth is the true mass of the moon; and this value we shall use.

Result given by Laplace.

THE DENSITIES OF BODIES.

Standard for density.

(173.) The density of a body is only a comparative term, and to find the comparison, some one body must be taken as the standard of measure. The earth is generally taken for that standard.

It is an axiom, in philosophy, that the same mass in a smaller volume, must be greater in density; and larger in volume, must be less in density; and, in short, the density must be directly proportional to the mass, and inversely proportional to the volume; and if the earth is taken for unity in *mass*, and unity in *volume*, then it will be unity in density also; and the density of any other planetary body will be *its mass divided by its volume;* and if its volume is not given, the density may be found by the following proportion, in which d represents the density sought, and r the radius of the body; *the radius of the earth being unity.* The proportion is drawn from the consideration that spheres are to one another as the cubes of their radii.

$$\frac{1}{1} \ : \ \frac{mass}{r^3} \ :: \ 1 \ : \ d; \text{ hence } d=\frac{mass}{r^3}.$$

Expression for the densities of

From this equation we readily find the density of the sun, for we have its mass (354945), and its semidiameter 111.6 times the semidiameter of the earth (Art. 156); therefore its

CHAP. III.

spheres compared to the density of the earth.

density must be $\frac{354945}{(111.6)^3}=0.254$, or a little more than $\frac{1}{4}$th the density of the earth.

The mass of Jupiter is 332 times that of the earth, and its volume is 1260 times the volume of the earth; therefore the density of Jupiter is $\frac{332}{1260}=0.264$; which is a little more than the density of the sun.

Densities of Jupiter, moon, &c.

The mass of the moon is $\frac{1}{75}$, and its volume $\frac{1}{49}$, therefore its density is $\frac{1}{75}$ divided by $\frac{1}{49}$, or $\frac{49}{75}=0.6533$; about $\frac{2}{3}$ the density of the earth.

From these examples the reader will understand how the densities were found, as expressed in table IV.

GRAVITY ON THE SURFACE OF SPHERES.

Gravity on the surfaces of the other planets, how found

(174.) The gravity on the surface of a sphere depends on the mass and volume. The attraction on the surface of a sphere is the same as if its whole mass were collected at its center; and the greater the distance from the center to the surface, the less the attraction, in proportion to the *square* of the distance: but here, as in the last article, some one sphere must be taken for the unit, and we take the earth, as before.

The mass of the sun is 354945, and the distance from its center to its surface is 111.6 times the semidiameter of the earth; therefore a *pound*, on the surface of the earth, is to the pressure of the same mass, if it were on the surface of the sun, as $\frac{1}{1}$ to $\frac{354945}{(111.6)^2}$, or as 1 to 28 nearly. That is, one pound on the surface of the earth would be nearly 28 pounds on the surface of the sun, if transported thither.

The mass of Jupiter is 332, and its radius, compared to that of the earth, is 11.1 (Art. 131); therefore one pound, on the surface of the earth, would be $\frac{332}{(11.1)^2}$, or 2.48 pounds on the surface of Jupiter; and by the same principle, we can compute the pressure on the surface of any other planet. **Results will be found in table IV.**

CHAPTER IV.

PROBLEM OF THE THREE BODIES.—LUNAR PERTURBATIONS

CHAP. IV.

The theory of gravity.

(175.) By the theory of universal gravitation, every body in the universe attracts every other body, in proportion to its mass; and inversely as the square of its distance; but simple and unexceptionable as the law really is, it produces very complicated results, in the motions of the heavenly bodies.

The complexity of results.

If there were but two bodies in the universe, their motions would be comparatively simple, and easily traced, for they would either fall together or circulate around each other in some one undeviating curve; but as it is, when two bodies circulate around each other, every other body causes a deviation or vibration from that primary curve that they would otherwise have.

The final result of a multitude of conflicting motions cannot be ascertained by considering the whole in mass; we must take the disturbance of one body at a time, and settle upon its results; then another and another, and so on; and the sum of the results will be the final result sought.

The problem of the three bodies.

We, then, consider two bodies in motion disturbed by a third body; and to find all its results, in general terms, is the *famous problem of "the three bodies;"* but its complete solution surpasses the power of analysis, and the most skillful mathematician is obliged to content himself with approximations and special cases. Happily, however, the masses of most of the planets are so small in comparison with the mass of the sun, and their distances so great, that their influences are insensible.

We shall make no attempt to give minute results; but we hope to show general principles in such a manner, that the reader may comprehend the common inequalities of planetary motions.

Abstract attraction.

Let m, Fig. 35, be the position of a body circulating around another body A, moving in the direction PmB, and disturbed by the attraction of some distant body D.

CHAP IV.

Fig. 35.

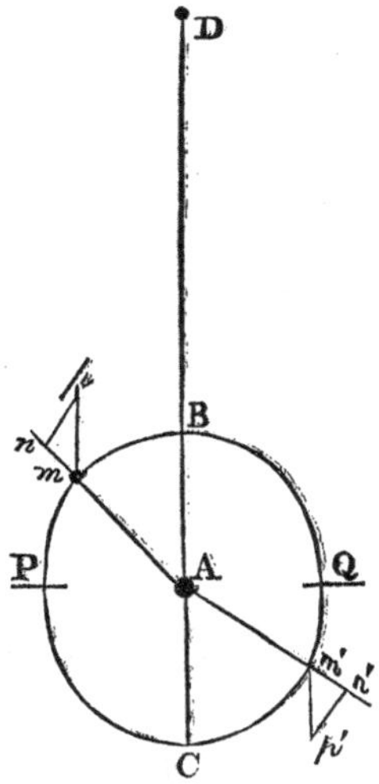

We now propose to show some of the most general effects of the action of D, *without paying the least regard to quantity.*

Two bodies equally attracted in parallel lines are not affected in their mutual relations

If A and m were equally attracted by D, and the attraction exerted in parallel lines, then D would not disturb the mutual relations of A and m. But while m is nearer to D than A is to D, it must be more strongly attracted, and let the line $m p$ represent this *excess of attraction.* Decompose this force (see Nat. Phil.) into two others, $m n$ and $n p$, the first along the line $A m$, the other at right angles to it.

The first is a *lifting force* (called by astronomers the radial force), the other is a *tangental force*, and affects the motion of m. It will accelerate the motion of m, while acting with it, from P to B; and retard its motion, while acting against it, from B to Q.

Wé must now examine the effect, when the revolving body is at m', a greater distance from D than A is from D.

Now A is more strongly attracted than m', and *the result* of this unequal attraction is the same as though A were not attracted at all, and m' attracted the other way by a force equal to the difference of the attractions of D on the two bodies A and m'. Let this difference be represented by the line $m' p'$, and decompose it into two other forces, $m' n'$ and $n' p'$, the first a lifting force, the other the tangental force.

The *rationale* of this last position may not be perceived by every reader; and to such we suggest, that they conceive A and m' joined together by an inflexible line $A m'$, and both A and m' drawn toward D, but A drawn a greater distance than m'. Then it is plain that the *position* of the line $A m'$ will be changed: the angle $D A m'$ will become greater, and the angle $C A m'$ less; that is, the motion of m' will be accelerated from Q to C, but from C to P it will be retarded.

In short, the motion of m will be accelerated when moving toward the line D B C, and retarded while moving from that line. That is, retarded from B to Q, accelerated from Q to C, retarded from C to P, and again accelerated from P to B.

The disturbing body constantly urges a revolving body to the line of syzigies.

If we conceive A to be the earth, m the moon, and D the sun; then $D B C$ is called the line of the *syzigies*, a term which means the plane in which conjunctions and oppositions take place. At the point B the moon falls in conjunction with the sun, and is new moon; at the point C it is in opposition, or full moon.

Fig. 36.

(176.) Conceive a ring of matter around a sphere, as represented in Fig. 36, and let it be either attached or detached from the sphere, and let D be *not* in the plane of the ring.

Action of an attracting body on a ring.

From what was explained in the last article, the particles of matter at m are constantly urged toward the line $D B C$, and the particles at m' are constantly urged toward the same line; that is, the attraction of D, on the ring, has a tendency to diminish its inclination to the line $D B C$; and its position would be changed by such attraction from what it would otherwise be; and if the ring is attached to the sphere, the sphere itself will have a slight motion in consequence of the action on the ring.

Now there is, in fact, a broad ring attached to the equatorial part of the earth, giving the whole a spheroidal form; and the plane of the equator is in the plane of the ring.

Cause of nutation

When the sun or moon is without the plane of this ring, that is, without the plane of the equator, their attraction has a tendency to draw the plane of the equator toward the attracting body, and actually does so draw it; which motion is called *nutation*. How this motion was discovered, and its amount ascertained, will be explained in a subsequent chapter.

(177.) We may conceive the line $D B C$ to be in the

CHAP. IV. plane of the ecliptic, D the sun, and the ring around the earth the moon's orbit, inclined to the plane of the ecliptic with an angle of about *five degrees;* then when the sun is out of the plane of the ring, or moon's orbit, the action of the sun has a constant tendency to bring the moon into the ecliptic, and by this tendency the moon does fall into the ecliptic from either side sooner than it otherwise would.

Application of the ring to the lunar orbit.

The moon's nodes retrograde.

The point where the moon falls into the ecliptic is called the *moon's node;* and by this external action of the sun the moon falls into the ecliptic from its greatest inclination before it describes 90°, and goes from node to node before it describes 180° — and hence we say that the moon's nodes fall *backward* on the ecliptic. The rate of retrogradation is 19° 19′ in a year, making a whole circle in about 18.6 years.

Fig. 37.

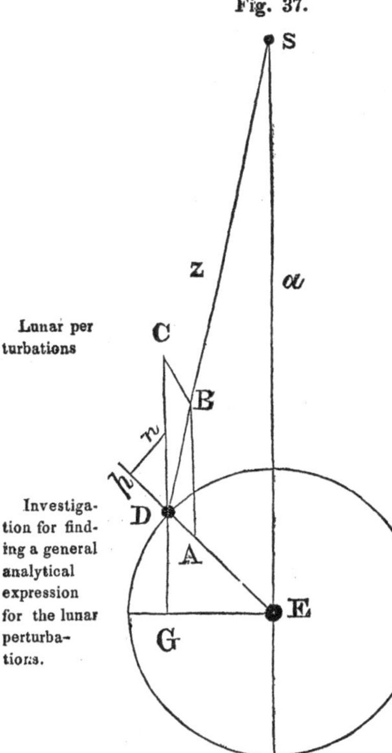

Lunar perturbations

(178.) We are now prepared to be a little more definite, and inquire as to the amount of some of the lunar irregularities.

Investigation for finding a general analytical expression for the lunar perturbations.

Let S be the mass of the sun, E that of the earth, and m the moon, situated at D. Let a be the mean distance between the earth and sun, z the distance between the sun and moon, and r the mean radius of the lunar orbit. Let the moon have any indefinite position in its orbit. (It is represented in the figure at D.)

The attraction of the sun on the earth is $\frac{S}{a^2}$, the attraction of the sun on the moon is $\frac{S}{z^2}$; and the attraction of the earth and moon, on the moon, is $\frac{E+m}{r^2}$. (Art. 156.)

Let the line DB, the diagonal of the parallelogram AC, be the attraction of the sun on the moon, and decompose it into the two forces DA and DC; the first along the lunar radius vector, the other parallel to SE.

The two triangles CDB and DSE are similar, and give the proportion $a : z :: CD : DB$. But $DB = \frac{S}{z^2}$;

Therefore $CD = \frac{aS}{z^3}$. By a similar proportion we find $DA = \frac{rS}{z^3}$.

Let the angle SED be represented by x, then DG will be expressed by $r \cos. x$, and SDG will be a right line *nearly*, for the angle DSE is never greater than 7′.

Now if the force DC, which is parallel to SE, is only equal to the force of the sun's attraction on the earth, it will not disturb the mutual relations of the earth and moon. The force of the sun's attraction on the earth is $\frac{S}{a^2}$; and as this must be less than the force of attraction on the moon when the moon is at D, conceive it represented by the line Cn, and subtracted from CD, will leave Dn the excess of the sun's attraction on the two bodies, the earth and the moon; and this alone constitutes the disturbing force of the moon's motion;

An expression for the whole disturbing force.

That is, $Dn = CD - Cn = \frac{aS}{z^3} - \frac{S}{a^2}$;

Or $Dn = aS\left(\frac{1}{z^3} - \frac{1}{a^3}\right)$, *the disturbing force.* Decompose this force (Dn) into two others, Dp and pn, by means of the right angled triangle Dpn; the angle pDn being equal to DES, which we represent by x.

CHAP. IV.

Whence $Dp = Sa\left(\frac{1}{z^3} - \frac{1}{a^3}\right)$ cos. x;

And $pn = Sa\left(\frac{1}{z^3} - \frac{1}{a^3}\right)$ sin. x.

The force DA, *i. e.* $\left(\frac{rS}{z^3}\right)$ is called the *additious* force;

The radial force.

the force Dp the *ablatitious* force. The difference of these two forces is called the radial force; that is

$Sa\left(\frac{1}{z^3} - \frac{1}{a^3}\right)$ cos. $x - \frac{rS}{z^3} =$ the radial force; pn is the tangental force.

Expression of the radial force at the quadratures.

When the angle x is equal to 90°, cos. $x = o$, $SD = SE$, or $z = a$; which values, substituted, give $-\frac{rS}{a^3}$ for the value of the radial force at the quadratures, and its tendency there is to increase the gravity of the moon to the earth. When the angle x is zero (the moon is in conjunction with the sun) the cos. $x = 1$, and the radial force becomes

$$\frac{Sa}{z^3} - \frac{Sa}{a^3} - \frac{rS}{z^3}; \text{ or } \frac{S\,(a-r)}{z^3} - \frac{Sa}{a^3}.$$

But at that point $z = (a - r)$, which value substituted, and rejecting the comparatively very small quantities in both numerator and denominator, we have, for the radial force at conjunction, $\frac{2r\,S}{a^3}$.

When the angle $x = 180°$ (the moon is in opposition to the sun), cos. $x = -1$, and the force becomes

$$\frac{Sa}{a^3} - \frac{Sa}{z^3} - \frac{rS}{z^3}; \text{ or } \frac{S}{a^2} - \frac{S\,(a+r)}{z^3}.$$

But at this point $z = a + r$, which, substituting as before, and we have for the radial force in opposition $\frac{2r\,S}{a^3}$, the same expression as at conjunction.

If we compare the radial force at the syzigies with the expression for it at the quadratures, we shall find it the same in form, but double in amount and opposite in sign, showing that it is opposite in effect.

(179.) As the radial force increases the gravity of the moon to the earth at the quadratures, and diminishes it at the syzigies, there must be points in the orbit symmetrically situated, in respect to the syzigies, where the radial force neither increases nor diminishes the gravity, and of course its expression for those points must be zero; and to find these points we must have the equation

Points where the radial force is zero.

How to find them.

$$Sa\left(\frac{1}{z^3}-\frac{1}{a^3}\right)\cos. x-\frac{rS}{z^3}=0 \quad . \quad . \quad (1)$$

By inspecting the figure we perceive that the line SDG is in value nearly equal to the line SE, and for all points in the orbit we have

$$z=a\pm r\cos. x. \quad . \quad . \quad . \quad . \quad . \quad . \quad (2)$$

Reducing equation (1), we have

$$(a^3-z^3)\cos. x=ra^2. \quad . \quad . \quad . \quad (3)$$

Cubing (2),

$$z^3=a^3\pm 3a^2 r\cos. x\mp 3ar^2\cos.^2 x\pm r^3\cos.^3 x.$$

As r is very small in relation to a, the terms containing the powers of r, after the first, may be rejected; we then have

$$(a^3-z^3)=\mp 3a^2 r\cos. x. \quad . \quad . \quad (4)$$

This value substituted in (3), and reduced, gives

$$\mp 3\cos.^2 x=1.$$

Result of the radial force at the quadratures and syzigies.

Hence $\cos. x=\sqrt{\tfrac{1}{3}}$ and $x=54^\circ\ 44'$, or the points are $35^\circ\ 16'$ from the quadratures.

This shows that at the quadratures, and about 35° on each side of them, the gravity of the moon is increased by the action of the sun, and at the syzigies, and about 54° on each side of them, the gravity is diminished; and the diminution in the one case is double the amount of increase in the other, and by the application of the differential calculus we learn that the mean result, for the entire revolution, is a diminution whose analytical expression is $\frac{rS}{2a^3}$, an expression which holds a very prominent place in the lunar theory; the

Mean radial force.

CHAP. IV. result of which we have used in Art. 171, and there stated it to be $\frac{1}{358}$th part of the force that retained the moon in its orbit.

Value of the mean radial force, and how found.

But how do we know this to be its numerical value, is a very serious inquiry of the critical student?

The force that retains the moon in its orbit is $\frac{E+m}{r^2}$ (Art. 156); and if the radial force can be rendered *homogeneous* with this, some numerical ratio must exist between them. Let x represent that ratio, and we must find some numerical value for x to satisfy the following equation:

$$\frac{rS}{2a^3}x = \frac{E+m}{r^2}. \quad . \quad . \quad . \quad . \quad . \quad (\mathrm{A})$$

Therefore $$x = \frac{2\,(E+m)a^3}{r^3\,S};$$

calling $E=1$, $m=\frac{1}{75}$ (Art. 172), or $E+m$ is 1.013. $S = 354945$ (Art. 169), and the relation between the mean distance to the sun, and the mean radius of the lunar orbit, is 397.3,* therefore

$$x = \frac{(2.026)(397.3)^3}{354945} = 358;$$

or the coefficient to x, in equation (A), is *one three hundredth and fifty-eighth part of the force which retains the moon in its orbit.*

General effect of the radial force.

(180.) The mean radial force causes the moon to circulate at $\frac{1}{358}$th part greater distance from the earth than it otherwise would have, and its periodical revolution is increased by its 179th part; but this would cause no variation or irregularity in its distance or angular motion, provided its orbit were circular, and the earth and moon always at the same mean distance from the sun.

The radial force variable.

But we perceive the expression $\frac{rS}{2a^3}$ contains two variable quantities, r and a, which are not always the same in value; and, therefore, the value of the expression itself must be va-

* This relation is found by dividing the horizontal parallax of the moon, 56′ 57″, by the horizontal parallax of the sun, 8″.6.

riable; and it will be least when the earth is at the greatest distance from the sun, and, of course, the moon's motion will then be increased. But the earth's variable distance from the sun depends on the eccentricity of the earth's orbit; and hence we perceive that the same cause which affects the apparent solar motion, affects also the motion of the moon, and gives rise to an equation called the *annual equation** of the moon's motion. It amounts to 11′ in its maximum, and varies by the same law as the equation of the sun's center.

The annual equation of the moon's motion.

(181.) If we take the general expressions for the radial force, $Sa\left(\frac{1}{z^3} - \frac{1}{a^3}\right)$ cos. $x - \frac{rs}{z^3}$, and banish the letter z from it by means of the equation

A general expression for the radial force at any point of the moon's orbit.

$$z = a \pm r \text{ cos. } x$$

Or, $$z^3 = a^3 \pm 3a^2\, r \text{ cos. } x,$$

(neglecting the powers of r) and we shall have,

$$\frac{rS\,(3 \text{ cos. }^2 x - 1)}{a^3}$$

for an expression of the radial force corresponding to any angle x from the syzigy.

If we take the general expression for the line pn, the tangental force, and banish z, as before, we have,

tangental force $$= \frac{3rs \text{ cos. } x \text{ sin. } x}{a^3}.$$

By doubling numerator and denominator, this fraction can take the following form:

Expression for the tangental force.

$$\frac{3rs\,(2 \text{ cos. } x \text{ sin. } x)}{2a^3}.$$

But, by trigonometry, 2 cos. x sin. x = sin. $2x$,

Therefore the tangental force $= \frac{3rs \text{ sin. } 2x}{2a^3}$.

This expression vanishes when $x = o$ and $x = 90°$; for then sin. $2x$ = sin. 180 = 0. Hence the tangental force vanishes at the syzigies and quadratures, attains its maximum

Its vanishing points.

* This is equation I, in the Lunar Tables.

CHAP. IV. value at the octants, and *varies as the sine of the double angular distance of the moon from the sun.*

The tangental force greatest when the earth is in perigee.

The *mean maximum* for this force must be determined by observation. It is known by the name of *variation*, and by mere inspection we can see that its amount must correspond to the variations of r and of a^3. Hence, to obtain the moon's place, we must have correction on correction.

The variation amounts to about 35′. It increases the velocity of the moon from the quadratures to the syzigies, and diminishes it from the syzigies to the quadratures; hence, in consequence of the variation, the velocity of the moon is greatest at the syzigies, and least at the quadratures.

Application of the radial force to an elliptical orbit.

(182.) Let us now examine the effect of the radial force on the lunar orbit, considered as elliptical.

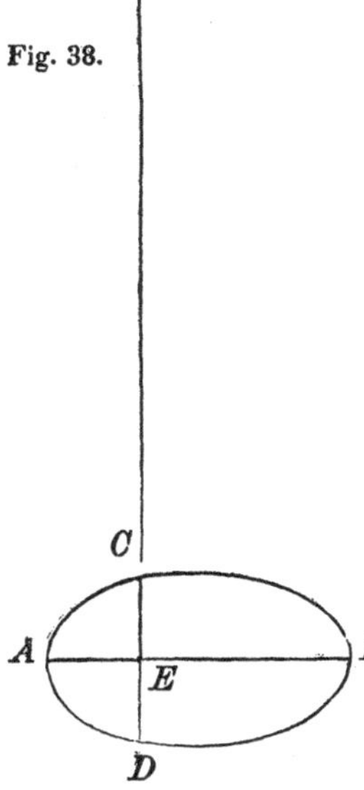

Fig. 38.

Let SE (Fig. 38) be at right angles to AB, the greater axis of the lunar orbit, and conceive ACB to represent the orbit that the moon would take if it were undisturbed by the sun.

But when the moon comes round to its perigee at A, it is in one of its quadratures, and the radial force then increases the gravity of the moon toward the earth by the expression $\frac{rs}{a^3}$. But here r *is less than its mean value*, and the expression is less than its mean, and therefore the moon is not *crowded* so near the earth as it otherwise would be, and, of course, at this point the moon will run farther from the earth.

At the point C, the radial force tends to increase the distance between the earth and moon, *and to widen the orbit.*

When the radial force

When the moon passes round to B, the radial force again increases the gravity of the moon, and r, in the expression

$\frac{rs}{a^3}$, *is greater than its mean value;* and, of course, *crowds* the moon nearer to the earth than it otherwise would go; and thus we perceive that the action of the radial force on an elliptical orbit has a tendency *to decrease the eccentricity of the ellipse, when the sun is at right angles to its greater axis.*

decreases the eccentricity of the lunar ellipse.

(183.) Now conceive the sun to be in a line, or nearly in a line, with the longer axis of the lunar orbit, as represented in Fig. 39.

When the radial force increases the eccentricity of the lunar orbit.

The radial force at the quadratures, *C* and *D*, has a tendency to press in the orbit, or *narrow* it. At the point *A*, the tendency, it is true, is to increase the distance between the earth and moon; but that tendency is not so strong as it would be if the moon were at its mean distance from the earth.

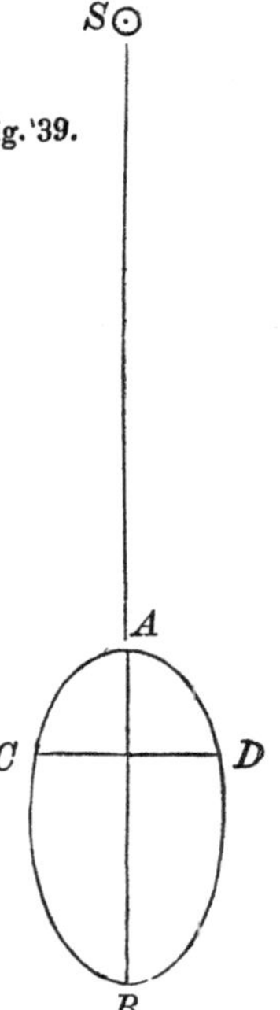

Fig. 39.

The tendency at *B* is to increase the distance, and it is a tendency greater than the medium. That is, the tendency at *A* is *less* than the medium; at *B*, greater than the medium; and at *C* and *D*, the compressed parts of the orbit, the tendency is to a still greater compression; therefore, the entire action of the radial force is *to increase the eccentricity of the lunar orbit, when the sun is in line, or nearly in line, with the longer axis.*

Thus, we perceive, that under the disturbing action of the sun, the eccentricity of the moon's orbit must be in a state of perpetual change, now more, now less, than its mean state.

Corresponding with this change of eccentricity there must be changes in the lunar motion; and to keep account of it, and allow for it, astronomers have formed a table called **EVECTION.**

CHAP. IV.

Effect of the radial force on the motion of the lunar perigee

(184.) Now let us examine the effect of the radial force on the position of the lunar apogee.

Fig. 40.

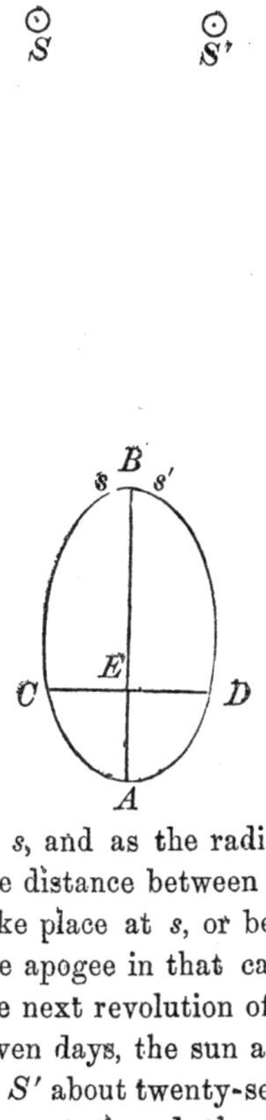

Let E (Fig. 40), be the earth, and, for the sake of simplicity, we conceive the earth to be stationary, and the sun and moon both to revolve about it with their apparent angular velocities; the moon in the orbit $A\ C\ B$, and in the direction $A\ C\ B$; the sun in a distant orbit, part of which is represented by $S\ S'$.

Let $A\ B$ be the greater axis of the moon's orbit, in its natural position, or as it would be if undisturbed by the sun; and being undisturbed, the perigee and apogee would remain constant at the points A and B; and the time from A to B, or from B to A, would be just equal to the mean time of half a revolution, as explained in a former part of this work.

Retrograde motion of the perigee and apogee.

Now let us conceive the sun to be in its orbit at S, then the moon will be in the syzigy when it comes round to s, and as the radial force at that point tends to increase the distance between the earth and the moon, the apogee will take place at s, or between s and B; and it is evident that the apogee in that case would recede or run back. But at the next revolution of the moon, in a little more than twenty-seven days, the sun at that time will, apparently, have moved to S' about twenty-seven degrees. Now the syzigy will take place at s', and the greatest distance between the earth and moon will now be between B and s', that is, the apogee will advance, in one revolution, from near s to near s'; and thus, in general, the longer axis of the moon's orbit is strongly inclined to follow the sun; and this is the source of its progressive motion. It makes a revolution in $3232\frac{1}{2}$ days; but its motion is very irregular, for, as we have just seen,

The major axis of the lunar orbit is inclined to follow the sun.

when the line which joins the earth and sun makes a very acute angle with the longer axis of the lunar orbit, and is approaching that axis, the motion of the apogee and perigee is retrograde; but, all of a sudden, when the sun passes the longer axis of the lunar orbit, the motion of the apogee becomes direct, and moves with considerable rapidity.

Under what position of the sun, the lunar perigee remains stationary.

When the sun is at right angles to the major axis of the moon's orbit, the tendency of the radial force is to diminish the eccentricity of the orbit, but it has no tendency to change the position of the axis.

From this investigation it follows, that when the sun has just passed the greater axis of the lunar orbit, the interval from apogee to apogee, or from perigee to perigee, will be greater than a revolution. Just before the sun arrives at the position of the longer axis, the time from one apogee to another is less than a revolution; and when the sun is at right angles to the longer axis, the time is just equal to a revolution in longitude.

Ancient eclipses compared with modern observations.

(185.) By comparing eclipses of the moon, observed by the ancient Egyptians and Chaldeans, with those of more modern times, Dr. Halley, and other astronomers, concluded that the periodic time of the moon is now a little shorter than at those remote periods; and to make these extreme observations agree with modern ones, it became necessary to conceive the moon's mean motion to be *accelerated about* 11 *seconds* per century.

The result.

For a long time this fact seriously perplexed astronomers: some were for condemning the theory of gravity as insufficient to explain the cause of the lunar perturbations, while others were for rejecting the facts, although as well established as any *mere historical* facts could be.

In this dilemma, says Herschel, "Laplace stepped in to rescue physical astronomy from reproach by pointing out the real cause of the phenomenon in question."

Although this subject troubled the greatest philosophers of the past age — the greatest mathematical philosophers the world ever saw — the problem is quite simple, now the solution is pointed out, and we are sure that every reader of or-

CHAP. IV. dinary capacity can understand it, provided he gives his serious attention to the subject.

A summary statement of the cause.

The secular acceleration of the moon's mean motion *is caused by a small change in the mean value of the radial force, occasioned by a change in the eccentricity of the earth's orbit.*

The expression $\frac{rS}{2a^3}$ is the mean radial force of the sun acting on the moon's orbit, dilating it and increasing the time of the lunar revolution.

When the moon's motion is increased.

If the earth's orbit had no eccentricity, $2a^3$, the denominator of the fraction, would always have the same value, and then regarding the numerator as constant, there would be no variation of the moon's motion arising from this cause. But in consequence of the earth and moon moving toward the apogee of the earth's orbit, a, of course, a^3 becomes greater, and the value of the radial force becomes less than its mean value, and in consequence of this, the moon's motion is *increased.* And when the earth and moon move toward the earth's perigee, a and a^3 become less, and the value of the radial force becomes greater than its mean; the moon's orbit is dilated to excess, and its motion is *diminished; and the orbit is more dilated when the earth is in perigee than it is contracted when the earth is in apogee.* In other words, the mean dilatation of the lunar orbit is greater, and the mean motion of the moon less, in proportion as the earth's orbit is more eccentric.

When diminished.

The expression for the mean radial force is not the true mean.

The less the value of $\frac{rS}{2a^3}$ the greater is the moon's mean motion, and that value is least when a is greatest. But a would have no variation of value if the earth's orbit were circular.

The earth's orbit, however, is eccentric, and in the course of a year the value of the radial force is exactly expressed by $\frac{rS}{2a^3}$ only at two instants of time, when the earth passes the extremities of the shorter axis of its orbit. At all other times a is either greater or less than its mean value, and the variations are equal on each side of it; that is, a becomes $(a-d)$ or $(a+d)$, and the radial force is really CHAP. IV.

$$\frac{rS}{2(a-d)^3} \text{ or } \frac{rS}{2(a+d)^3};$$

which expressions correspond to equal distances on each side of the mean distance, and d may have all values from 0 to ae, the eccentricity. *The mean value of the radial force corresponding to the whole year, is equal to*

The true mean value of the radial force.

$$\frac{1}{2}\left(\frac{\frac{1}{2}rS}{(a-d)^3}+\frac{\frac{1}{2}rS}{(a+d)^3}\right);$$

Or,

$$\frac{rS}{4}\left(\frac{1}{(a-d)^3}+\frac{1}{(a+d)^3}\right).$$

But this expression is *always greater* than $\frac{rS}{2a^3}$, except when $d=0$; then it is the same, as any algebraist can verify. Hence the mean radial force for the whole year is greater as the earth's orbit is more eccentric, and it will be least of all when that orbit becomes a circle; and then, and then only, it will be accurately represented by $\frac{rS}{2a^3}$.

The mean value of the radial force will be least of all when the earth's orbit is a circle.

But when the radial force is least, the mean motion must be greatest, and that force is less and less as the eccentricity of the earth's orbit becomes less and less; and corresponding thereto the moon's motion becomes greater and greater, *as has been the case for more than* 4000 *years.*

(186.) The mean distance between the earth and sun remains constant. It must be so from the nature of motion, force, action, and reaction; but by the attraction of the planets the eccentricity of the earth's orbit is in a state of perpetual change; the change, however, is *excessively slow.* From the earliest ages the eccentricity of the orbit has been diminishing; and this diminution will probably continue until it is annihilated altogether, and the orbit becomes a circle; after which it will open out in another direction, again become eccentric, and increase in eccentricity to a certain moderate amount, and then again decrease.

The cause of the change of eccentricity of the earth's orbit.

CHAP. IV.

The immense period corresponding to these changes.

The period for these *vibrations*, "though calculable, *has never been calculated* further than to satisfy us that it is not to be reckoned by hundreds or even by thousands of years." It is a period so long that the history of astronomy, and of the whole human race, is but a point in comparison.

The moon's mean motion will continue to increase until the earth's orbit becomes a circle; after which it will again decrease, corresponding with the increase of a new eccentricity.

The inclination of the lunar orbit taken into account.

(187.) For the sake of simplicity, we have thus far considered the moon's orbit to be in the same plane as the earth's orbit; but this is not true; the mean inclination of the lunar orbit to the ecliptic is 5° 8′, varying about 9′ each way, according to the position of the sun.

Owing to this inclination of the lunar orbit, the expressions which we have obtained for the tangental force need correction, by *multiplying them by the cosine of the inclination;* and for the effect of the same forces in a perpendicular direction to the moon's longitude, multiply them *by the sine of the inclination of the orbit.*

The position of the moon's orbit, in relation to the sun, is strictly analogous to the ring in relation to the disturbing body *D* (Art. 176); the sun is constantly urging the moon into the plane of the ecliptic, which has a constant tendency to diminish the inclination of the lunar orbit (except when the sun is in the positions of the moon's nodes); and this constant force urging the moon to the ecliptic, causes the moon's nodes to retrograde.

We conclude this chapter by a brief summary of the principal causes which affect the moon's motion.

A summary statement of the lunar irregularities.

1. The eccentricity of the earth's orbit; which gives rise to the annual equation of the moon in longitude.

2. The eccentricity of the lunar orbit; producing the equation of the center.

3. The tangental force; giving rise to the equation called variation.

4. The position of the sun in respect to the greater axis of the lunar orbit; giving rise to the inequality called *evection.*

5. The inclination of the moon's orbit.

6. The combination of the first cause, when differing from its mean state, augments or diminishes the result of every other — thus making many additional small equations. CHAP. IV.

7. The ellipsoidal form of the earth.

CHAPTER V.

THE TIDES.

CHAP. V.

Definition of the term tide.

(188.) THE alternate rise and fall of the surface of the sea, as observed at all places directly connected with the waters of the ocean, is called tide; and before its cause was definitely known, it was recognized as having some *hidden and mysterious connection with the moon*, for it rose and fell twice in every lunar day. High water and low water had no connection with the hour of the day, but it always occurred in *about such an interval* of time after the moon had passed the meridian.

Connection with the moon.

High tides.

When the sun and moon were in conjunction, or in opposition, the tides were observed to be higher than usual.

When the moon was nearest the earth, in her perigee, other circumstances being equal, the tides were observed to be higher than when, under the same circumstances, the moon was in her apogee.

The space of time from one tide to another, or from high water to high water (when undisturbed by wind), is 12 hours and about 24 minutes, thus making two tides in one lunar day; showing high water on opposite sides of the earth at the same time.

Tides affected by the declination of the moon.

The declination of the moon, also, has a very sensible influence on the tides. When the declination is high in the north, the tide in the northern hemisphere, which is next to the moon, is greater than the opposite tide; and when the declination of the moon is south, the tide opposite to the moon is greatest.

A difficulty which meets a superficial reasoner.

It is considered mysterious, by most persons, that the moon by its attraction should be able to raise a tide on the opposite side of the earth.

CHAP. V.

That the moon should attract the water on the side of the earth next to her, and thereby raise a tide, seems rational and natural, but that the same simple action also raises the opposite tide, is not readily admitted; and, in the absence of clear illustration, it has often excited mental rebellion — and not a few popular lecturers have attempted explanations from false and inadequate causes.

The true cause.

But the true cause is the sun and moon's attraction; and until this is clearly and decidedly understood — not merely assented to, but fully comprehended — it is impossible to understand the common results of the theory of gravity, which are constantly exemplified in the solar system.

Fig. 41.

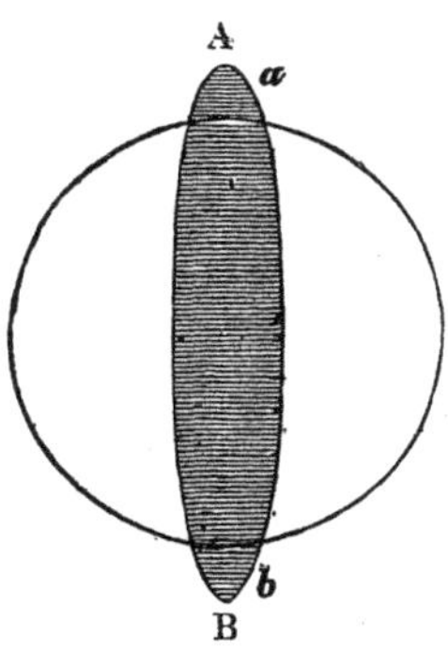

We now give a rude, but striking, and, we hope, a satisfactory explanation.

A summary illustration of the tides.

Conceive the frame-work of the earth to be an inflexible solid, as it really is, composed of rock, and incapable of changing its form under any degree of attraction; conceive also that this solid protuberates out of the sea, at opposite points of the earth, at A and B, as represented in Fig. 41, A being on the side of the earth next to the moon, m, and B opposite to it. Now in connection with this solid conceive a great portion of the earth to be composed of water, whose particles are inert, but readily move among themselves.

The solid AB cannot expand under the moon's attraction, and if it move, the whole mass moves together, in virtue of the moon's attraction *on its center of gravity*. But the particles of water at a, being free to move, and being under a more powerful attraction than the solid, rise toward A, producing a tide.

The particles of water at b being *less* attracted toward m than the solid, will not move toward m as fast as the solid, and being *inert*, they will be, as it were, left behind. The solid is drawn toward the moon more powerfully than the particles of water at b, and sinks in part into the water, but the observer at B, of course, conceives it the water rising up on the shore (which in effect it is), thereby producing a tide.

Analogy between the lunar perturbations and the perturbations of the ocean.

(189.) The mathematical astronomer perceives a strict analogy between the analytical expressions for the tides and the expressions for the perturbations of the lunar motion.

What we have called the *radial force*, in treating of the lunar irregularities, is the same *in its nature* as the force that raises the tides; the tide force is a radial force, which diminishes the pressure of the water toward the center of the earth under and opposite to the moon, in the same manner as the radial force diminishes the gravity of the moon toward the earth in her syzigies.

The radial force as applied to the moon.

In Art. 179 we found that the radial force for the moon, at the syzigies, is expressed by $\frac{2rS}{a^3}$; in which expression S is the mass of the sun, a its distance from the earth, and r the radius of the lunar orbit.

Converted into an expression for the tides.

The same expression is true for the tides, if we change S to m, the mass of the moon, and conceive a to represent the distance to the moon, and r the radius of the earth. For the tides, then, we have $\frac{2rm}{a^3}$, and as the numerator is always constant, the variation of the tides must correspond to the cube of the inverse distance to the moon.

Sun's attraction considered.

(190.) The sun's attraction on the earth is vastly greater than that of the moon; but by reason of the great distance to the sun, that body attracts every part of the earth nearly alike, and, therefore, it has much less influence in raising a tide than the moon.

CHAP. V.

Observations at Brest.

From a long course of observations made at Brest, in France, it has been decided that the medium high tides, when the sun and moon act together in the syzigies, is 19.317 feet; and when they act against each other (the moon in quadrature), the tides are only 9.151 feet. Hence the efficacy of the moon, in producing the tides, is to that of the sun, as the number 14.23 to 5.08.

Comparative influences of the sun and moon.

Among the islands in the Pacific ocean, observations give the proportion of 5 to 2.2, for the relative influences of these two bodies; and, as this locality is more favorable to accuracy than that of Brest, it is the proportion generally taken.

Having the relative influences of two bodies in raising the tides, we have the relative masses of those two bodies, provided they are at the *same* distance. But by the expression for the tides, as we have just seen, the variation for distance corresponds with the *inverse cube* of the distance, and the distance to the sun is 397.2 times the mean distance to the moon. Hence, to have the influence of the moon on the tides, when that body is removed to the distance of the sun, we must divide its observed influence by the cube of 397.2.

Mass of the moon computed.

That is, the mass of the moon is, to the mass of the sun, as the number $\frac{5}{(397.2)^3}$ to the number 2.2.

In all preceding computations we have called the mass of the earth *unity*, and in relation thereto, the mass of the sun is 354945 (Art. 169). Let us represent the mass of the moon by m, then we have the following proportion:

The result.

$$m : 354945 :: \frac{5}{(397.2)^3} : 2.2.$$

This proportion makes the mass of the moon a little less than $\frac{1}{77}$; but I have little confidence in the accuracy of the result, as the data, from their very nature, must be vague and indefinite.

The times of high water different in different localities.

(191.) The time of high water at any given point is not commonly at the time the moon is on the meridian, but two or three hours after, owing to the inertia of the water; and places, not far from each other, have high water at very different times on the same day, according to the distance and direction that the tide wave has to undulate from the main ocean.

The interval between the meridian passage of the moon and the time of high water, is nearly constant at the same place. It is about fifteen minutes less at the syzigies than at the quadratures; but whatever the mean interval is at any place, it is called the *establishment of the port.*

The tides would not instantly cease on the removal of their causes.

It is high water at Hudson, on the Hudson river, before it is high water at New York, on the same day; but the tide wave that makes high water one day at Hudson, made high water at New York the day before; and the tide waves that make high water now, were, probably, raised in the ocean several days ago; and the tides would not instantly cease on the annihilation of the sun and moon.

Tides very much affected by local circumstances.

The actual rise of the tide is very different in different places, being greatly influenced by local circumstances, such as the distance and direction to the main ocean, the shape of the bay or river, &c., &c.

In the Bay of Fundy the tide is sometimes fifty and sixty feet; in the Pacific ocean it is about two feet; and in some places in the West Indies, it is scarcely fifteen inches. In inland seas and lakes there are no tides, because the moon's attraction is equal over their whole extent of surface.

The following table shows the hight of the tides at the most important points along the coast of the United States, as ascertained by recent observation.

	Feet.
Annapolis (Bay of Fundy),	60
Apple River,	50
Chicneito Bay (north part of the Bay of Fundy),	60
Passamaquoddy River,	25
Penobscot River,	10
Boston,	11
Providence, R. I.,	5
New Bedford,	5
New Haven,	8
New York,	5
Cape May,	6
Cape Henry,	4½

CHAPTER VI.

PLANETARY PERTURBATIONS.

CHAP. VI.

Planetary and lunar perturbations analogous.

(192.) The perturbations of a planet, produced by the attractions of another planet, are precisely analogous to the perturbations of the moon, produced by the action of the sun. The disturbing forces are of the same kind, and they are subject to similar variations from precisely the same causes. But the amount of the disturbances is, in most cases, very trifling, on account of the small mass of the disturbing planet compared with the mass of the sun, or its great distance from the body disturbed.

Action and reaction among the planets reciprocal.

As action and reaction are everywhere equal, the planets mutually disturb each other, and if one is accelerated in its motion, the other must be retarded; if the tendency of one toward the sun is diminished, that of the other must be increased.

Examine Fig. 23, and conceive *V*, Venus, to be disturbed by the attraction of the earth at *E*, and if the motion of the planets is in the direction of *VB*, it is perfectly clear that Venus will be accelerated by the earth, and the earth will be retarded by Venus.

One planet is accelerated while another is retarded.

But Venus will be more accelerated in its motion than the earth will be retarded, for the disturbance at this point is in a line with the motion of Venus, and not in a line with the motion of the earth.

When the action changes.

After Venus passes conjunction, that is, passes the varying line *S E*, her motion becomes retarded, and the earth's is accelerated; *but every motion of the earth we ascribe to the sun;* and in all modern solar tables, the corrections of the sun's longitude corresponding to the action of *Venus, Mars, Jupiter, the moon,* &c., are simply the effect that these bodies have on the motion of the earth.

What is meant by solar perturbations.

The direct effect of any of these bodies on the position of the sun is absolutely insensible.

The relative disturbances of two planets are reciprocal to their masses; for if one is double in mass of another, the greater mass will move but half as far as the smaller, under their mutual action. But when the amount of disturbance is referred to angular motion for its measure, regard must be had to the distances of each planet from the sun; for the same distance on a larger orbit corresponds to a less angle.* Also, the whole amount of the disturbing force of a superior planet on an inferior will, at times, be a tangental force (Fig. 23); but the reaction of the inferior planet on the superior can *never* be in a tangent directly with, or opposed to, the motion of the superior.

Angular irregularities indicate the amount of planetary disturbance after certain reductions.

If observations can give the mutual disturbance of any two planets, then these circumstances being taken into consideration, an easy computation will give the relative masses of the planets.

(193.) As a general result, the attraction of a superior planet on an inferior, is to increase the time of revolution of the inferior, and to maintain it at a greater distance from the sun than it would otherwise have. The action of the inferior is to diminish the time of revolution of the superior; and the general effect *is greater* than it would be, if the inferior planet were constantly situated at the distance of the sun. (Art. 185.)

The general results in respect to the times of revolution.

As an illustration of this truth, we say, that if Venus were annihilated, the length of our year, and the times of revolution of all its superior planets, would be a little increased, and the revolution of Mercury, its inferior planet, would be a little diminished. If Jupiter were annihilated, the times of revolution of all its inferior planets would be a little diminished; for it acts as a *radial force* to keep them all a little farther from the sun.

(194.) If the orbits of all the planets were circular, the acceleration in one part of an orbit would be exactly compen-

Inequalities in circular orbits.

* Geometry demonstrates, that, on the average of each revolution, the proportion in which this reaction will affect the longitudes of the two planets, is that of their masses multiplied by the square roots of the major axes of their orbits, inversely; and this result of a very intricate and curious calculation is fully confirmed by observation.—HERSCHEL.

sated by the retardation in another; and in the course of a whole revolution, the mean motions of both planets (the disturber and the disturbed) would be restored, and the errors in longitude would destroy each other. But the orbits are not circles, and it is only in certain very rare occurrences that symmetry on each side of the line of conjunctions takes place; and hence, in a single revolution the acceleration of one part cannot be exactly counterbalanced by the retardation of the other; and, therefore, there is commonly left a certain outstanding error, which increases during every synodical revolution of the two planets, until the conjunctions take place in opposite parts of the orbits, then it attains its maximum, which is as gradually frittered away as the line of conjunctions works round to the same point as at first.

Long periods of inequalities depending on conjunctions in the same parts of the orbits.

Some of these inequalities too minute to be noticed.

Hence, between every two disturbing planets there is a common inequality depending on their mutual conjunctions, in the same, or nearly in the same, parts of their orbits. But it would be folly to compute the inequalities for *every* two planets, by reason of the extreme minuteness of the amounts; for instance, Mercury is not sensibly disturbed by Saturn or Uranus; and Mars, and Mercury, and Uranus, practically speaking, do not disturb each other; but Jupiter and Saturn have very considerable mutual perturbations, on account of their orbits being near each other, and both bodies far away from the sun.

The effect of commensurate revolutions of the planets.

(195.) Again, if the revolutions of two planets are exactly commensurate with each other, or, what is the same thing, the mean motion of both exactly commensurate with the circle, then the conjunctions of those two planets will always occur at the same points of the orbits (just as the conjunctions of the two hands of a clock always occur at the same points on the dial plate), and, in that case, the conjunctions will not revolve and distribute themselves around the orbits, so that in time, the *radial* and *tangental forces* will have an opportunity to accelerate on one side of the line of conjunctions as much as they retard on the other; and, therefore, a permanent *derangement* would then take place.

A supposed case for illustration.

For instance, if three times the mean angular motion of one planet were exactly equal to twice the mean angular motion of another, then three revolutions of the one would exactly correspond to two of the other, and every second conjunction of the two would take place in the same points of the orbits; and the orbits, not being circular, the portions of them on each side of the line of conjunctions cannot be symmetrical, unless the longer axes of the two orbits are in the same line, and the conjunctions also taking place on that line.

Here, then, is a case showing that the disturbing force may *constantly differ in amount* on each side of the line of conjunctions, and, of course, could never compensate each other, and a permanent derangement of these two planets would be the result.

Stability of the solar system.

Hence, we perceive, that, to preserve the solar system, it is necessary that the orbits should be circles, or their times of revolution *incommensurable;* but we do not pretend to say that the converse of this is true: we do say, however, that *no* natural cause of destruction has thus far been found.

(196.) The times of the planetary revolutions are *incommensurable;* but, nevertheless, there are instances that approach commensurability, and, in consequence, approach a derangement in motion, which, when followed out, produce very long periods of inequality, called *secular* variation. The most remarkable of these, and one *which very much perplexed the astronomers of the last century,* is known by the term of "*the great inequality*" of Jupiter and Saturn.

The great inequalities of Jupiter and Saturn.

"It had long been remarked by astronomers that, on comparing together ancient with modern observations of Jupiter and Saturn, their mean motions could not be uniform." The period of Saturn appeared to have been increased throughout the whole of the *seventeenth century,* and that of Jupiter shortened. Saturn was constantly lagging behind its calculated place, and Jupiter was as constantly in advance of his. On the other hand, in the *eighteenth century,* a process precisely the reverse was going on.

The perplexity given to the philosophers.

The amount of retardations and accelerations, corresponding to one, two, or three revolutions were not very great; but, as they went on accumulating, material differences, at length, existed between the observed and calculated places of both

CHAP. VI. these planets; and, as such differences could not then be accounted for, they excited a high degree of attention, and formed the subject of prize problems of several philosophical societies.

Laplace solved the mystery.

For a long time these astonishing facts baffled every endeavor to account for them, and some were on the point of declaring the doctrine of universal gravity overthrown; but, at length, the immortal Laplace came forward, and showed the cause of these discrepancies to be in the near commensurability of the mean motions of Jupiter and Saturn; which cause we now endeavor to bring to the mind of the reader in a clear and emphatic manner.

(197.) The orbits of both Jupiter and Saturn are elliptical, and their perihelion points have different longitudes, and, therefore, their different points of conjunction are at different distances from each other, and no line* of conjunction cuts the two orbits into two equal or symmetrical parts; hence, the inequalities of a single synodical revolution will not destroy each other; and, to bring about an equality of perturbations, requires a certain period or succession of conjunctions, as we are about to explain.

The revolutions of Jupiter and Saturn compared.

Five revolutions of Jupiter require 21663 days, and two of Saturn, 21518 days. So that, in a period of two revolutions of Saturn (about sixty of our years), after any conjunction of these two planets, they will be in conjunction again not many degrees from where the former took place.

Their synodical revolution determined.

To determine definitely where the third mean conjunction will take place, we compute the synodical revolution of these two planets by dividing the circumference of the circle in seconds (1296000) by the difference of the mean daily motion of the planets in seconds (178″.6),† and the quotient is 7253.4 days; three times this period is 21760 days. In this period Jupiter performs five revolutions and 8° 6′ over; Saturn makes two revolutions and 8° 6′ over; showing that the line

* Line of conjunction, an imaginary line drawn from the sun through the two planets when in conjunction.

† See problem of the two couriers, Robinson's Algebra.

of conjunction advances 8° 6′ in longitude during the period of 21760 days. CHAP. VI.

In the year 1800, the longitude of Jupiter's perihelion point was 11° 8′, and that of Saturn 89° 9′; the inclination of the greater axis of the orbits, therefore, was 78° 1′.

Fig. 42

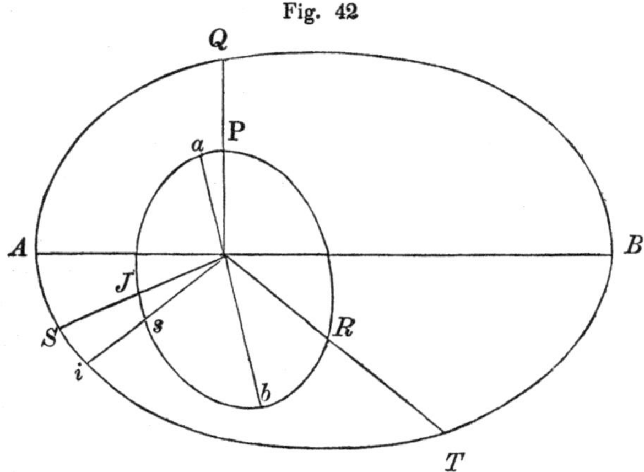

Let *A B* (Fig. 42) represent the major axis of Saturn's orbit, and *a b* that of Jupiter; the two are placed at an angle of 78°.* The series of conjunctions explained.

Suppose any conjunction to take place in any part of the orbits, as at *J S* (the line *J S* we call the line of conjunction); in 7253.4 days afterward another conjunction will take place. In this interval, however, Saturn will describe about 243° in its orbit, at a mean rate, and Jupiter will describe one revolution and about 243° over, and it will take place as represented in the figure, at *P Q* (*S T B* being the direction of the motion). The next conjunction will be 243° from *P Q*, or at *R T*. From *R T* the next conjunction will be at *s i*, 8° 6′ in advance of *J S*, and thus the conjunction *J S* (so to speak) will gradually advance along on the orbit from *S* to *T*. Line of conjunction explained.

But, as we perceive, by inspecting the figure, there is a

* We have very much exaggerated the eccentricities of these ellipses, for the purpose of magnifying the principle under consideration.

CHAP. VI.

Certain conjunctions bring the planets nearer together than most others.

certain portion of the orbits, between *S* and *T*, where the two planets would come nearer together in their conjunction, than they do at conjunctions generally, and, of course, while any one of the three conjunctions is passing through that portion of the orbits — Jupiter disturbs Saturn, and Saturn reacts on Jupiter more powerfully than at other conjunctions; and this is the cause of "*the great inequality of Jupiter and Saturn.*"

The period of this remarkable inequality computed, and the computation confirmed by observation.

(198.) To obtain the period of this inequality, we compute the time requisite for one of these lines of conjunction to make a third of a revolution, that is, divide 120° by 8° 6′, and we shall find a quotient of $14\frac{2}{27}$, showing the period to be $14\frac{2}{27}$ times 21760 days, or nearly 883 years; which would be the actual period, provided the elements of the orbits remained unchanged during that time. But in so long a period the relative position of the perigee points will undergo considerable variation; which causes the period to lengthen to about 918 years.

Fig. 43.

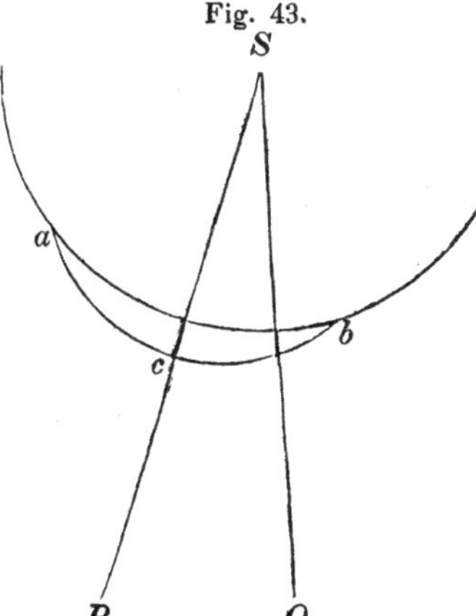

The maximum amount of this inequality, for the longitude of Saturn, is 49′, and for Jupiter 21′, always opposite in effect, on the principle of action and reaction.

An explanation of the principle that led to the discovery of Neptune.

(199.) The last great achievement of the powers of mind in the solar system, was the discovery of the new planet *Neptune*, by Leverrier and Adams analyzing the inequalities of the motion of Uranus. To give a rude explanation of the possibility of this problem, we present Figure 43. Let *S* be the sun, and the regular curve the orbit of Uranus, as corresponding to all known perturbations; but at *a* it departs from its computed track and runs out in the protuberance *a c b*. This indicated that some attracting body must be somewhere in the direction

SP, although no such body was ever seen or known to exist. The next time the planet comes round into the same portions of its orbit,* *suppose* the center of the protuberance to have changed to the line SQ. This would indicate that the unknown and unseen body was now in the line SQ, and that since the former observations it had changed positions by the angle PSQ; and, by this angle, and the time of its description, something like a *guess* could be made of the time of its revolution.

CHAP. VI.

How computations could be made for the revolution of an unseen planet.

With the approximate time of revolution, and the help of Kepler's third law, its corresponding distance from the sun can be known. With the distance of the unseen body, and the amount that Uranus is drawn from its orbit by it, we can approximate to its mass.

Thus, we perceive, that it is possible to know much about an existing planet, although so distant as never to be seen. But the body that disturbed the motion of Uranus has been *seen*, and is called *Neptune*.

CHAPTER VII.

ABERRATION, NUTATION, AND PRECESSION OF THE EQUINOXES.

CHAP. VII.

Dr. Bradley's observations on the fixed stars for the purpose of finding their parallax. Unexpected results.

(200.) ABOUT the year 1725 Dr. Bradley, of the Greenwich observatory, commenced a very rigid course of observations on the fixed stars, with the hope of detecting their parallax. These observations disclosed the fact, that all the stars which come to the upper meridian near midnight, have an increase of longitude of about 20″; while those opposite, near the meridian of the sun, have a decrease of longitude of 20″; thus making an annual *displacement* of 40″. These observations were continued for several years, and found to be the same at the same time each year; and, what was most

---

* Leverrier and Adams had not the advantage of a complete revolution of Uranus.

CHAP VII. perplexing, the results were directly opposite from such as would arise from parallax.

These facts were thrown to the world as a problem demanding solution, and, for some time, it baffled all attempts at explanation; but it finally occurred to the mind of the Doctor, that it might be an effect produced by the progressive motion of light combined with the motion of the earth; and, on strict examination, this was found to be a satisfactory solution.

Aberration illustrated.

Fig. 44.

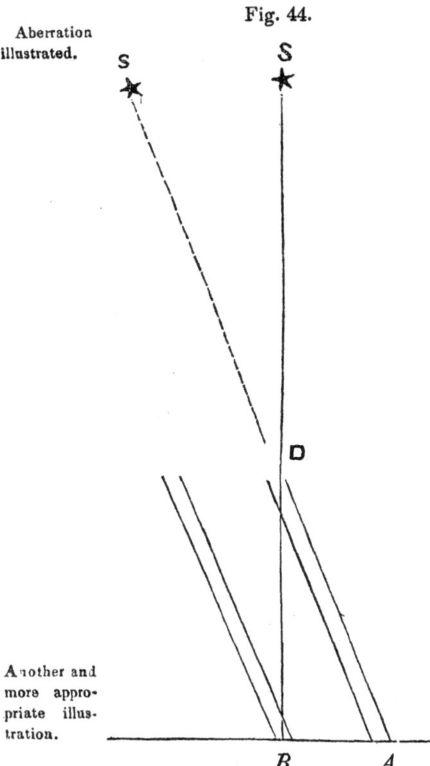

(201.) A person standing still in a rain shower, when the rain falls perpendicularly, the drops will strike directly on the top of his head; but if he starts and runs in *any direction*, the drops will strike him in the face; and the effect would be the same, in relation to the *direction* of the drops, as if the person stood still and the rain came inclined from the direction he ran.

This is a full illustration of the principle of these changes in the positions of the stars, which is called *aberration;* but the following explanation is more appropriate.

Another and more appropriate illustration.

Conceive the rays of light to be of a material substance, and its particles progressive, passing from the star S (Fig. 44) to the earth at B; passing directly through the telescope, while the telescope itself moves from A to B by the motion of the earth. And if $D B$ is the motion of light, and $A B$ the motion of the earth, then the tele-

CHAP. VII.

scope must be inclined in the direction of AD, to receive the light of the star, and the apparent place of the star would be at S', and its true place at S and the angle ADB is $20''.36$, at its maximum, called the angle of aberration.

By the known motion of the earth in its orbit, we have the value of AB corresponding to one second of time: we have the angle ADB by observation: the angle at B is a right angle, and (from these data) computing the side BD we have the velocity of light, corresponding to one second of time. To make the computation, we have

$$DB : BA :: Rad. : tan. 20''.36.*$$

The velocity of light computed by means of aberration.

But BA, the distance which the earth moves in its orbit

Fig. 45.

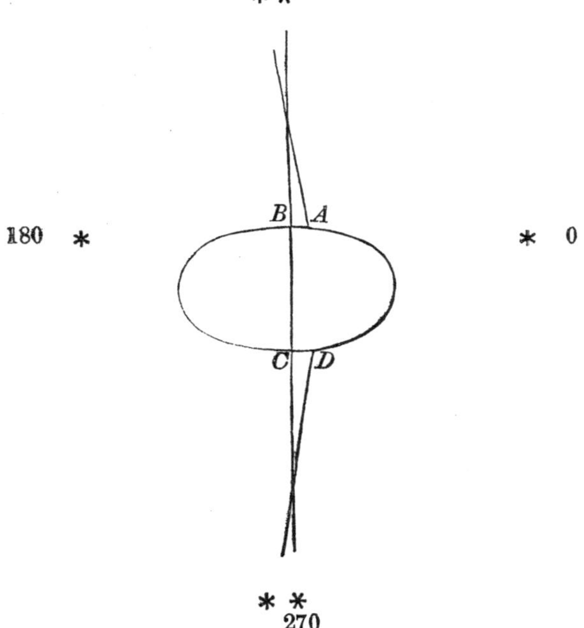

*To obtain the logarithmetic tangent of $20''.36$ see note on page 128.

 in one second of time, is within a very small fraction of 19 miles; the logarithm of the distance is 1.278802, and, from this, we find that $B D$ must be 192600 miles, the velocity of light in a second; a result very nearly the same as before deduced from observations on the eclipses of Jupiter's moons. (Art. 143.)

The agreement of these two methods, so disconnected and so widely different, in disclosing such a far-hidden and remarkable truth, is a striking illustration of the power of science, and the order, harmony, and sublimity that pervades the universe.

A comprehensive view of the effects of aberration.

To show the effects of aberration on the whole starry heavens, we give figure 45. Conceive the earth to be moving in its orbit from A to B. The stars in the line AB, whether at 0 or 180, are not affected by aberration. The stars, at right angles to the line $A B$, are most affected by aberration, and it is obvious that the general effect of aberration is to give the stars an apparent inclination to that part of the heavens, toward which the earth is moving. Thus the star at 90 has its longitude increased, and the star opposite to it, at 270, has its longitude decreased, by the effect of aberration; both being thrown more toward 180. The effect on each star is 20″.36. But when the earth is in the opposite part of its orbit, and moving the other way, from C to D, then the star at 90 is apparently thrown nearer to 0; so also is the star at 270, and the whole annual variation of each star, in respect to longitude, is 40″.72.

Proof of the annual motion of the earth.

(202.) The supposition of the earth's annual motion fully explains aberration; conversely, then, the observed variations of the stars, called aberration, *are decided proofs of the earth's annual motion.*

In consequence of aberration, each star appears to describe a small ellipse in the heavens, whose semi-major axis is 20″.36, and semi-minor axis is 20″.36 multiplied by the sine of the latitude of the star. The true place of the star is the center of the ellipse. If the star is on the ecliptic, the ellipse, just mentioned, becomes a straight line of 40″.72 in length

If the star is at either pole of the ecliptic, the ellipse becomes a circle of 40″.72 in diameter, in respect to a great circle; but a circle, however small, around the pole, will include all degrees of longitude; hence it is possible for stars very near either pole of the ecliptic, to change longitude very considerably, each year, by the effect of aberration; but no star is sufficiently near the pole to cause an apparent revolution round the pole by aberration; and the same is true in relation to the pole of the celestial equator.

All these ellipses have their longer axis parallel to the ecliptic, and for this reason it is easy to compute the aberration of a star in latitude and longitude,* but it is a far more complex problem to compute the effects in respect to right ascension and declination.

Aberration of the sun.

(203.) The aberration of the sun varies but a very little, because the distance to the sun varies but little, and without material error, it may be always taken at 20″.2, subtractive. The apparent place of the sun is always behind its true place by the whole amount of aberration; but the solar tables give its apparent place, which is the position generally wanted.

In computing the effect of aberration on a planet, regard must be had to the apparent motion of the planet while light is passing from it to the earth.

The moon not affected by aberration.

The effects of aberration on the moon are too small to be noticed, as light passes that distance in about one second of time.

Other inequalities observed by Dr. Bradley.

(204.) While Dr. Bradley was continuing his observations to verify his theory of aberration, he observed other small variations, in the latitudes and declinations of the stars, that could not be accounted for on the principle of aberration.

The period of these variations was observed to be about

---

*Aber. in Lon. $= \dfrac{-20''.36 \cos. (S-s)}{\cos. l}$;

Aber. in Lat. $= 20''.36 \sin. (S-s) \sin. l$.

In these expressions S represents the longitude of the sun, s the longitude of the star, and l its latitude.

 the same as the revolution of the moon's node, and the amount of the variation corresponded with particular situations of the node; and, in short, it was soon discovered that the cause of these variations was a slight vibration in the earth's axis, caused by the action and reaction of the sun and moon on the protuberant mass of matter about the equator, which gives the earth its spheroidal form, and the effect itself is called Nutation.

Fig. 46.

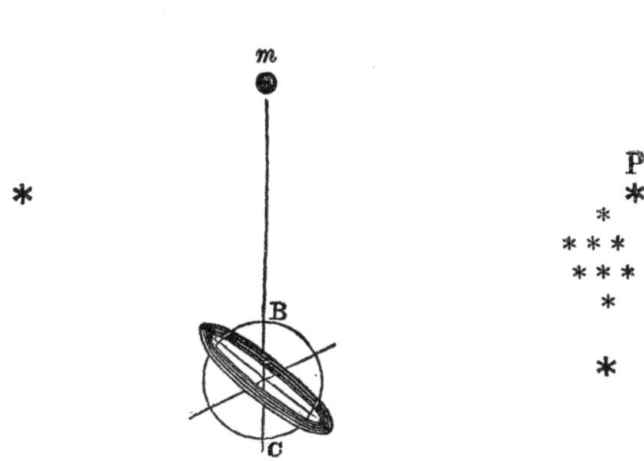

Nutation fully explained by the theory of gravity.

(205.) We have shown, in Art. 176, that the attraction of a body, m, on a ring of matter around a sphere, has the effect of making the plane of the ring incline toward the attracting body.

Let $B\,C$, Fig. 46, represent the plane of the equator; and conceive the protuberant mass of matter, around the equator, to be represented by a ring, as in the figure. Let m be the moon at its greatest declination, and, of course, without the plane of the ring.

Let P be the polar star. The attraction of m on the ring inclines it to the moon, and causes it to have a slight motion on its center; but the motion of this ring is the motion of the whole earth, which must cause the earth's axis to change its position in relation to the star P, and in relation to all the stars.

When the moon is on the other side of the ring, that is, opposite in declination, the effect is to incline the equator to the opposite direction, which must be, and is, indicated by *an apparent motion of all the stars.*

A slight alternate motion of all the stars *in declination, corresponding to the declinations* of the sun and moon, was carefully noted by Dr. Bradley, and since his time has been fully verified and definitely settled: this vibratory motion is known by the name of *nutation,* and it is fully and satisfactorily explained on the principles of universal gravity; and conversely, these minute and delicate facts, so accurately and completely conforming to the theory of gravity, served as one of the many strong points of evidence to establish the truth of that theory.

The general effect of nutation illustrated by Fig. 46.

(206.) By inspecting Fig. 46, it will be perceived that when the sun and moon have their greatest northern declinations, all the stars north of the equator and in the *same hemisphere* as these bodies, will incline toward the equator; or all the stars in that hemisphere will incline southward, and those in the opposite hemisphere will incline northward; the amount of vibration of the axis of the earth is only 9″.6 (as is shown by the motion of the stars), and its period is 18.6, or about nineteen years, the time corresponding to the revolution of the moon's node. When the moon is in the plane of the equator, its attraction can have no influence in changing the position of that plane; and it is evident that the greatest effect must be when the declination is greatest.

Where the node must be to correspond to the moon's greatest declination.

The moon's declination is greatest when the longitude of the moon's *ascending* node is 0, or at the first point of Aries. The greatest declination is then 28° on each side of the equator; but when the *descending* node is in the same point, the moon's greatest declination is only 18°. Hence there will be times, *a succession of years,* when the moon's action on the *protuberant matter* about the equator must be greater than in an opposite succession of years, when the node is in an opposite position. Hence, the amount of *lunar* nutation depends on the position of the moon's nodes.

Monthly nutation, effect small.

It is very natural to suppose that the period of *lunar* nutation would be simply the time of the revolution of the moon; *and so, in fact, it is;* but the corresponding amount is very small, *only about one-tenth of a second.* This is because half a lunar revolution, about 13½ days, while the moon is on one side of the equator, is not a sufficient length of time for the moon to effect much more than to overcome the inertia of the earth; but, in the space of nine years, effecting a little more than a mean result at every revolution, the amount can rise to 9″.6, a perceptible and measurable quantity.

The mean effect of the moon on the mass of matter around the equator.

(207.) The *mean* course of the moon is along the ecliptic: its variation from that line is only about five degrees on each side; hence, the *medium* effect of the moon on the protuberant mass of matter at the equator is the same as though the moon was all the while in the ecliptic. But, in that case, its effect would be the same at every revolution of the moon; and the earth's equator and axis would then have an equilibrium of *position,* and there would be *no nutation,* save the slight monthly nutation just mentioned, which is too small to be sensible to observation; and the nutation that we observe, is only an *inequality* of the moon's attraction on the protuberant equatorial ring; and, however great that attraction might be, it would cause no vibration in the position of the earth, if it were constantly the same.

Solar nutation.

We have, thus far, made particular mention of the moon, but there is also a *solar nutation:* its period is, of course, a year; and it is very trifling in amount, because the sun attracts all parts of the earth nearly alike; and the short period of one year, or half a year (which is the time that the unequal attraction tends to change the plane of the ring in one direction), is too short a time to have any great effect on the inertia of the earth.

The solar nutation, in respect to declination, is *only one second.*

(208.) Hitherto we have considered only one effect of nutation—that which changes the *position* of the plane of the equator—or, what is the same thing, that which changes the position of the earth's axis; but there is another effect, of greater magnitude, earlier discovered, and better known, resulting from the same physical cause, we mean the

PRECESSION OF THE EQUINOXES.

First principles again examined.

We again return to first principles, and consider the mutual attraction between a ring of matter and a body situated out of the plane of the ring; the effect, as we have several times shown, is to incline the ring to the body, or (which is the same in respect to relative positions), the body inclines to run to the plane of the ring.

The mean attractions of the sun and moon are in one plane, the ecliptic.

The mean attraction of the moon is in the plane of the ecliptic. The sun is all the while in the ecliptic. Hence, the *mean* attraction of both sun and moon is in one plane, the ecliptic; but the equator, considered as a ring of matter surrounding a sphere, is inclined to the plane of the ecliptic by an angle of $23\frac{1}{2}$ degrees, and hence the sun and moon have a constant tendency to draw the equator to the ecliptic, and actually do draw it to that plane; and the visible effect is, to make both sun and moon, in revolutions, cross the equator sooner than they otherwise would, and thus the equinox falls back on the ecliptic, called the precession of the equinoxes.

The precession of the equinoxes.

The annual mean precession of the equinoxes is $50''.1$ of arc, as is shown by the sun coming into the equinox, or crossing the equator at a point $50''.1$ before it makes a revolution in respect to the stars.

Natural mode of expression.

Perhaps it is clearer to the mind to say, that the sun is drawn to the equator by the protuberant mass of matter around the earth, and, in consequence, arrives at the equator, in its apparent revolutions, sooner than it otherwise would. But the truth is, that the ecliptic is stationary in position,

CHAP. VII. and the equator, by a slight motion, meets the ecliptic; which motion is caused by the attractions of the sun and moon, as has been several times explained.

The true physical cause of the precession of the equinoxes.

If the moon were all the while in the ecliptic, the precession of the equinoxes would then be a *constantly flowing quantity*, equal to 50″.1 for each year; but, for a succession of about nine years, the moon runs out to a greater declination than the ecliptic, and, during that time, its action on the equatorial matter is greater than the *mean* action, and then comes a succession of about nine years, when its action is less than its mean; hence, for nine years, the precession of the equinoxes will be *more* than 50″.1 per year, and, for the nine years following, the precession will be *less* than 50″.1 for each year; *and the whole amount of variation, for this inequality, in respect to longitude*, is 17″.3, and its period is half a revolution of the moon's nodes. This inequality is called the equation of the equinoxes, and varies as the sine of the longitude of the moon's nodes.

Equation of the equinoxes.

The equation of the equinoxes, of course, affects the length of the tropical year, and slightly, *very slightly*, affects sidereal time.

Mean and true sidereal time.

There is a *true* equinox and a *mean* equinox; and, as sidereal time is measured from the meridian transit of the equinox, there must be a *true* sidereal and a *mean* sidereal time; but the difference is never more than 1.1 s. in time, and, generally, it is much less.

Explanation of Fig. 47.

(209.) In the hope of being more clear than some authors have been, in explaining the results of precession, we present Fig. 47. E represents the pole of the ecliptic, and the great circle around it is the ecliptic itself. P is the pole of the earth, 23° 27′ from the pole E, and around P, as a center, we have attempted to represent the equator, but this, of course, is a little distorted; ♈ and ♎ are the two opposite points where the ecliptic and equator intersect; ♈E is the first meridian of longitude; ♈P is the first meridian of right ascension. The angle E♈P is 23° 27′, and $E\,P$, produced, is the meridian passing through the solstitial points. To obtain a clear conception of the precession of the equinoxes, the stars the ecliptic, and its pole E, must be considered as FIXED, and the line ♈ ♎ as having a slow motion of $50''.1$ per an-

Fig. 47.

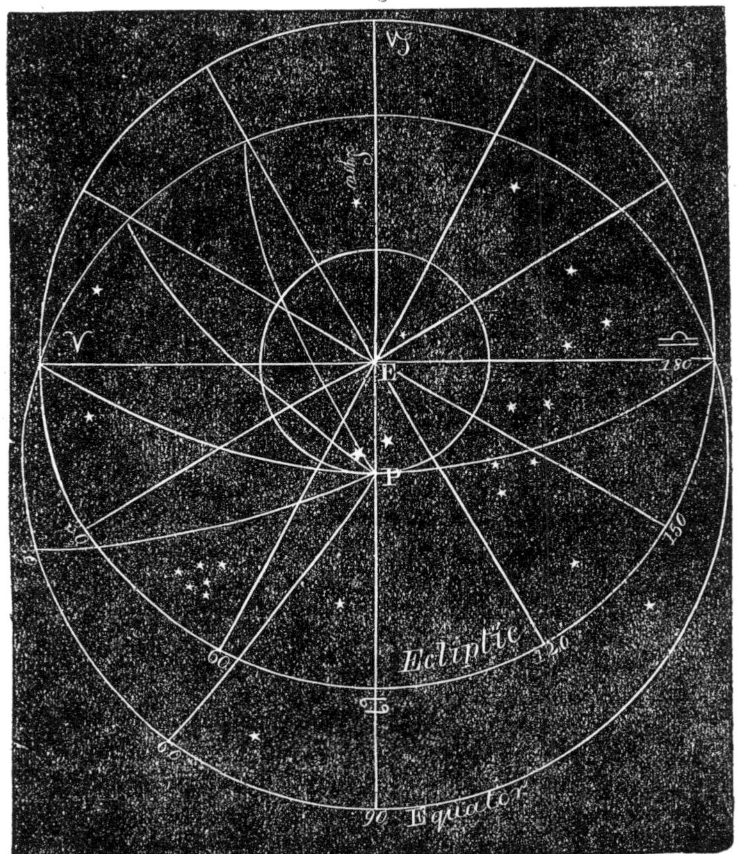

From the fixed position of the ecliptic, and also of the stars, the stars never change latitude.

num, on the ecliptic, in a retrograde direction; and this must carry the pole P, around the point E, as a center, carrying also the solstitial points backward on the ecliptic. Some of the stars have proper motions; but, putting that circumstance out of the question, the stars are fixed, and the ecliptic is fixed; therefore, the stars never change latitude, but

CHAP. VII.

the whole frame-work of meridians from the pole P, the pole itself, and the equator, revolve over the stars; and, in respect to that motion of the meridian and the equator, the stars change *right ascension, declination, and longitude*, but do not change latitude. The stars change longitude, simply because the first meridian of longitude, ♈ E, moves backward; they change right ascension, because the meridian, ♈ P, and all the meridians of right ascension, revolve backward.

One hemisphere of stars approaches the north pole, the other recedes from it.

By inspecting the figure, we readily perceive that all the stars near ♈ must, apparently, approach the north pole, because the pole, in its revolution round E, is approaching toward that part of the ecliptic; for the same reason, all the stars near ♎ are, apparently, moving southward, because the equator is being drawn over them. In short, all the stars, from the eighteenth hour of right ascension through ♈, to the sixth hour of right ascension, must diminish in north polar distance, and all the stars, from the six hours through ♎, to the eighteenth hour of right ascension, must increase in north polar distance, in consequence of the precession of the equinoxes.

Inspection Table II.

These observations may be confirmed by inspecting Table II, in which is registered the positions of the principal fixed stars, with their annual variations. The column of annual *variation* of declination changes sign at the point corresponding to six hours, and eighteen hours of right ascension; and the rapidity of this *variation* is greater as the star is nearer to 0 hours, or twelve hours of right ascension.

Annual variation in declination, how computed.

When the right ascension of a star is 0 hours, or twelve hours, it is easy to compute its annual variation in declination, corresponding to its precession along the ecliptic of 50″.1. Conceive a small plane triangle whose hypothenuse is 50″.1, the angle at the base 23° 27′ 40″ (*i. e.* the obliquity of the ecliptic), the side opposite to this angle will be found to be a little over 20″, corresponding to the figures in the table.

Proper motions, how discovered.

It is thus, by the motion of these imaginary lines over the whole concave of the heavens, that the annual variation of both right ascension and declination of each individual star in the catalogue is computed and put down; and if any particular star does not correspond with this, it is said to have *proper* motion; *and it is thus that proper motions are detected.*

Final effect of precession.

As P must circulate round E by the slow motion of 50″.1 in a year, it will require 25868 years to perform a revolution; and the reader can perceive, by inspecting the figure, why the pole star is in apparent motion in respect to the pole, and why that star will cease to be the polar star, and why, at the expiration of about 12000 years, the bright star, Lyra, will be the polar star.

Comparative effect of sun and moon.

(210.) The mean effect of the moon in producing the precession of the equinoxes is, to the mean effect of the sun, as five to two. The sun's action is nearly constant, because the sun is always in the ecliptic; a small annual variation, however, is observed. The great inequality of 17″.3, corresponding to about nineteen years, is caused entirely by the unequal action of the moon, depending on the longitude of the moon's ascending node.

Undulatory motion of the earth's axis around the pole of the ecliptic.

In consequence of this inequality, the pole, P, does not move round the pole of the ecliptic, E, in an even circumference of a circle, but it has a waving or undulating motion, as represented in this figure; each wave corresponding to nineteen years; and, therefore, there must be as many of them in the whole circle as 19 is contained in 25868. From this, we perceive, that the undulations in the figure are much exaggerated, and vastly too few in number; an exact linear representation of them would be impossible.

Mean and apparent place of a star.

(211.) From the foregoing, we learn that the positions of all the stars are affected by *aberration*, *precession*, and *nutation*: the amount for each cause is very trifling in itself, yet, in most cases, too great to be neglected, when accuracy is required; and it is as difficult to make computations for a small quantity as for a large one, and often greater; and to reduce the apparent place of a fixed star from its mean place,

CHAP. VII.

and its mean place from its apparent place, is one of the most troublesome problems in practical astronomy.

General formulæ, where found.

The mean place of a fixed star, reduced to the time of observation, is sufficiently near its apparent place to be considered the same. The practical astronomer, however, who requires the star as a point of reference, or uses it for the adjustment of his instruments, must not omit any cause of variation; but such persons will always have the aid of a *Nautical Almanac*, where general formulæ and tables will be found, to direct and facilitate all the requisite reductions.

Importance of physical astronomy.

(212.) Physical astronomy brings many things to light that would otherwise escape observation, and some of these developments, at first, strike the learner with surprise, and he is not always ready to yield his assent. For instance, as a general student, he learns that the anomalistic year, the time that the earth moves from its perigee to its perigee again, is 365 d. 6 h. 14 m.; that the perigee is very slow in its motion, moves only about 12″ in a year, and is subject to but few fluctuations. He has also learned that the earth, in its orbit, describes equal areas in equal times; hence, he concludes, that the time from perigee to perigee, or from apogee to apogee, must be very nearly a constant quantity; but, on consulting and comparing the predictions to be found in the English nautical almanacs, he will find these periods to be (in comparison to his anticipations) very fluctuating. They differ from the stated mean times, not only by minutes and seconds, but by hours, and even days. The investigator is, at first, surprised, and fancies a mistake; at least, a misprint; but, on examining concurrent facts, such as the logarithms of the distance from the sun, and the sun's true motion at the time, he finds that, if a mistake has been made, it is a very harmonious one, and every other circumstance has been adapted to it.

The latitude of the sun explained.

But let us turn a moment from these facts, and examine the first page of our Tables. There it will be found, that the sun has latitude; that it deviates to the north and south of the ecliptic, by a quantity *too small ever to be observed:* it is, therefore, a quantity wholly determined by theory, and, as the sun's latitude changes with the latitude of the moon, we must seek for its cause in the lunar motions.

Fig. 48.

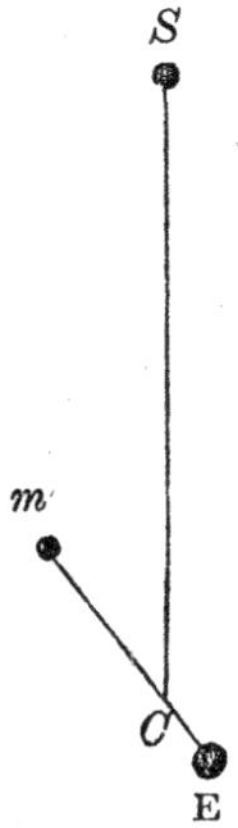

To understand the fact of the sun having latitude, we must admit that it is the center of gravity between the earth and moon, that moves in an elliptical orbit round the sun; and that center is always in the ecliptic; and the sun, viewed from that point, would have no latitude. But when the moon, *m*, (Fig. 48), is on one side of the plane of the ecliptic, *Sc*, the earth, *E* would be on the other side, and the sun, seen from the center of the earth, would appear to lie on the same side of the ecliptic as the moon. *Hence, the sun will change his latitude, when the moon changes her latitude.*

Longitude of the sun affected by the position of the moon.

If the moon were all the while in the plane of the ecliptic, the sun would have no latitude (save some *extremely minute* quantities, from the action of the planets, when not in the plane of the ecliptic); but the moon does not deviate more than 5° 20 from the ecliptic, and, of course, the earth makes but a proportional deviation on the other side; but, in longitude, the moon deviates to a right angle on both sides, in respect to the sun, and when the moon is in advance in respect to longitude, the sun appears to be in advance also; and when the moon is at her third quarter, the longitude of the sun is apparently thrown back by her influence:—the greatest variation in the sun's longitude, arising from the motion of the earth and moon about their center of gravity, is about 6″ each side of the mean. Now it is this motion of the earth around the common center of gravity of the earth and moon, that chiefly affects the time when the earth comes to its apogee and perigee. When the moon is in conjunction with the sun, the center of the earth is farther from the sun than it otherwise would be; and when the moon is in opposition to the sun, the earth is about 3200 miles nearer the sun than it would be in its mean orbit; and thus, we perceive, that the longitude of the moon has a great influence in

Longitude of the moon affects the time that the earth comes to its apogee and perigee.

CHAP. VII. bringing the earth into, or preventing it from coming into, its perigee or apogee; but the perigee and apogee points, *for the center of gravity*, are quite uniform, agreeably to the views expressed in the first part of this article. These explanations will give a general insight into some of the apparent intricacies of physical astronomy.

Small equations of the sun's center explained.

The small equations of the sun's center are computed on the principle explained by Fig. 48, the sun having a motion round the center of gravity between itself and each of the planets. For example, the perturbation produced by Jupiter is greatest when Jupiter is in longitude 90° from the sun, as seen from the earth; the greatest effect is then about 8″, and varies very nearly as the sine of Jupiter's elongation from the sun.

When Jupiter is in conjunction with the sun, the sun is nearer the earth than it otherwise would be; and, on this account, we have a small table to correct the sun's distance from the earth, called the perturbations of the sun's distance.

The same remarks apply to other planets but, to avoid confusion, the effects of each one must be computed separately.

SECTION IV.

PRACTICAL ASTRONOMY.

PREPARATORY REMARKS.

WE have now done with general demonstrations, and with minute and consecutive explanations; but we shall give all necessary elucidation in relation to the particular problems under consideration. To go through this part of astronomy with success and satisfaction, the reader *must have* a passable understanding of plane and spherical trigonometry; and if to these he adds a general knowledge of the solar system, as taught in the foregoing pages, he will have a full comprehension of all we design to embrace in this section. TRIG.

To prompt the student in his knowledge of trigonometry, we give the following *formulæ:*

I. Relative to a single arc or angle.

1. - - - $\sin. a = \tan. a \cos. a.$*

2. - - - $\sin. a = \dfrac{\tan. a}{\sqrt{1+\tan.^2 a.}}$

3. - - - $\cos. a = \dfrac{1}{\sqrt{1+\tan.^2 a.}}$

4. - - - $\cos. a = 2 \cos.^2 \tfrac{1}{2} a - 1.$

5. - - - $\tan. \tfrac{1}{2}a = \dfrac{\sin. a}{1+\cos. a}$

6. - - - $\tan.^2 \tfrac{1}{2}a = \dfrac{1-\cos. a}{1+\cos. a}.$

7. - - - $\sin. 2a = 2 \sin. a \cos. a.$

* Radius is unity in all these equations.

TRIG.

8. - - $\cos. 2a = 2\cos.^2 a - 1 = 1 - 2\sin.^2 a$.

II. Relative to two arcs, a and b, of which a is supposed to be the greater.

9. - $\sin. (a+b) = \sin. a \cos. b + \sin. b \cos. a$.

10. - $\cos. (a+b) = \cos. a \cos. b - \sin. a \sin. b$.

11. - $\sin. (a-b) = \sin. a \cos. b - \sin. b \cos. a$.

12. - $\cos. (a-b) = \cos. a \cos. b + \sin. a \sin. b$.

Sum of (9) and (11) gives 13, diff. gives 14.

13. - $\sin. (a+b) + \sin. (a-b) = 2\sin. a \cos. b$.

14. - $\sin. (a+b) - \sin. (a-b) = 2\cos. a \sin. b$.

15. - $\tan. (a+b) \quad - \quad = \dfrac{\tan. a + \tan. b}{1 - \tan. a \tan. b}$.

16. - $\tan. (a-b) \quad - \quad = \dfrac{\tan. a - \tan. b}{1 + \tan. a \tan. b}$.

17. - $\dfrac{\sin. a + \sin. b}{\sin. a - \sin. b} \quad - \quad = \dfrac{\tan. \frac{1}{2}(a+b)}{\tan. \frac{1}{2}(a-b)}$.

18. - $\dfrac{\tan. a + \tan. b}{\tan. a - \tan. b} \quad - \quad = \dfrac{\sin. (a+b)}{\sin. (a-b)}$.

19. $\begin{cases} \dfrac{1+\tan. b}{1-\tan. b} \quad - \quad = \tan. (45^\circ + b). \\ \dfrac{1-\tan. b}{1+\tan. b} \quad - \quad = \tan. (45^\circ - b). \end{cases}$

We shall, probably, make an application of the following theorem; it applies to finding the unknown angles of a triangle, when the *logarithms* of two sides (not the sides themselves) and the angle included between the sides are given.

The greater of two sides of a plane triangle is, to the less, as radius to the tangent of a certain angle. Take this angle from 45°, *and call the difference a. Lastly, radius is to the tangent, a, as the tangent of the half sum of the angles at the base is to the tangent of half their difference.*

III. *Resolution of right-angled spherical triangles.*

In the following equations, h is the hypothenuse, s a given side, a a given angle, and x the quantity sought. (The right angle is unity, and always given.) TRIG.

Given,	Required,	Solution.
h and a	side op. a	20. $\sin. x = \sin. h \sin. a$.
	side adj. a	21. $\tan. x = \tan. h \cos. a$.
	the other angle	22. $\cot. x = \cos. h \tan. a$.
h and s	the other side	23. $\cos. x = \frac{\cos. h}{\cos. s}$
	ang. adj. s	24. $\cos. x = \tan. s \cot. h$
	ang. op. s	25. $\sin. x = \frac{\sin. s}{\sin. h}$
s and a opposite,	h	26. $\sin. x = \frac{\sin. s}{\sin. a}$
	the other side,	27. $\sin. x = \tan. s \cot. a$
	the other ang.	28. $\sin. x = \frac{\cos. a}{\cos. s}$
s and a adjacent,	h	29. $\cot. x = \cos. a \cot. s$
	the other side,	30. $\tan. x = \tan. a \sin. s$
	the other ang.	31. $\cos. x = \sin. a \cos. s$.
The two sides.	h	32. $\cos. x = \cos. s$ cos. other side
	the angles,	33. $\cot. x = \sin.$ adj. side $\times$ cot. opp. side.

IV. Resolution of oblique angled spherical triangles. Let A B and C be the three angles of any spherical triangle, and a b and C the sides opposite to them respectively, that is, the side a is opposite to A, &c.

In spherical trigonometry, the sines of the angles are proportional to the sines of the opposite sides.

Therefore 34. $\frac{\sin. A}{\sin. a} = \frac{\sin. B}{\sin. b} = \frac{\sin. C}{\sin. c}$.

Given the three sides a b c;

Required one of the angles, A.

35. - - $\text{Sin.}^2 \frac{1}{2} A = \frac{\sin. (s-b) \sin. (s-c)}{\sin. b \sin. c}$.

TRIG.

$$36. \quad \text{Cos.}^2 \tfrac{1}{2} A = \frac{\sin. S \sin. (s-a)}{\sin. b \sin. c}$$

In 35 and 36, $2S = a + b + c$.

CHAPTER I.

ASTRONOMICAL PROBLEMS.

PROBLEM I.

CHAP. I.

Given the right ascension and declination of any heavenly body, to find its latitude and longitude; or conversely, given the latitude and longitude of a body to find its corresponding right ascension and declination.

A general projection for connecting right ascension, declination, longitude, and latitude.

Fig. 49.

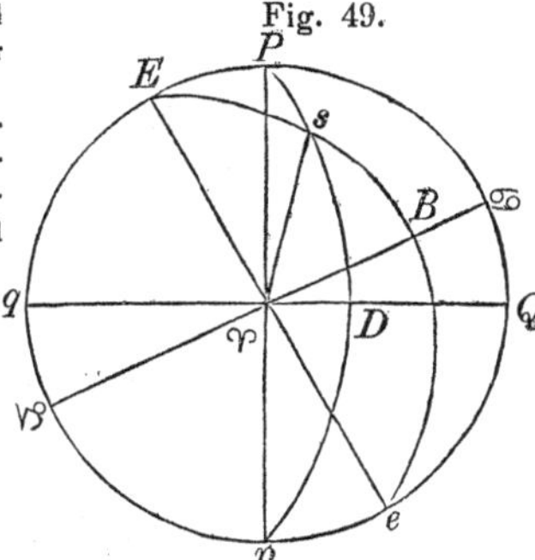

From any point as a center (Fig. 49) describe a circle Q EP♋, &c. Let this circle represent the meridian, which passes through the pole of the ecliptic E, the pole of the earth's axis P, and through the solstitial points ♋ and ♑. Then the point Aries (♈) will be at the center of the circle, and ♑ ♈ ♋ and Q ♈ q will be lines crossing each other by an angle equal to the obliquity of the ecliptic. Pp is the celestial meridian which passes through the equinoctial points, and is the first meridian of right ascension. E ♈ e is the first meridian of longitude, and, of course, the angle E ♈ P is equal to the obliquity of the ecliptic.

The figure is considered transparent, and both sides of it are represented.

Let s be the position of any celestial body, and draw the meridian of right ascension Psp; also draw the meridian of longitude Ese draw also ♈ s. We have now two right-angled spherical triangles sD♈ and ♈ Bs, having a common hypothenuse ♈ s; the first is the right ascension triangle, the second is the longitude triangle. Let the student observe that the line Qq represents a circle, the whole equator; and the point ♈ represents, in fact, two points, the 0 degree of right ascension and the 180th degree. So the point s represents two points, and ♈ D is the right ascension from 0 degree, or from 180 degrees. **CHAP. I.**

In our figure, the point s is north of both ecliptic and equator; but it might have been between the two, or south of both; hence, to meet every case, the judgment of the operator must be called into exercise to perceive a general solution.

Now, having the right ascension and declination of s, we find its latitude and longitude thus:

In the triangle ♈ Ds, ♈ D and Ds are given, and equation 32 gives ♈ s (h); 33 gives the angle s ♈ D. From s ♈ D subtract B ♈ D, the obliquity of the ecliptic, and there remains the angle s ♈ B.*

With the angle s ♈ B and the side ♈ s, equation 20 gives sB the latitude, and 21 gives ♈B the longitude.

EXAMPLES.

1. *The right ascension of a certain point in the heavens is* 5 h. 7 m. 50 s., *or in arc* 76° 57′ 30″; *and its declination is* 26° 11′ 36″ N.:

What is the latitude and longitude of the same point? Four equations contained in one operation.

	(32.)		(33.)
♈D 76° 57′ 30″	cos. 9.353454	- - -	sin. 9.988651
sD 26° 11′ 36″	cos. 9.952952	- - -	cot. 10.308104
♈ s 78° 19′ 3″	cos. 9.306406	26° 47′ 27″	cot. 10.296755
	B♈D - - - -	23 27 32	
	s♈B - - - -	3 19 55 = a	

* In general, take the difference between the angle s ♈ D and the obliquity of the ecliptic; and if the angle s ♈ D is the greater quantity, the body is north of the ecliptic, otherwise it is south of it. When the declination is south, the angle s ♈ D must be added to the obliquity of the ecliptic in the first and second quadrants, and subtracted in the third and fourth. Hence the judgment of the operator must be called in to decide the particulars of the case; or he must have a general formula that will give no exercise to the mind.

CHAP. I.

		(20.)		(21.)
(*h*)	78° 19′ 3″	sin. 9.990911		tan. 10.684611
(*a*)	3 19 55	sin. 8.763965		cos. 9.999265
	3° 15′ 36″	sin. 8.754876	78 18 6	tan. 10.683876

Thus we determine that the longitude must be 78° 18′ 6″, and the latitude 3° 15′ 36″ N.

2. *The longitude of the moon, at a certain time, according to computation, was* 102° 7′; *and latitude* 5° 14′ 15″ S.:

*What was the corresponding right ascension and declination?**

From these examples we might form a general rule; but rules thus formed seldom reflect principles; therefore for educational purposes, we fall back on the primary equations.

		(32.)		(33.)
♈*B*	77° 53′	cos. 9.322019		sin. 9.990215
sB	5° 14′ 15″	cos. 9.998183		cot. 11.037780
♈ *s*	77° 56′ 12″	cos. 9.320202	5° 21′ 27″	cot. 11.027995
		B♈*D* - -	23 27 42	
			18 6 15	

		(20.)		(21.)
(*h*)	77° 56′ 12″	sin. 9.990302		tan. 10.670170
(*a*)	18 6 15	sin. 9.492400		cos. 9.977948
	17 41 22	sin. 9.482702	77° 19′ 41″	tan. 10.648118

Thus we find that the right ascension distance on the equator, from the 180th degree, was 77° 19′ 41″; or its right ascension in arc was 102° 40′ 19″, or in time, 6h. 50m. 41s.

3. *By meridian observations on the moon, at a certain time, its right ascension was found to be* 16h. 53m. 33s., *and its declination* 17° 51′ 36″ S.: *what was its longitude and latitude?*

Ans. Lon. 254° 9′ 14″, Lat. 4° 41′ 12″ N.

Any number of the like examples can be found.

In the following examples either right ascension and declination may be taken for the data, and the longitude and latitude the sought terms, or conversely; the longitude and latitude may be the given data, and the right ascension and

* As the longitude is more than 90° and less than 180°, the moon is in the second quadrant of right ascension, and 77° 53′ in longitude from the equator; and as her latitude is south, it does not correspond to *Bs* in the figure, and we give the example to exercise the judgment of the learner.

declination the required terms. A Nautical Almanac will furnish any number of similar examples.

	R. A.	Dec.	Lon.	Lat.
	h. m. s.	° ′ ″	° ′ ″	° ′ ″
4	15 47 36	15 58 15 south,	238 14 48	4 30 17 north,
5	6 13 22	18 23 2 north,	93 10 55	5 4 23 south,
6	11 24 44	1 45 28 north,	171 12 40	1 52 51 south,
7	20 23 33	14 11 9 south,	304 47 15	5 2 23 north.

PROBLEM II.

Given the latitude of the place, and the declination of the sun or star; to find the semidiurnal arc, or the time the sun or star would remain above the horizon; and to find its amplitude, or the number of degrees from the east and west points of the horizon, where it will rise and set.

Tables for the semidiurnal arc and amplitudes are computed by this problem.

To illustrate this problem we draw Figure 50. Let PZH, &c., represent the celestial meridian passing through the place. Make the arc QZ equal to the latitude, then ZP will equal the co-latitude. The line Hh is everywhere 90° from Z, and represents the horizon. Pp represents the earth's axis, and the meridian 90° distant from the meridian of the place; Qq is the equator. From the points Q and q set off d and d', equal to the declination (north or south, as the case may be) and describe the small circle of declination, $d \odot d'$, where this circle crosses the circle of the horizon Hh is the point where the body (sun, moon, or star) will rise or set (rise on one side of the meridian and set on the other, both are represented by the same point in the projection). Through $P \odot p$ describe the meridian as in the figure, and the right-angled spherical triangle $R \odot C$ appears; right angled at R.

These examples do not take refraction into account.

Fig. 50.

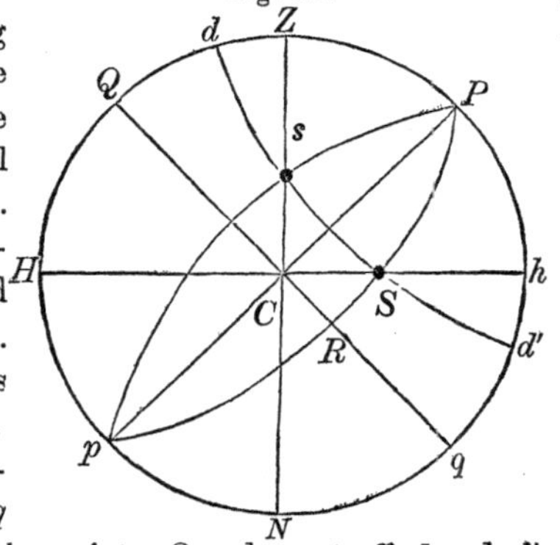

CHAP. I.

In the triangle $R \odot C$, there is given the side $R \odot$, the declination, and the angle opposite $R C \odot$, which is equal to the co-latitude. $R C$, expressed in time, at the rate of 15° to one hour, will be the time before and after 6 hours, from the time the body is on the meridian to the time it is in the horizon; and the arc $C \odot$ is the amplitude. The triangle is immediately resolved by equations 26 and 27.

$$(27.) \qquad \text{Sin. } R\,C = \text{tan. declin.} \times \text{tan. lat.}$$

$$(26.) \qquad \text{Sin. } C \odot = \frac{\text{sin. declin.}}{\text{cos. lat.}};$$

Observing that the tangent of the latitude is the same as the cotangent of the angle $R C \odot$, and the cosine of the lati- tude is the same as the sine of $R C \odot$, corresponding to a in the equation.

EXAMPLE.

The time in all these examples is, of course, apparent, because it refers directly to the sun, and not to a clock.

In the latitude of 40° N., *when the sun's declination is* 20° N., *what time before and after six will it rise and set, and what will be its amplitude?*

	(27.)		(26.)
20°	tan. 9.561066		sin. 9.534052
40	tan. 9.923813		cos. 9.884254
17° 47′	sin. 9.484879	26° 31′	sin. 9.649798

Thus we find that the arc called the *ascensional difference*, is 17° 47′, or, in time, 1h. 11m. 8s., showing that the sun or heavenly body, whatever it may be (when not affected by parallax or refraction), will be found in the horizon 7h. 11m. 8s. before and after it comes to the meridian.

Its amplitude for that latitude and declination is 26° 31′ north of east, or north of west, and, if observed by a compass, the apparent deviation would be the *variation of the compass.*

2. *At London, in Lat.* 51° 32′ N., *the sun's amplitude was observed to be* 39° 48′ *toward the north; what was its declination, and what was the apparent time of its rising and setting?*

Ans. Sun's declination, 23° 27′ 59″ N.

Sun's rising, 3h. 47m. 32s.; sun's setting, 8h. 12m. 28s.

The amplitude of the sun is frequently observed, at sea, to discover the variation of the compass; but, by reason of refraction, the results are not perfectly accurate.

CHAP. I

Refraction not taken into account, if it were, the time that the sun would remain above the horizon would be increased while it rose in altitude 33′ of arc.

From the right-angled spherical triangle (Fig. 50) $PZ\odot$, we can compute the time when the sun is east or west in position, and the altitude it must have, when in that position. The angle Z is a right angle, PZ is the co-latitude, and $P\odot$ is the co-declination.

Equation (23) gives the cosine of $Z\odot$, or the sine of the altitude of the sun when it is east or west — the latitude and declination being given — and equation (24) will give the angle or time from noon.

We may also find the altitude and azimuth of the sun, at 6 o'clock, by making use of a triangle formed by drawing a vertical through $Z\,s\,N$: $C\,S$, the given declination, will be its hypothenuse and $P\,Ch$, the latitude, will be the arc of its angles.

By means of right-angled spherical trigonometry, as comprised in the equations from 20 to 33, we can resolve all possible problems that can occur in astronomy, pertaining to the sphere; but, for the sake of brevity, mathematicians, in some cases, use oblique-angled spherical trigonometry, which is nothing more than right-angled trigonometry *combined and condensed.*

PROBLEM III.

Given, the latitude of the place of observation, the sun's declination, and its altitude above the horizon, to find its meridian distance, or the time from apparent noon.

The sun's distance from the meridian, as measured from the pole as a center, and on the equator as a circumference, is the measure of time from apparent noon.

There is no problem more important in astronomy than that of time. No astronomer puts implicit faith in any chronometer or clock, however good and faithful it may have been; and even to suppose that a chronometer runs true, it can only show time corresponding to some particular meridian; and hence, to obtain *local* time, we must have some method, directly or indirectly, of finding the sun's distance from the meridian.

When the center of the sun is on any meridian, it is *then* and *there* apparent noon; and the equation of time will be the

CHAP. I.

interval *to* or *from* mean noon; but none, save an astronomer in an observatory, can define the instant when the sun is on the meridian; no one else has a meridian line sufficiently definite and accurate, and with him it is the result of great care, combined with a multitude of nice observations.

Great importance of this problem.

To define the time, then (when anything like accuracy is required), we must resort to observations on the sun's altitude.

Direct meridian observations not generally accurate.

It is evident that the altitude of the sun is greater and greater from sunrise to noon, and from noon to sunset the altitude is continually becoming less. If we could determine, by observation, exactly when the sun had the greatest altitude, that moment would be apparent noon; but there is a considerable interval, *some minutes*, before and after noon, that it is difficult to determine, without the nicest observations, whether the sun is rising or falling; therefore, meridian observations are not the most proper to determine the time.

Proper times of observation.

From two to four hours before and after noon (depending in some respects on the latitude), the sun rises and falls most rapidly; and, of course, that must be the best time to fix upon some definite instant; for every minute and second of altitude has its corresponding time from noon; and thus the time and altitude have a scientific connection, which can only be disentangled by spherical trigonometry. But we proceed to the problem.

Fig. 51.

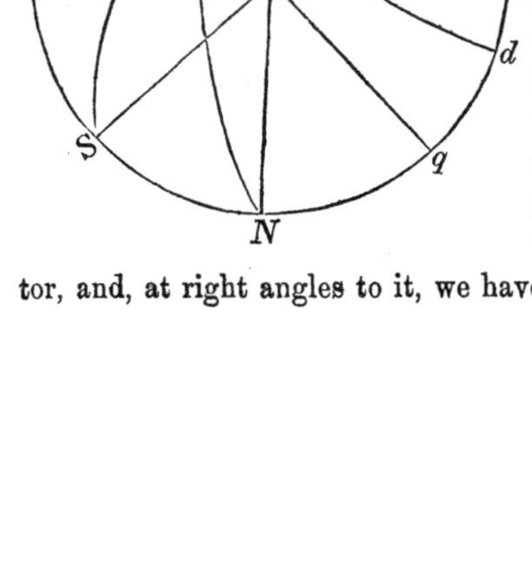

Description of the figure.

Draw a circle, P Z Q H, &c., (Fig. 51), representing the meridian; Z is the zenith, and Z N is the prime vertical; Hh is the horizon; Z Q is an arc equal to the given latitude; Q q is the equator, and, at right angles to it, we have the earth's axis, P S.

Take $H a$, $h a$, equal to the observed altitude of the sun, and draw the small circle, $a a$, parallel to the horizon, $H h$. From the equator take $Q d$, $q d$, equal to the declination of the sun, and draw the small circle, $d d$, parallel to $Q q$. Where these two small circles, $a a$, $d d$, intersect, is the position of the sun at the time.

From Z draw the vertical, $Z \odot N$, and from P draw the meridian through the sun, $P \odot S$. The triangle $P Z \odot$ has all its sides given, from which the angle $Z P \odot$ can be computed; which angle, changed into time at the rate of 15° to one hour, will give the time from noon, when the altitude was taken. If the time, per watch, should agree with the time thus computed, the watch is right, and as much as it differs is the error of the watch.

The observation defines and points out a triangle.

The side $Z \odot$ is the complement of the altitude, $P \odot$ is the complement of the declination, and $P Z$ is the complement of the latitude, and equation (35) or (36) will solve the problem; that is, find the angle P which can be made to correspond to A, in the equation. But, in place of using the complement of the latitude, we may use the latitude itself; and, in place of using the complement of the altitude, we may use the altitude itself; provided we take the cosine, when the side of the triangle calls for the sine; for it would be the same thing. By thus taking advantage of every circumstance, ingenious mathematicians have found a less troublesome practical formula than either (35) or (36) would be; but we cannot stop to explain the modifications and changes in a work like this; we contemplate doing so in a work more appropriate to such a purpose: the student must be content with the following practical rule, *to find the time of day, from the observed altitude of the sun, together with its polar distance, and the latitude of the observer.*

Mathematicians make great exertions to abbreviate practial operations.

Practical rule used at sea.

RULE 1.—*Add together the altitude, latitude, and polar distance, and divide the sum by two. From this half sum subtract the altitude, reserving the remainder.*

2.—*Take the arithmetical complement of the cosine of the latitude, the arithmetical complement of the sine of the polar distance, the cosine of the half sum, and the sine of the remainder. Add*

CHAP. I.

these four logarithms together, and divide the sum by two; the result is the logarithmetic sine of half the hourly angle.

3.—This angle, taken from the Tables, and converted into time at the rate of four minutes to one degree, will be the time from apparent noon; the equation of time applied, will give the mean time when the observation was made.*

The quadrant and sextant and reflecting circle essentially the same instrument.

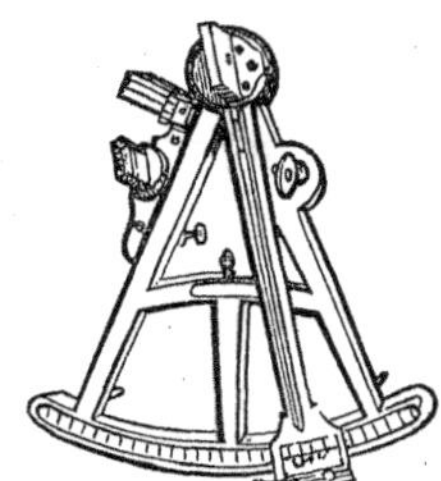

* The instrument for taking altitudes at sea, or wherever the observer may happen to be, is a quadrant or sextant, according to the number of degrees of the arc. It is made on the principle of reflecting the image of one body to another, by means of a small mirror revolving on a center of motion, carrying an index with it over the arch. Nearly opposite to the *index mirror* is another mirror, one half silvered, the other half transparent, called the horizon glass. Directly opposite to the horizon glass is the line of sight, in which line there is sometimes placed a small telescope. The line of sight must be *parallel* to the plane of the instrument. The two mirrors must be *perpendicular* to the plane of the instrument. To be in adjustment, the two mirrors, namely the index glass and horizon glass, *must be parallel*, when the index stands at 0.

To examine whether a sextant is in adjustment or not, proceed as follows:

1. *Is the index mirror perpendicular to the plane of the instrument?*

Put the index in about the middle of the arch, and look into the index mirror, and you will see part of the arch reflected, and the same part direct; and if the arch appears perfect, the mirror is in adjustment; but if the arch appears broken, the mirror is not in adjustment, and must be put so by a screw behind it, adapted to this purpose.

2. *Are the mirrors parallel when the index is at* 0?

Place the index at 0, and clamp it fast; then look at some well-defined, distant object, like an even portion of the dis-

EXAMPLE.

In latitude 39° 46′ north, when the sun's declination was 3° 27′ north, the altitude of the sun's center, corrected for refraction, index error, &c., was 32° 20′, *rising;* what was the apparent time?

Altitude,	32	20				
Latitude,	39	46	-	cos. comple.	-	.114268
Polar dis.,	86	33	-	sine comple.	-	.000788
	2)158	39				
	79	19	30	cosine	-	9 .267652
	32	20				
	46	59	30	sine	-	9 .864090
						2)19 .246798
$\frac{1}{2}$ *ZP* ⊙	24	50	30	sine		9 .623399
			2			

The hourly angle is 49 41 0, which, converted into time, gives 3 h. 18 m. 44 s., the time from apparent noon, and, as

---

tant horizon, and see part of it in the mirror of the horizon glass, and the other part through the transparent part of the glass; and, if the whole has a natural appearance, the same as without the instrument, the mirrors are parallel; but, if the object appears broken and distorted, the mirrors are not parallel, and must be made so, by means of the lever and screws attached to the *horizon glass.*

3. *Is the horizon glass perpendicular to the plane of the instrument?*

The former adjustments being made, place the index at 0, and clamp it; look at some smooth line of the distant horizon, while holding the instrument perpendicular; a continued, unbroken line will be seen in both parts of the horizon glass; and if, on turning the instrument from the perpendicular, the horizontal line *continues unbroken,* the horizon glass is in full adjustment; but, if a break in the line is observed, the glass is not perpendicular to the plane of the instrument, and must be made so, by the screw adapted to that purpose.

After an instrument has been examined according to these

CHAP. I. the sun was rising, it was before noon, and the apparent time was 8 h. 41 m. 16 s.

An arc may be measured by the quadrant within one minute.

A good observer, with a good instrument, in favorable circumstances, can define the time, from the sun's altitude, to within three or four seconds.

An artificial horizon.

At sea, the observer brings the reflected image of the sun to the horizon, and allows for the dip (Tables p.25). On shore, where no natural horizon can be depended upon, resort is had to an *artificial horizon,* which is commonly a little mercury turned out into a shallow vessel, and protected from the wind by a glass roof. The sun, or any other object, may be seen reflected from the surface of the mercury (which, of course, is horizontal), and the image, thus reflected, appears as much below the natural horizon as the real object is above the horizon; and, therefore, if we measure, by the instrument, the angle between the object and its image in the artificial horizon, that angle will be double the altitude.

When mercury is not at hand, a plate of molasses will do very well; and in still, calm weather, any little standing pool of water may be used for an artificial horizon.

Observations taken in an artificial horizon are not affected by *dip,* but they must be corrected for refraction and index error, and, if the two limbs of the sun are brought together, its semidiameter must be added after dividing by two.

A practical example.

The following example is from a sailor's note book:

"On the 18th of May, 1848, at sea, in latitude 36° 21′ north, longitude 54° 10′ west, by account, at 7 h. 43 m. per watch; the altitude of the sun's lower limb was 32° 51′ rising; the hight of the eye was eighteen feet, and the index

---

directions, it may be considered as in an approximate adjustment—a re-examination will render it more perfect—and, finally, we may find its *index error* as follows:—measure the sun's diameter both on and off the arch—that is, both ways from 0, and if it measures the same, there is *no index error;* but if there is a difference, half that difference will be the index error, *additive,* if the greatest measure is off the arch, subtractive, if on the arch.

error of the sextant was 2′ 12″ additive. What was the error of the watch?"

CHAP. I.

PREPARATION.

Time, per watch, - - - 7 h. 43 m., morning.

Longitude, 54° 10′, in time, - 3 38

Estimated mean time at Greenwich, 11 h. 21 m.

Preparations to be made according to circumstances.

The declination of the sun at mean noon, Greenwich time, was 19° 38′ 29″ increasing, the daily variation being 13′; the variation, therefore, for 39′, the time before noon, was 21″ subtractive. Hence, the declination of the sun, at the time of observation, was 19° 38′ 8″ north, and the polar distance 70° 21′ 52″.

Observed altitude, - - - - -		32° 51′ 00″
Index error, - - - - - -	+	2 12
Semidiameter, - - - - - -	+	15 49
Refraction, - - - - - - -	—	1 28
Dip of the horizon, - - - - -	—	4 13
True altitude of sun's center, - -		33° 3′ 20″

Altitude,	33°	3′	20″		
Latitude,	36	21		cos. complement,	.093982
Polar dis.,	70	21	52	sin. complement,	.026013
2)139		46	12		
	69	53	6	cosine, - -	9.536470
	33	3	20		
	36	49	46	sine, - -	9.777770
					2)19.434235
½ hourly angle,	31	25	30	sine, - -	9.717117

This angle corresponds to 4 h. 11 m. 24 s., or, in reference to the forenoon, 7 h. 48 m. 36 s. apparent time.

On the 18th of May, noon, Greenwich time, the equation of time was 3 m. 54 s. subtractive; therefore, the true mean time, at ship, was - - - 7 h. 44 m. 42 s.

Time, per watch, - - - 7 43

Watch slow, - - - 1 42

By observations thus taken at different times at the same place, the rate of the watch can be determined.

A short time before this observation was taken, the watch

CHAP. I. was compared with a chronometer in the cabin, which was too fast for mean Greenwich time, 19 m. 12.5 s., according to estimation from its rate of motion. The chronometer was fast of watch by 3 h. 56 m. 39 s. What was the longitude of the ship?

	h. m. s.
Time of observation, per watch,	7 43 00
Diff. between watch and chron.,	3 56 39
Time, per ch., at observation,	11 39 39
Chron. fast of Greenwich time,	19 12
Greenwich mean time, -	11 20 27
Mean time at ship, - -	7 44 42
Longitude in time, - -	3 35 45=53° 56′ west.

How to decide from the observations whether the longitude is east or west.

The longitude is west, because it is later in the day, at Greenwich, than at the ship. This example explains all the details of finding the longitude by a chronometer.

How to determine and mark out a meridian line.

By taking advantage of the observations for time on shore, we may draw a meridian line with considerable exactness; for instance, in the last observation (if it had been on land), in 4 h. 11 m. 24 s. after the observation was taken, the sun would be exactly on the meridian; and if the watch could be depended upon to measure that interval with tolerable accuracy, the direction from any point toward the sun's center, at the end of that interval, would be a meridian line. Several such meridians, drawn from the same point, from time to time, and the mean of them taken, will give as true a meridian as it is practical to find; although, for such a purpose, a prominent fixed star would be better than the sun.

Absolute and local time.

The problem of time includes that of longitude, and finding the difference of longitude between two places always resolves itself into the comparison of the *local times*, at the same instant of absolute time. When any definite thing occurs, wherever it may be, that is absolute time. For instance, the explosion of a cannon is at a certain instant of absolute time, wherever the cannon may be, or whoever may note the event; but if the instant of its occurrence could be known at far distant places, the local clocks would mark very different hours and minutes of time; but such difference would be occasioned entirely by difference of longitude: the event is the same for all places — it is a *point* in absolute time.

Absolute time defined by means of events.

Thus any single event marks a *point* in absolute time. If the same event is observed from different localities, the difference in local time will give the difference in longitude. But a *perfect* clock is a noter of events, it marks the event of noon, the event of sunrise, the event of one hour after noon, &c.; and if we could have perfect confidence in this marker of events, nothing more would be necessary to give us the local time at a distant place. The time, at the place where we are, can be determined by the altitude of the sun, *or a star*, as we have just seen. But, unfortunately, we cannot have perfect confidence in any chronometer or clock; and therefore we must look for some event that distant observers can see at the same time.

A clock is a noter of events, when it runs true, but not otherwise.

Eclipses are events, which mark absolute time, but for common purposes they are of little value.

The passage of the moon into the earth's shadow is such an event, but it occurs so seldom as to amount to no practical value. The eclipses of Jupiter's satellites are such events, but they cannot be observed without a telescope of considerable power, and a large telescope cannot be used at sea. Hence these events are serviceable to the local astronomer only; the sailor and the practical traveler can be little benefited by them. The moon has comparatively a rapid motion among the stars (about 13° in a day), and when it comes to any definite distance to or from any particular star, that circumstance may be called an event, and it is an event that can be observed from half the globe at once.

The motion of the moon among the stars, may be considered as an index moving round a circle marking absolute time.

Thus, if we observe that the moon is 30° from a particular star, that event must correspond to some instant of *absolute time;* and if we are sufficiently acquainted with the moon, and its motion, so as to know exactly how far it will be from certain definite points (*stars*) at the times, when it is noon, 3, 6, 9, &c., hours at Greenwich, then, if we observe these events from any other meridian, we thereby know the Greenwich time, and, of course, our longitude.

Finding the Greenwich time by means of the moon's angular distance from the sun or stars, is called *taking a lunar;*

CHAP. I. and it is probably the only reliable method for long voyages at sea.

If the motion of our moon had been very slow, or if the earth had not been blessed with a moon, then the only methods, for sea purposes, would have been chronometers and *dead reckoning*. For a practical illustration of the theory of lunars, we mention the following facts.

Lunar observations illustrated by an example.

In a sea journal of 1823, it is stated that the distance of the moon from the star Antares was found to be 66° 37′ 8″, *when the observation was properly reduced to the center of the earth*, and the time of observation at ship was September 16th, at 7h. 24m. 44s. P. M. apparent time.

By comparing this with the Nautical Almanac, it was found that at 9 P. M., apparent time at Greenwich, the distance between the moon and Antares was 66° 5′ 2″, and at midnight it was 67° 35′ 31″; but the observed distance was between these two distances, therefore the Greenwich time was between 9 and 12 P. M., and the time must fall between 9 and 12 hours in the same proportion as 66° 37′ 8″ falls between the distances in the Nautical Almanac; and thus an observer, with a good instrument, can at any moment determine the Greenwich time, whenever the moon and stars are in full view before him.

The moon, in connection with the stars in the heavens, may be considered a public clock (quite an enlargement of the town-clock), by which certain persons, who understand the dial plate and the motion of the index, and who have the necessary instrument, can read the Greenwich time, or the time corresponding to any other meridian to which the computations may be adapted.

Observed distances, and distances as seen from the center of the earth.

The angular distances from the moon to the sun, stars, and planets, as put down in the Nautical Almanac, corresponding to every third hour, are distances as seen from the center of the earth, and when observations are taken on the surface the distance is a little different; the position of the moon is affected by parallax and refraction, the sun or star by refraction alone; and therefore a reduction is necessary, which is called *clearing the distance*. This is done by spheri-

Clearing the distance.

cal trigonometry. The distance between the moon and star is observed, the altitudes of the two bodies are also observed. The co-altitudes come to the zenith, and the co-altitudes, with the distance, form three sides of a spherical triangle, from which the angle at the zenith can be computed. Then correct the altitude of the moon for parallax and refraction, and the star for refraction, and find the true altitudes and co-altitudes, and the true co-altitudes and angle at the zenith give two sides and the included angle of a spherical triangle, and the third side, computed, is the *true distance.*

An immense amount of labor has been expended by mathematicians, to bring in artifices to abbreviate the computation of clearing lunar distances; and they have been in a measure successful, and many special rules have been given, but they would be out of place in a work of this kind.

PROPORTIONAL LOGARITHMS.

Proportional logarithms — an explanation of the construction of the table given.

In every part of practical astronomy there are many proportional problems to be resolved, and as the terms are mostly incommensurable, it would be very tedious, in most cases, to proceed arithmetically, we therefore resort to logarithms, and to a prepared scale of logarithms, very appropriately called *proportional logarithms.*

The tables of proportional logarithms commonly correspond to one hour of time, or 60′ of arc, or to three hours of time. The table in this book corresponds to one hour of time, or 3600 seconds of either time or arc. To explain the construction and use of a table of proportional logarithms, we propose the following problem :

At a certain time, the moon's hourly motion in longitude was 33′ 17″ ; *how much would it change its longitude in* 13m. 23s.?

Put x to represent the required result, then we have the following proportion :

$$\begin{array}{lcccc} & \text{m.} & \text{m. s.} & ' \ '' & \\ & 60 : & 13\ 23 :: & 33\ 17 : & x; \\ \text{Or} & 3600 : & 13\ 23 :: & 33\ 17 : & x. \end{array}$$

Divide the first and second terms of this proportion by the

CHAP. I.

second, and the third and fourth by the third, then we have

$$\frac{3600}{13.23} : 1 :: 1 : \frac{x}{33.17}$$

Divide the third and fourth terms by x, and multiply the same terms by 3600, and the proportion becomes

$$\frac{3600}{13.23} : 1 :: \frac{3600}{x} : \frac{3600}{33.17}.$$

Multiplying extremes and means, using logarithms, and remembering that the addition of logarithms performs multiplication,

Then we have $\log. \frac{3600}{x} = \log. \left(\frac{3600}{13.23}\right) + \log. \left(\frac{3600}{33.17}\right).$

By the construction of the table, the *proportional* logarithm of 1″ is the *common* logarithm of $\frac{3600}{1}$; of 2″ is the common logarithm of $\frac{3600}{2}$; of 3″ is $\frac{3600}{3}$, &c., to $\frac{3600}{3600}$; hence the proportional logarithm of 3600 is 0.

We now work the problem:

	′ ″			
	13 23	- - -	P. L.	6516
	33 17	- - -	P. L.	2559
Result, - -	7 25½	- - -	P. L.	9075

Examples given to illustrate the practical utility of proportional logarithms.

EXAMPLES FOR PRACTICE.

1. When the sun's hourly motion in longitude is 2′ 29″, what is its change of longitude in 37 m. 12 s.?

Ans. 1′ 32″.5.

2. When the moon's declination changes 57″.2 in one hour, what will it change in 17 m. 31 s.? Ans. 16″.8.

3. When the moon changes longitude 27′ 31″ in an hour, how much will it change in 7 m. 19 s.? Ans. 3′ 21″.

4. When the moon changes her right ascension 1 m. 58 s. in one hour, how much will it change in 13 m. 7 s.?

Ans. 25″.8.

N. B. This table of proportional logarithms will solve any proportion, *provided* the first term is 60, or 3600; therefore, when the first term is not 60, reduce it to 60, and then use the table.

CHAP. I.

EXAMPLES.

1. If the sun's declination changes 16′ 33″ in twenty-four hours, what will be the change in 14 h. 18 m.?

Examples given to illustrate the practical utility of proportional logarithms.

Statement,	24	:	14.18	::	16′ 33″
Or,	12	:	7.09		
Or,	60	:	35.45	::	16′ 33″

	16′ 33″	P. L.	5594
	35′ 45″	P. L.	2249
Ans.	9′ 51″.5	P. L.	7843

2. If the moon changes her declination 1° 31′ in twelve hours, what will be the change in 7 h. 42 m.? Ans. 58′.

Conceive degrees and minutes to be minutes and seconds, and hours and minutes to be minutes and seconds.

When 60 minutes or 3600 seconds are not the first term of a proportion, the result is found by taking the difference of the proportional logarithms of the other term for the P. L. of the sought term, as in the following example:

The moon's hourly motion from the sun is 26′ 30″, *what time will it require to gain* 30″?

Statement, 26′ 30″ : 60m. : 30″ : x

Other examples.

	30″	P. L.	2.0792
	60 m.	P. L.	0.0000
Product of extremes,			2.0792
	26′ 30″	P. L. sub.	3549
Result,	1 m. 7 s.	P. L.	1.7243

3. The equation of time for noon, Greenwich, on a certain day, was 6 m. 21 s.; the next day, at noon, it was 6 m. 43 s.: what was it corresponding to 3 h. 42 m. P. M., in longitude 74° west, on the same day? Ans. 6 m. 29 s.

CHAPTER II.

GENERAL PROBLEM.

CHAP. II.

A general problem preparatory to the computations of eclipses.

Given, the motions of sun and moon, to determine their apparent positions at any given time; from which results their apparent relative situations, and the eclipses of the sun and moon.

This problem covers two-thirds of the science of astronomy, and includes many minor problems; therefore a brief or hasty solution must not be expected.

From the foregoing portions of this work, the reader is supposed to have acquired a good general knowledge of the solar and lunar motions, and the tables give all the particulars of such motions; and all the artifices and ingenuity that astronomers could devise, have been employed in forming and arranging these tables, for the double purpose of facilitating the computations and giving accuracy to the results.

The tables, in general, must be left to explain themselves, and the mere heading, combined with the good judgment of the reader, will furnish sufficient explanation, in most instances; but some of them require special mention. *All the tables are adapted to mean time at Greenwich.*

EXPLANATION OF TABLES.

A very general and comprehensive explanation of the tables.

Table IV contains the sun's mean longitude, the longitude of its perigee (each diminished by 2°), and the *Arguments* * for some of the small inequalities of the sun's apparent motion.

Explanation of the term argument.

* The term, ARGUMENT, in astronomy, means nothing more than a correspondence in quantities. Thus, each and every degree of the sun's longitude corresponds with a particular amount of declination; and therefore, a table could be formed for the declination, and the argument would be the sun's longitude.

The moon's horizontal parallax and semidiameter vary together, and each minute of parallax corresponds to a particular amount of semidiameter; hence, a table can be made for finding the semidiameter, and the arguments would be the horizontal parallax. But the hori-

Argument I, corresponds to the action of the moon; Argument II, to the action of Jupiter; Argument III, to Venus; and Argument N, is for the equation of the equinoxes, and corresponds with the position of the moon's node; and, by inspecting the column in the table, it will be perceived that the *argument* runs round the circle in a little more than eighteen years, as it should; and thus, by inspection, we can obtain an insight as to the period of any argument in the solar or lunar tables. CHAP. II.

Explanation of the solar tables.

The object of diminishing the mean longitude and perigee of the sun by 2°, is to render the equation of the center always additive; for if 2° are taken from the longitude, and 2° added to the equation of the center, the combination of the two quantities will be the same as before; and, as the equation of the center is always less than 2°, therefore, 2° added to its greatest *minus value,* will give a positive result. By the same artifice all equations may be rendered always positive. The 2°, taken from the mean longitude, are restored by adding 1° 59′ 30″ to the equation of the center, and 10″ to each of the other equations; hence, to find the real equation of the center corresponding to any degree of the anomaly, subtract 1° 59′ 3″ from the quantity found in the table.

Table XII, shows the time of the mean new moon, &c., in January, diminished by fifteen hours, to render the corrections always additive. The fifteen hours are restored by adding 4 h. 20 m. to the first equation, 10 h. 10 m. to the second, 10 m. to the third, and 20 m. to the fourth.

Argument I, corrects for the action of the sun on the lunar

---

zontal parallax and semidiameter of the moon depend (not solely) on the moon's distance from its perigee; hence, a table can be formed giving both horizontal parallax and semidiameter; which ARGUMENTS are the anomaly. In other words, an argument may be called an INDEX, and when the arguments correspond to points in a circle, or to the difference of points in a circle, the circle may be considered as divided into 1000 or 100 parts, then 500, or 50, as the case may be, would correspond to half a circle, and so on in proportion. This mode of dividing the circle has been adopted, with certain limitations, to avoid the greater labor of computing by denominate numbers.

CHAP. II.

orbit; Argument II, corrects for the mean eccentricity of the lunar orbit; Argument III, corrects for the different combinations of the solar and lunar perigee; and Argument IV, corrects for the variation occasioned by the inclination of the lunar orbit to the ecliptic; N. shows the distance from or to the nodes.

Tables adapted to the synodical motion of the moon, by which new and full moons can be computed.

New and full moons, calculated by these tables, can be depended upon within *four minutes*, and commonly much nearer; but when great accuracy is required, the more circuitous and elaborate method of computing the longitudes of both sun and moon must be employed.

Tables XIII, XIV, and XV, are used in connection with Table XII.

Explanation of the lunar table.

Table XVI, shows the reduction of the latitude, and also of the moon's horizontal parallax, corresponding to the latitude, occasioned by the peculiar shape of the earth, and the diminution of its diameter as we approach the poles. *The table is put in this place because of the convenient space in the page.*

Table XVII, and the following tables to No. XXX, contain the arguments and epochs of the moon's mean longitude, evection, &c., necessary in computing the moon's true place in the heavens.

The method of computing the true longitude of the sun.

The argument for evection is diminished by 29′; the anomaly by 1° 59′, the variation by 8° 59′, and the longitude by 9° 44′, and the balances are restored by adding the same amounts to the various equations, which, at the same time, renders the equation affirmative, as explained in the solar tables.

The arguments in Table XXXII, are also arguments for polar distance, or latitude, in Table XXVIII. Anything like a minute explanation of these tables would lead us too far, and not comport with the design of this work. The use of the tables will be shown by the examples.

We have carried the mean motions of the sun and moon only to five minutes of time — and this is sufficient for all practical purposes — for we can proportion to any intermediate minute or second, by means of the hourly motions.

PROBLEM I.

From the solar tables find the sun's longitude, hourly motion in longitude, declination, semidiameter and equation of time; and for a specific example, find these elements corresponding to mean time, at Greenwich, 1854, *May* 26 d. 8 h. 40 m.

To find the sun's declination, spherical trigonometry gives us the following proportion: (Eq. 20, page 231.)

As radius - - - - -	10.000000
Is to sin. of ☉'s lon. (65° 12′ 15″) - -	9.957994
So is sin. of obliq. of the eclip. (23° 27′ 32″)	9.599900
To sin. declination N., 21° 10′ 54″ - -	9.557894

In nearly all astronomical problems, time is reckoned from noon to noon—from 0 hour to 24 hours.

When the given time is apparent, reduce it to mean time, and when not at Greenwich, reduce it to Greenwich time, by applying the longitude in time.—(This is necessary because the tables are adapted to Greenwich mean time.)

From Table IV, and opposite the given year, take out the whole horizontal line of numbers (headed as in the table), and from Tables V, VII, VIII, take out the numbers corresponding to the month—day of the month—hour and minute of the day, as in the following example.

Add up the perpendicular columns, as in compound numbers, rejecting *entire circles* in every column, and the sums or surplus, as the case may be, will give the mean values of all the quantities for the given instant.

The sun's distances from its perigee point is called its mean anomaly.

Subtract the longitude of the perigee from the mean longitude, and the remainder will be the mean anomaly; which is the argument for the equation of the center.

With the respective arguments take out the corresponding equations, all of which add to the mean longitude, and the true longitude of the sun from the *mean* equinox will be found.

With the argument N* take out the equation of the equi-

* The reason why N is not applied with the other equations is because it is sometimes negative.

CHAP. II. noxes from Table X, and apply it according to its sign, and the result will be the true longitude from the true equinox.

		M. Lon.	Lon. Perig.	I.	II.	III.	N.
		s. ° ′ ″	s. ° ′ ″				
1854		9 8 48 48	9 8 25 29	073	998	902	809
	May	3 28 16 40	20	59	301	206	18
	26 d	24 38 28	4	844	63	43	4
	8 h	19 43		11	0	0	0
	40m	1 39		987	362	151	831

	2 2 5 18	9 8 25 53
Eq. of center	3 6 42	2 2 5 18
I	10	4 23 39 25 = Mean anomaly.
II	13	
III	8	
	2 5 12 31	
Eq. of the equinox	—16	Sun's hourly motion in lon. 2′ 24″
True lon.	2 5 12 15	" semidiameter, 15′ 49′

To find the equation of time to great accuracy.

These principles were explained on pages 94 and 95.

	° ′ ″
By equation 21, page 231, we find the sun's R. A., - - - - -	63 16 10
Subtract this from the sun's lon., - -	65 12 15
Equatorial point is *west* of mean eastward motion by - - - - -	1° 56′ 5″ (*a*)
From the equation of the center, as just found, - - - -	3 6 42
Subtract the constant of the table, -	1 59 30
The sun east of its mean place, - -	1 7 12 (*b*)
Subtract (*b*) from (*a*) because one is east, the other west, and we have the arc - -	48′ 53″

This *arc*, converted into time, gives 3 m. 15.5 s. for the equation of time at this instant, and the sun will not come to the meridian at mean noon, but 3 m. 15½ s. afterward, Hence, to convert mean into apparent time, in the month of May, add the equation of time.

Thus, in general, we can determine the exact amount of the equation of time, by means of the two arcs (a) and (b) (which are roughly tabulated on page 95), and, without strictly scrutinizing each particular case, we can determine whether we are to take the *sum* or *difference* of the arcs by inspecting the table on page 95, or by referring our results to some respectable calendar.

EXAMPLE.

2. What will be the sun's longitude, declination, right ascension, hourly motion in longitude, semidiameter of the sun, and equation of time corresponding to 20 minutes past 9, mean time at Albany, N. Y., on the 17th of July, 1860?

N. B. At this time the sun will be eclipsed.

Ans. Lon. 214° 38′ 21″; Dec. 21° 12′ 48″.

R. A., in time, 7h. 46m. 15s.; Eq. of time to add to apparent time, 5m. 46.2s.; hourly motion in lon., 2′ 23″; S. D., 15′ 45.6″.

PROBLEM II.

From Tables XI, XII, and XIII, to find the approximate time of new and full moons.

Take the time of new moon, and its arguments, from Table XI, corresponding to January of the given year, and take as many lunations, from the following table, as correspond to the number of the months after January, for which the new moon is required; add the sums, rejecting the sums corresponding to whole circles, in the arguments, and in the column of days, rejecting the number corresponding to the expired months, as indicated by Table XIII; the sums will be the mean new moon and arguments for the required month.

When a full moon is required, add or subtract half a lunation. Sometimes one more lunation than the number of the month after January, will be required to bring the time to the required month, as it occasionally happens that two lunations occur in the same month.

Add the number of lunations necessary to bring the result to the required time of year.

Apply the equations corresponding to the different arguments taken from Table XIV, and their sum, added to the mean time of new or full moon, will give the true mean time of new or full moon for the meridian of Greenwich, within four minutes, and generally within two minutes.

CHAP. II.

For the time at any other meridian apply the time corresponding to the longitude.

EXAMPLES.

1. Required the approximate time of new moon, in May, 1854, corresponding to the day of the month, and the time of the day, at Greenwich, England, Boston, Mass., and Cincinnati, Ohio.

January.	Mean N. Moon.			I.	II.	III.	IV.	N.
1854,	27d.	18h.	14m.	0761	1168	19	04	668
Four Luna.	118	2	56	3234	2869	61	96	341
	145	21	10	3995	4037	80	00	009
Table XIII.	120							
May,	25	21	10	N shows an eclipse of the sun—visible in the United States.				
I.		6	46					
II.		4	14					
III.			17					
IV.			20					
May,	26	8	47					

New ◉ mean time at Greenwich, -	8 h.	47 m., P. M.
Boston, Longitude, - - -	4	44
New ◉ Boston time, - - -	4	3
Cincinnati, Longitude from Boston,		53
New ◉ Cincinnati time, - -	3	10

2. Required the approximate time of full moon, in July, 1852, for the meridian of Greenwich, and for Albany time, New York.

January.	Mean N. Moon.			I.	II.	III.	IV.	N.
1852,	20d.	11h.	53m.	0549	3239	38	27	538
Five Luna.	147	15	40	4042	3586	76	95	426
Half Luna.	14	18	22	404	5359	58	50	43
	182	21	55	4995	2184	72	72	007
Tab. 13. Bis.	182			The column N shows that the moon is very near her node. There will be a total eclipse of the moon—invisible in the United States.				
July,	0	21	55					
I.		4	21					
II.			42					
III.			17					
IV.			10					
July,	1	3	25	Mean time at Greenwich.				

Full ◉ Greenwich time, - -	3 h.	25 m.	P. M.
Albany, Longitude, - - -	4	55	
Full ◉ Albany time, - - -	10	30	A. M.

Thus we can compute the time of new or full moon for any month in any year; but, as the numbers for the arguments correspond to mean or average motions, and cannot, without immense care and labor, be corrected for the true, variable motions, the results are but approximate, as before observed.

ECLIPSES.

Eclipses take place at new and full moons; an eclipse of the sun at new moon, and an eclipse of the moon at full moon; but eclipses do not happen at every new and full moon; and the reason of this must be most clearly comprehended by the student before it will be of any avail for him to prosecute the further investigation of eclipses.

When eclipses take place.

If the moon's orbit coincided with the ecliptic, that is, if the moon's motion was along the ecliptic, there would be an eclipse of the sun at every new moon, and an eclipse of the moon at every full moon; but the moon's path along the celestial arch does not coincide with the sun's path, the ecliptic; but is inclined to it by an angle whose average value is 5° 8′, crossing the ecliptic at two opposite points on the apparent celestial sphere, which are called the moon's nodes.

Why eclipses do not take place every month

If the moon's path were *less* inclined to the ecliptic, there would be *more* eclipses in any given number of years than now take place. If the moon's path were *more* inclined to the ecliptic than it now is, there would be *fewer* eclipses.

What would be essential for more and what for fewer eclipses.

The time of the year in which eclipses happen, depends on the *position* of the moon's nodes on the ecliptic; and if that position were always the same, the eclipses would always happen in the same months of the year. For instance, if the longitude of one node was 30°, the other would be in longitude 30+180, or 210°; and, as the sun is at the first of these points about the 20th of April, and at the second about the 20th of October, the moon could not pass the sun in these months without coming very nearly in range with it, of course, producing eclipses in April and October.

Why an eclipse should take place in any particular month.

CHAP. II.

Fig. 52.

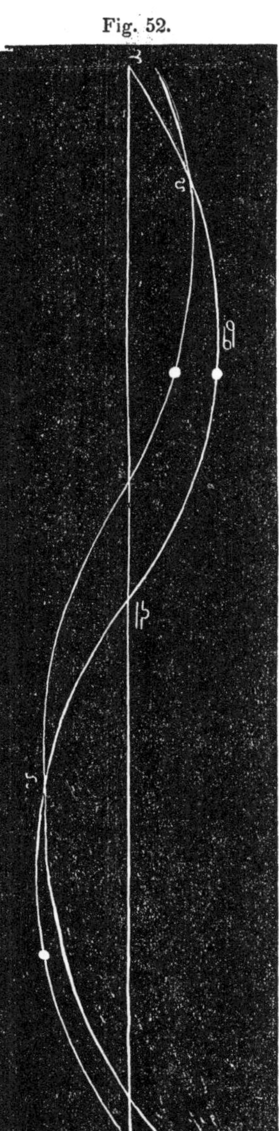

The figure represents the particular paths of the sun and moon through the heavens.

For a clearer illustration, we present Fig. 52: the right line through the center of the figure, represents the equator, the curved line ♈ ♋ ♎, crossing the equator at two opposite points, represents the ecliptic; and the curved line ☋ ● ☊ represents the path of the moon crossing the ecliptic at the points ☋ and ☊; the first of these points is the descending, the other, the ascending node.

As here represented, the ascending node is in longitude about 210°, and the descending node in about 30°; which was about the situation of the nodes in the year 1846, and, of course, the eclipses of that year must have been, and really were, in April and October.

The sun and moon at conjunction are represented in the figure a little after the sun has passed the northern tropic, which must be about the first of August; and it is perfectly evident that no eclipse can then take place, the moon running past the sun, at a distance of about *five degrees* south; and at the opposite longitude, the moon must pass about *five degrees* north.

The moon's nodes move backward at the mean rate of 19° 19′ per year; but the sun moves over 19° in about twenty days; therefore, the eclipses, on an average, must take place about twenty days earlier each year, or at intervals of about 346 days.

In May, 1846, the moon's ascending node was in longitude 216°; in eight years, at the rate of 19° 19′ per year, it would bring the same node to longitude 61° 28′. The sun attains this longitude each year on the 23d of May; therefore, the eclipses for 1854 must happen in May, and in the opposite month, November.

The meaning of the columns N, in the tables

In computing the time of new and full moons, as illustrated by the preceding examples, the columns marked N, not hitherto used, indicate the distance of the sun and moon from the moon's node at the time of conjunction or opposition.

Eclipses are limited to a certain space along the ecliptic.

The circle is conceived to be divided into 1000 parts, commencing at the ascending node; the other node then must be at 500; and when the moon changes within 37 of 0, or 500, that is, 37 of either node, there must be an eclipse of the sun, seen from some portion of the earth. When the distance to the node is greater than 37, and less than 53, there may be an eclipse, but it is doubtful: we shall explain how to remove the doubt in the next chapter.

Comparative number of eclipses of the sun and moon.

When the moon fulls within 25 divisions of either node, there must be an eclipse of the moon: when the distance is greater than 25, and less than 35, the case is doubtful; but, like the limits to the new moon, the doubts are easily removed. *We repeat, the ecliptic limits for eclipses of the sun are* 53 *and* 37; *for eclipses of the moon, the limits are* 35 *and* 25. Hence, in any long period of time, the number of eclipses of the sun is, to the number of eclipses of the moon, as 53 to 35.

In the same period of time, say in one hundred years, there will be more visible eclipses of the moon than of the sun; for every eclipse of the moon is visible over half the world at once, while an eclipse of the sun is visible only over a very small portion of the earth; therefore, as seen from any one place, there are more eclipses of the moon than of the sun.

In the preceding examples the columns N are far within the limits, and, of course, there must be an eclipse of the sun on the 26th of May, 1854, and an eclipse of the moon in July, 1852.

How we know that an eclipse of the sun will happen on the 26th of May, 1854, and from what circumstance we learn that it will be an eclipse to some northern portion of the earth.

As N is in value 9, at the time of new moon, in May, 1854, it shows that the moon will then have passed the ascending node, and be north of the ecliptic, and the eclipse must be visible on the northern portions of the earth, and *not* on the southern.

When the moon changes in south latitude, which will be shown by N being a little more than 500, or a little less than 1000, the corresponding eclipse, if of the sun, will be visible on some southern portion of the earth, and not visible in the northern portion; and if of the moon, the moon will run through the southern portion of the earth's shadow.

What indicates that a solar eclipse will be visible on some southern portion of the earth.

Table B, p. 31, shows the moon's latitude, approximately corresponding to the column N; or N is the argument for the latitude, and the heading of the argument columns will show whether the moon is *ascending* to the northward, or *descending* to the southward.

The tables from XVI to XVIII, together with the solar tables, will give approximate values of the elements necessary for the calculation of eclipses; *and if accurate results are not expected,* these tables are sufficient to present general principles, and give primary exercises to the student in calculating eclipses; but he who aspires to be an astronomer, must continue the subject, and compute from the lunar tables, farther on.

The times, and the intervals of time, between eclipses, depend on the relative motion of the sun and moon, and the motion of the moon's nodes. The relative motion of the sun and moon is such as to bring the two bodies in conjunction, or in opposition, at the average interval of 29 d. 12 h. 44 m. 3 s., and the retrograde motion of the node is such as to bring the sun to the same node at intervals of 346 d. 14 h. 52 m. 16 s. Neglecting the seconds, and conceiving the *sun, moon,* and *node* to be together at any *point* of time, and after an unknown interval of time, which we represent by P, suppose them together again. Then $\frac{P}{29\ 12'\ 44}$ represents the number of returns of the lunation to the node in the time P, and the expression $\frac{P}{346\ 14\ 52}$ represents the number of returns of the sun to the node in the same time. Each return of either body to the node is unity; therefore, these expressions are to each other as *two whole numbers;* say as m to n; that is, $\frac{P}{29\ 12\ 44} : \frac{P}{346\ 14\ 52} :: m : n$;

The motions of the sun and moon in relation to moon's node investigated.

$$\text{Or,} \quad \frac{n}{(29\ 12\ 44)} = \frac{m}{(346\ 14\ 52)};$$

$$\text{Or,} \quad (346\ 14\ 52)n = (29\ 12\ 44)m \quad - \ - \ - \quad (a)$$

$$\text{Or,} \quad \frac{n}{m} = \frac{29\ 12\ 44}{346\ 14\ 52}.$$

Reducing to minutes, and dividing numerator and denominator by 4, we have $\frac{n}{m} = \frac{10631}{124783}$. As this last fraction is irreducible, and as m and n must be whole numbers *to answer the assumed condition,* therefore, the smallest whole number for m is 124783, and for n is 10631; that is, as we see by equation (a), the *sun, moon,* and *node* will not be exactly together a second time, until a lapse of 124783 lunations, or 10631 returns of the sun to the same node; which require a period of no less than 10088 years and about 197 days. We say about, because we neglected seconds in the computation, and because the mean motions will change, in some slight degree, through a period of so long a duration.

This period, however, contemplates an exact return to the same positions of the *sun, moon, and earth,* so that a line drawn from the center of the sun to the center of the moon would strike the earth's axis in exactly the same point; but to produce an eclipse, it is not necessary that an *exact return* to former position should be attained; a greater or less approximation to former circumstances will produce a greater or less approximation to a former eclipse: but exact coincidences, in all particulars, can never take place, however long the period.

This period contemplates practical impossibilities.

Exact coincidences never happen.

To determine the time when a return of eclipses may hap-

CHAP. II.

How to find the successive return of eclipses.

pen (particularly if we reckon from the most favorable positions — that is, commence with the *supposition* that the sun, moon, and node are together), it is sufficient to find the first approximate values of the fraction $\frac{10631}{124783}$. If we find the successive approximate fractions, by the rule of continued fractions,* we shall have the successive periods of eclipses, which happen about the same node of the moon.

The approximating fractions are

$$\frac{1}{11} \quad \frac{1}{12} \quad \frac{3}{35} \quad \frac{4}{47} \quad \frac{19}{223} \quad \frac{156}{1831}.$$

A series of fractions showing the periods at which eclipses occur.

These fractions show that 11 lunations from the time an eclipse occurs, we may look for another; but if not at 11, it will be at 12, and it may be at both 11 and 12 lunations; and at five or six lunations, we shall find eclipses at the other node, and the same succession of periods occurs at both nodes.

To be more certain of the time when an eclipse will occur, we must take 35 lunations from a preceding eclipse, which period is 1033 days 13 h. 40 m., and the sun at that time is about 6° 40′ farther from, or nearer to, the node, than before — and, if the count is from the ascending node, the moon's latitude is about 32′ farther south than before; and if from the descending node, the moon is about the same distance farther north.

The double of 11, 12, and 35 lunations, from any eclipse, may also bring an eclipse.

If an eclipse occurs within 10° of either node, it is certain that eclipses will again happen after the lapse of 47 lunations.

A brief examination of the periodical return of eclipses.

The period of 47 lunations includes 1387 d. 22 h. 31 m., and 4 revolutions of the sun to the node include 1386 d. 11 h. 29 m.; the difference is 1 day 11 h. 29 m.; but in this time the sun will move, in respect to the node, 1° 32 and some seconds; therefore, if the first eclipse *were exactly* at the node, the one which follows at the expiration of 47 lunations,

*See Robinson's Arithmetic.

or 3 years and nearly 11 months afterward, would take place 1° 32′ short of the same node; and if it were the ascending node, the moon's latitude would be about 5′ 40″ south, and, if the descending node, about 5′ 40″ more to the north.

The period, however, which is most known, and the most remarkable, appears in the next term of the series, which shows that 223 lunations have a very close approximate value to 19 revolutions of the sun to the node.

The Chaldæan astronomers called this period Saros.

The period of 223 lunations includes 6585.32 days, and 19 returns of the sun to the same node require 6585.78 days, showing a difference of only a fraction of a day; and if the sun and moon were at the node, in the first place, they would be only about 20′ from the node, at the expiration of this period, and the difference in the moon's latitude would be less than 2′, and therefore the eclipse, at the close of this period, must be nearly the same in magnitude as the eclipse at the beginning; and hence the expression "*a return of the eclipse,*" as though the same eclipse could occur twice.

By this period we can make a summary prediction of eclipses.

This period was discovered by the Chaldæan astronomers, and enabled them to give general and indefinite predictions of the eclipses that were to happen; and by it any learner, however crude his mathematical knowledge, can designate the day on which an eclipse will occur from simply knowing the date of some former eclipse. The period of 6585 days is 18 years, including 4 leap years, and 11 days over; therefore from any eclipse, if we add 18 years and 11 days, we shall come within one day of the time of an eclipse, and it will be an eclipse of about the same magnitude as the one we reckon from.

A summary mode of computing the time when an eclipse must occur.

For the purpose of illustrating the method of computing lunar eclipses, we wish to find the time when some future eclipse of the moon will take place; and from the American Almanac of 1833, we find that an eclipse of the moon took place on the 1st day of July of that year, therefore "*a return of this eclipse*" must take place on the 12th of July 1851.

By a simple glance into the American Almanac for the year 1834, we find a total eclipse of the moon on the 21st of

CHAP. II. June — therefore, on the first of July 1852, or at the time that the moon fulls on or about the first of July, there must be a large eclipse of the moon, visible to all places from where the moon will then be above the horizon; and furthermore, 18 years and 11 days after this, that is, in the year 1870, on the 12th day of July, the moon will be again eclipsed; and, in this way, we might go on for several hundred years, but in time the small variations, which occur at each period, will gradually wear the eclipse away, and another eclipse will as gradually come on and take its place.

In the same manner we may look at the calendar for any year, take any eclipse, that is anywhere near either node, and run it on, forward or backward.

Let us now return to the eclipse of July 12th, 1851.

Elements for the computation of lunar eclipses.

To decide all the particulars concerning a lunar eclipse we must have the following data, commonly called elements of the eclipse:

1. The time of full moon.
2. The semidiameter of the earth's shadow.
3. The angle of the moon's visible path with the ecliptic.
4. Moon's latitude.
5. Moon's hourly motion.
6. Moon's semidiameter.
7. The semidiameter of the moon and earth's shadow.

General directions to obtain the elements of eclipses.

To find these elements, the approximate time of full moon is found from Table XI, and the tables immediately connected. For the time thus found, compute the longitude of the sun from Table IV, and the tables immediately connected, as illustrated by examples on page 254.

Compute, also, the latitude, longitude, horizontal parallax semidiameter, and hourly motion, in latitude and longitude, from the lunar tables, commencing with Table XVI, and following out the computation by a strict inspection of the examples we have given (*rules, aside from the examples, would be of no avail*); and, if the longitude of the moon is exactly 180° *in advance* of the sun, it is then just the time of full moon; if not 180°, it is not full moon; if more than 180°, it is past full moon.

It will rarely, if ever, happen that the longitude of the moon will be exactly 180° in advance of the longitude of the sun; but the difference will always be very small, and, by means of the hourly motions of the sun and moon, the time of full moon can be determined by the *problem of the couriers*.*

The moon's latitude must be corrected for its variation, corresponding to the variation in time between the approximate and true time of full moon.

Rule to find the semidiameter of the earth's shadow.

To find the semidiameter of the earth's shadow, where the moon runs through it, we have the following rule: 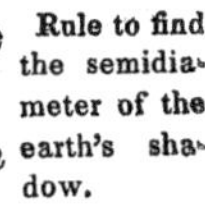

To the moon's horizontal parallax, add the sun's, and, from the sum, subtract the sun's semidiameter.

This rule requires demonstration. Let S (Fig. 53) be

Fig. 53.

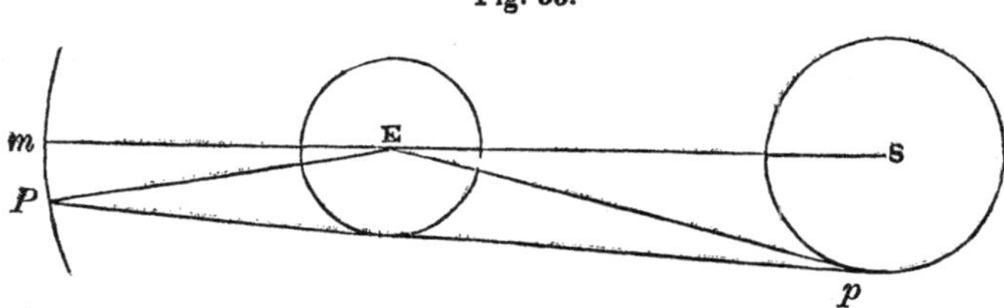

the center of the sun, E the center of the earth, and Pm a small portion of the moon's orbit. Draw $p\,P$, a tangent to both the earth and sun; from p and P, draw PE and pE, forming the triangle $p\,E\,P$.

Demonstration of the rule.

By inspecting the figure, we perceive that the three angles:

$$SEp + pEP + mEP = 180°.$$

Also, the three angles of the triangle, PEp, are, together, equal to 180°;

Therefore, $SEp + pEP + mEP = P + p + pEP$;

Drop the angle, pEP, from both members of the equation, and transpose the angle SEp, we then have

$$mEP = P + p - SEp.$$

---

* Robinson's Algebra—problem of the couriers.

CHAP. II. But the angle, mEp, is the semidiameter of the earth's shadow at the distance of the moon; SEp is the semidiameter of the sun; P, that is, the angle EPp, is the moon's horizontal parallax; and p is the horizontal parallax of the sun; therefore, the equation is the rule just given.*

What is meant by the angle of the moon's visible path with the ecliptic.

The angle of the moon's *visible* path with the ecliptic is always greater than its *real* path with the ecliptic, and depends, in some measure, on the relative motions of the sun and moon.

To explain why the real and visible paths of the moon are different, let AB (Fig. 54) be a portion of the ecliptic, and Am a portion of the moon's orbit; then the angle, mAB,

Fig. 54.

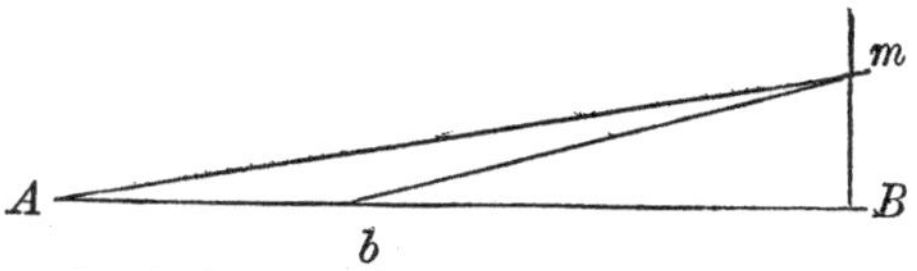

is the angle of the moon's real path with the ecliptic. Conceive the sun and moon to depart from the node, A, at the same time, the moon to move from A to m in one hour, and the sun to move from A to b in the same time; join b and m, and the angle mbB is the angle of the moon's visible path with the ecliptic, which is greater than the angle mAB; which is the angle of the moon's real path with the ecliptic. On this principle we determine the angle in question.

All the other elements are given directly from the tables.

* Some writers have directed us to increase this value of the shadow by its one-sixtieth part, but we emphatically deny the propriety of the direction.

CHAPTER III.

PREPARATION FOR THE COMPUTATION OF ECLIPSES.

CHAP. III.

Computation of a lunar eclipse.

WE shall now go through the computation in full, that it may serve for an example to guide the student in computing other eclipses.

The approximate time of full moon computed.

	Mean N. Moon.	I.	II.	III.	IV.	N.
1851,	1d. 14h. 21m.	0038	3916	40	39	431
Six Luna.	177 4 24	4851	4303	92	95	511
Half Luna.	14 18 22	404	5359	58	50	43
	193 13 7	5293	3578	90	84	985
	181					
July,	12 13 7	As N is within 25 of 1000, or 0, there must be an eclipse. The sun is 15 short of the ascending node, and the moon at full, being opposite, must be 15 short of the descending node, and therefore, in north latitude, descending.				
I.	3 35					
II.	2 9					
III.	14					
IV.	11					
Full ●	12 19 16					

Sun's longitude computed, corresponding to the approximate time of full moon.

We now compute the sun's longitude, hourly motion, and semidiameter for 1851, July 12, 19 h. 15 m. mean Greenwich time, as follows:

		☉ M. Lon.	Lon. Peri.	I.	II.	III.	N.
		s. ° ′ ″	s. ° ′ ″				
1851		9 8 32 39	9 8 22 24	958	250	025	648
	July	5 28 24 8	31	129	454	310	27
	12 d	10 50 32	2	371	28	19	2
	19 h	46 49		27	0	0	0
	15m	0 37		485	732	151	677
		3 18 34 45	9 8 22 57				
Eq. of center		1 39 38	3 18 34 45				
I.		10	6 10 11 48 = Mean anomaly.				
II.		18					
III.		20					
		3 20 15 11	☉'s hourly motion, 2′ 23″				
Eq. of equinox		—16	☉'s semidiameter, 15′ 46″				
☉ lon.		3 20 14 55					

CHAP. III.

Direction for computing the moon's true longitude.

We now compute the moon's longitude, latitude, semidiameter, horizontal parallax, and hourly motions for the same, mean Greenwich time, as follows:

FOR THE LONGITUDE.

1. Write out the arguments for the first twenty equations, and find their separate sums. With these arguments enter the proper tables (as shown by the numbers), and take out the corresponding equations, and find their sum.

2. Write out the evection, anomaly, variation, longitude, supplement to node, and the several arguments for latitude, in separate columns, corresponding to the given time, *and write the sum of the twenty preceding equations in the column of evection.*

3. Add up the column of evection first; its sum will be the corrected argument of evection, with which, take out the equation of evection (Table XXIV), and write it under the sum of the first twenty equations; their sum will be the correction to put in the column of anomaly.

4. Add up the column of anomaly, and the sum will be the moon's corrected anomaly, which is the argument for the equation of the center. With this argument take out the equation of the center from Table XXV, and write it under the sum of the preceding equations, and find the sum of all, *thus far.* Write this last sum in the column of variation, *and then* add up the column of variation; which sum is the correct argument of variation, and with it take out the equation for variation from Table XXVI.

5. Add the equation for variation to the sum of all the preceding equations, and the sum will be the correction for longitude, which, put in the column of longitude, and the whole added up, will give the moon's longitude *in her orbit,* reckoned from the *mean equinox.*

Equation of the equinox is sometimes called nutation in longitude.

6. Add the orbit longitude to the supplement of the node, and the sum is the argument of reduction to the ecliptic; it is also the first argument for polar distance.

With the argument of reduction take out the reduction from Table XXVII, and add it to the longitude.

With argument 19, *which is the same as* N *in the solar tables*, take out the equation of the equinox, and apply it according to its sign; the result will be the moon's *true longitude reckoned on the ecliptic from the true equinox.*

FOR THE LATITUDE.

General directions for finding the moon's latitude.

Add the same correction (to its nearest minute) to column II, as was added to the column of longitude, and add its value, expressed in the 1000th part of a circle, to all the following columns, except column X. Add up these columns, rejecting thousands (or full circles), and the sums will be the 5th, 6th, 7th, 8th, 9th, and 10th arguments of latitude.

The sum of the moon's orbit longitude, and supplement to node, is the first argument of latitude. The sum of column II is the second argument of latitude; the moon's true longitude is the third argument, and the twentieth of longitude is the fourth argument. Then follow 5, 6, &c., up to 10. With these arguments enter the proper Tables, and take out the corresponding equations, and their sum will be the moon's true distance from the *north pole of the ecliptic*, and, of course, will be in north latitude if the sun is less than 90°, otherwise in south latitude.

N. B. *When the first argument of latitude is nearer* 6 *signs than* 12 *signs, the moon is tending south; when nearer* 12 *signs, or* 0 *sign, than* 6 *signs, it is tending north.*

Equatorial parallax and semidiameter depend upon each other.

For the equatorial horizontal parallax.—The arguments for Evection, Anomaly, and Variation are also arguments for horizontal parallax, and with these arguments take out the corresponding equations from the tables adapted to this purpose.

For the semidiameter.—The equatorial parallax is the argument for semidiameter, Table XXXIV.

General directions for finding the hourly motion of the moon.

For the hourly motion in longitude.—Arguments 2, 3, 4, and 5 of longitude sensibly affect the moon's motion; they are, therefore, arguments for hourly motion, Table 36 (the units and tens in the arguments are rejected). Take out these equations from table, also take out the equation corresponding to the argument of evection, Table XXXVII. With the sum of the preceding equations, at the top, and the corrected anomaly at the side, take out the equations from Table XXXVIII. Also, with the correct anomaly, take out the equation from Table XXXIX. With the sum of all the preceding equations at top, and the argument of variation at the side, take out the equation from Table XL. Also with the variation, take the equation from Table XLI. With the argument of reduction take out the equation from Table XLII. These equations, all added together, will give the true hourly motion in longitude.

In this proportion the first term is the mean motion of the moon.

For the hourly motion in latitude.—With the 1st and 2d arguments of latitude, take out the corresponding quantities from Tables XLIII, and XLIV, and find their algebraic sum, noting the sign; call the result l.

Then make the following proportion:

$$32' \; 56'' \; : \; L \; :: \; l \; : \; \frac{L\,l}{32' \; 56''};$$

the true hourly motion in latitude, tending north, if the sign is *plus*, and south, if *minus*. In this proportion L is the true motion of the moon in longitude, and the first term is the moon's mean motion; and the proportion is founded on the principle that the true motion in latitude must vary by the same ratio as the motion in longitude.

N. B. In computing the moon's latitude we caution the pupil against omitting to add to the arguments II, V, VI, VII, VIII, and IX, the same correction as to the column of longitude; its value must be changed into the decimal division of the circle for all the columns except column II.

In the following example the correction for longitude is added to column II, and its value to all the following columns except column X.

We find the value in question thus:

$$360^\circ \; : \; 13^\circ \, 46' \; :: \; 1000 \; : \; x.$$

The proportion resolved gives $x =$ the number added to the several columns.

But to avoid the formality of resolving a proportion for every example, we give the following skeleton of a table that may be filled out to any extent to suit the convenience and taste of the operator.

Degrees	=	*decimal parts*	*Degrees*	=	*parts.*
° ′			° ′		
1 5	=	.003	5 24	=	.015
1 26	=	.004	7 12	=	.020
1 48	=	.205	9 0	=	.025
2 10	=	.006	10 48	=	.030
2 31	=	.007	12 36	=	.035
2 53	=	.008	14 24	=	.040
3 14	=	.009	16 12	=	.045
3 36	=	.010			

To make use of this table, we will suppose that the correction for longitude, in a particular example is, 11° 31′ 25″; what is the corresponding decimal or numeral part?

Thus			
	9°	=	.030
	2 31	=	7
	11 31	=	.037

We now continue the examples, hoping to follow these precepts.

	1	2	3	4	5	6	7	8	9	10	11	12	13	14	15	16	17	18	19	20
1851.	0005	9167	2644	5695	3467	4239	8539	3477	3885	306	958	918	570	487	493	960	201	569	648	038
July,	4955	7629	8273	1942	0732	7341	0444	0643	4396	698	634	754	185	948	525	625	613	699	27	83
D. 12,	301	7149	1442	3157	3691	4093	635	4293	267	772	342	775	376	91	336	403	463	286	2	5
h. 19,	22	515	823	227	266	295	46	309	19	*56	25	56	27	78	24	29	33	21	0	0
m. 15,	0	7	11	3	3	4	1	4	0	1	0	1	0	1	0	0	0	0	0	0
	5283	4467	3193	1024	8159	5972	9665	8726	8567	833	959	504	158	605	378	017	310	575	677	126

	Evection. s. ° ′ ″	Anomaly. s. ° ′ ″	Variation. s. ° ′ ″	Longitude. s. ° ′ ″	Sup. of N. s. ° ′ ″	II. s. ° ′	V.	VI.	VII.	VIII	IX.	X.
1851.	6 24 41 35	4 3 1 38	11 6 6 36	8 15 54 25	7 23 15 51	6 1 45	358	359	504	506	282	816
July,	8 8 17 16	6 24 45 48	1 16 31 32	7 14 55 40	9 35 5	7 8 32	156	147	112	103	486	962
Day 12,	4 4 28 54	4 23 42 54	4 14 5 54	4 24 56 25	34 57	4 2 40	374	434	311	371	394	58
h. 19,	8 57 32	10 20 35	9 39 3	10 25 53	2 31	8 50	27	31	22	27	28	4
m. 15,	7 4	8 10	7 37	8 14	2	7	0	0	0	0	0	0
Sum of eq.	31 34	1 2 57	13 8 19	13 45 54		13 46	23	23	23	23	23	
	7 17 23 55	4 3 2 2	5 29 39 1	9 20 6 31	8 2 27 6	6 4 40	938	994	972	030	213	840
			Reduction,	8 54	9 20 6 31							
				9 20 15 25	5 22 34 27	Argument I of Latitude.						
			Equation of Equinox,	—16								
			Moon's true Longitude,	9 20 15 9								

*From 10 to 20, the numbers are found by proportioning from Table XVIII; thus, 19 hours is $\frac{19}{24}$ of one day.

Arg.	Moon's Longitude.	
		′ ″
1		14 34
2		3 21
3		8
4		5 14
5		10
6		2 18
7		1 12
8		2 11
9		5
10		16
11		6
12		21
13		4
14		27
15		3
16		20
17		4
18		7
19		16
20		17
	Sum,	31 34
	Evection,	31 23
	Sum,	1 2 57
	Anom.	12 5 22
	Sum,	13 8 19
	Varia.	37 35
	Sum,	13 45 54

Arg.	Eq. Moon's Polar Dis.
	° ′ ″
I.	89 10 40
II.	9 41
◐ lon.	1
20 lon.	4
V.	20
VI.	31
VII.	29
VIII.	17
IX.	4
X.	41
◐ Polar Dis.	89 22 48
	90
◐ Lat. N.	37 12 tending S

For the Equatorial Parallax.

Arg.	Moon's Equation.
Evection,	0′ 25″
Anomaly,	53 54
Variation,	57
Parallax,	55 16
S. D.	15 4

Moon's hourly motion in Longitude.

Arguments.	Equations.
2 of Longitude,	1″
3 of Longitude,	3
4 of Longitude,	3
Evection,	13
	20
Anom. and Sum of Eq.	12
Anomaly,	28 58
	29 30
Variation and Sum,	3
Variation,	1 19
Reduction,	2
Hourly motion in Lon.	30 54

Moon's hourly motion in Latitude.

Argument I.	—2′ 56″
II.	— 4
	—3 0

32′ 56″ : 30′ 54″ : : —3′ : —2′ 49″. The result of this proportion gives —2′ 49″ for the hourly motion in Latitude.

CHAP. III.

	s	°	′	″
The moon's longitude, as just computed, will be	9	20	15	9
The sun's longitude, at the same time, will be	3	20	14	55
The difference will be - - -	6	0	0	14.

Therefore, at the time for which these longitudes were computed, the moon will be *past her full* by 14″ of arc: to correct the time, then, we must find how much time will be required for the moon to gain 14″; which, by the problem of the couriers, is

$$t=\frac{14}{(30.54)-(2.23)}=\frac{14''}{28'\,31''}=\frac{14}{1711}.$$

The correction is subtractive because the moon is past conjunction, otherwise it would be additive.

The unit for t is one hour, and the denominator of the fraction is the difference of the hourly motions of the sun and moon, as determined by the tables; the result is 29 seconds of time to be subtracted.

	d.	h.	m.	s.
The Greenwich time will be, 1851, July	12d.	19h.	15m.	0s.
Subtract - - -				29
True time of full moon - -	12	19	14	31

But the time given by the lunation table was 19 h. 14 m., differing only 31 seconds from the true time; the approximate and true time, however, do not commonly coincide as near as this: if they did, none but the most rigid astronomer would use the lunar tables for the time of conjunction or opposition.

To be *very exact* we must correct the moon's latitude for what it will vary in 31 seconds; that is, in this case, increase it 4″.5. The moon's latitude, at the time of full moon, is, therefore, 37′ 17″.4.

We have now all the elements necessary for computing the eclipse, or, at least, we have all the materials for finding them, and, for convenience, we collect the elements together:

	d.	h.	m.	s.
1. True time of full moon, July, - -	12	19	14	31
2. Semidiameter of earth's shadow (page 265), - - - - -		°	39′	39″
3. Angle of the moon's visible path with the ecliptic,* - - -		5	38	26

* This is the angle of the base of a right-angled triangle, whose base

CHAP. III.

	′	″
4. Moon's latitude N. descending, - -	37	17.4
5. Moon's hourly motion from the sun, -	28	31
6. Moon's semidiameter, - - -	15	4
7. Semidiameter of ☉ and earth's shadow,	54	43

Whenever the moon's latitude, at the time of full moon, is less than this last element, the moon must be more or less eclipsed; and it is by computing and comparing these two elements, viz., 4 and 7, that all *doubtful cases* are decided.

TO CONSTRUCT A LUNAR ECLIPSE.

From any convenient scale of equal parts, take the 7th element in your dividers (54 43) $= 54\frac{3}{4}$, and from C, as a center with that distance, describe the semicircle $BDHE$ (Fig. 55). Take $CA =$ the 2d element, and describe the semidiameter of the earth's shadow. From C the center of the shadow, draw Cn at right angles to BE the ecliptic, above BE when the latitude is north, as in the present example, but below, if south.

When the moon has very little latitude describe a full circle.

When large south latitude, describe only the lower semicircle.

Fig. 55.

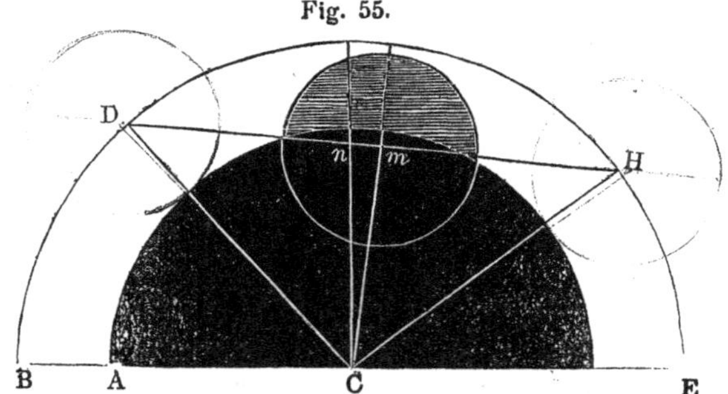

Take the moon's latitude from the scale of equal parts, and set it off from C to n. Through n draw DnH, the moon's path, so that the line shall incline to BE, the ecliptic, by an angle equal to the 3d element. Conceive the moon's

is the hourly motion of the moon from the sun (28′ 31″), and the perpendicular, the moon's hourly motion in latitude (2′ 49″). See page 266, figure 54.

CHAP. III center to run along the line from D to H, and from C draw Cm perpendicular to DH.

When the moon is *ascending* in her orbit, DH must incline the other way, and Cm must lie on the other side of Cn.

The eclipse commences when the moon arrives at D. It is the time of full moon when it arrives at n; the greatest obscuration occurs when it arrives at m, and the eclipse ends at H. The duration is the time employed in passing from D to H; and to find the duration apply DH to the scale, and thus find its measure. Divide this measure by the 5th element, and we shall have the hours and decimal parts of an hour in the duration. Also apply Dn to the scale and find its measure. Divide this measure by the 5th element, for the time of describing Dn, also divide the measure nH for the time of describing nH.

The 5th element is the moon's angular motion from the sun.

The time of describing Dn, subtracted from the time of full moon, will give the time of the beginning of the eclipse; and the time of describing nH, added to the time of full moon, will give the time when the eclipse ends.

With lunar eclipses the time of greatest obscuration is the instant of the middle of the eclipse, provided the moon's motion from the sun, for this short period of time, is taken as uniform, as it may be without sensible error.

In reference to this example $Dn = 36'$ and $nH = 44'$. These distances, divided by 28′ 31″, give 1 h. 14 m. 16 s. for the time of describing Dn, and 1 h. 32 m. 40 s. for nH: whole time, or duration, 2 h. 27 m. 20 s.

Astronomical time converted into civil time.

	h. m. s.
Therefore from the time of full ●	19 14 31
Subtract - - -	1 14 16
Eclipse begins - - -	18 0 15
Add the duration - -	2 47 20
Eclipse ends - - -	20 47 35

This eclipse not visible in Europe, and why.

That is, in 1851, July 12 d. 18 h. 0 m. 15 s. mean astronomical time, the eclipse begins; but this time corresponds with July 13, at 6 h. 0 m. in the morning; and at this time, the sun will be above the horizon of Greenwich, and, of course, the full moon, which is always opposite to the sun, will be below the horizon, and the eclipse will be invisible to all Europe.

Visible in the U. S.

In the United States, however, the eclipse will be visible; for, at these points of absolute time, the sun will not have risen nor the moon have gone down; but, to be more definite, we demand the times of the beginning, middle, and end of the eclipse, as seen from Albany, N. Y. To answer this demand, all we have to do, is to subtract from the Greenwich time the difference of meridians between the two places, which, in this case, is 4 h. 55 m.; and the result is,

Beginning of the eclipse	13 d.	1 h.	5 m.	morning,
Middle - - - -		2	30	"
End of the eclipse - -		3	52	"

In the same manner we would compute the time for any other place.

The quantity of the eclipse how found.

For the quantity of the eclipse we take the portion of the moon's diameter, which is immersed in the shadow, at the time of greatest obscuration, and compare it with the whole diameter of the moon; and in the present example, we perceive, that more than half of the diameter is eclipsed — about 7 digits when the whole is called 12, or 0.6 when the diameter is 1.

All these results, however, except the time of full moon, are approximate, because we cannot, nor do we pretend to *construct* to accuracy; *but any mathematician can obtain accurate results* by means of the triangles DCH and Cnm, and the relative motion of the moon from the sun.

The exact computation of the duration of the eclipse.

In the right-angled triangle Cnm, right-angled at m, Cn is the latitude of the moon $= 37' \; 17''.4 = 2237''.4$, and the angle $nCm = 5^\circ \; 38' \; 26''$; with these data we find $mn = 220''$, and $Cm = 2212''$

In the right-angled triangle CDm, or its equal CmH, we have - - $Cm^2 + mH^2 = CH^2$;

Or, - - $mH^2 = CH^2 - Cm^2$;

Or, - - $mH^2 = (CH + Cm)\,(CH - Cm)$.

CH is the 7th element $= 3283''$, and $Cm = 2212''.6$.

Therefore, $mH = \sqrt{(5495)\;(1071)} = 2426''$ This

X

CHAP. III. divided by 1711″, the 5th element, gives the time of half the duration of the eclipse 1 h. 25 m.; therefore the whole duration is 2 h. 50 m., which is 2 m. 40 s. more than the time we obtained by the *rough construction.*

The distance *nm*, as just determined, is 220″, and the time of describing this space, at the rate of 1711″ per hour, requires 7 m. 52 s., which taken from and added to the semi-duration, gives 1 h. 17 m. 8 s. from the beginning of the eclipse to full moon, and 1 h. 32 m. 52 s. from the full moon to the end of the eclipse.

The trigonometrical computation of the magnitude of the eclipse.

For the magnitude of the eclipse, we add the moon's semi-diameter in seconds (904″) to *Cm* (2212″), and from the sum subtract the semidiameter of the shadow in seconds (2379), and the remainder is the portion of the moon's diameter *not eclipsed.* Subtract this quantity from the moon's diameter, and we shall have the part eclipsed. Divide this by the whole diameter, and the quotient is the magnitude of the eclipse, the moon's diameter being unity.

Following these directions, we find the magnitude of this eclipse must be 0.587.

The construction a sufficient guide to carry out the trigonometrical computations.

In all these computations we were guided by the construction; *which will always prove a sufficient index, and all that should be required.*

We may determine, in any case, whether the eclipse will or will not be total, by the following operation:

Subtract the ☾'s semidiameter from the semidiameter of the shadow, and if the moon's latitude, at the time of full moon, is less than the remainder, the eclipse will be total, otherwise not.

To find the duration of total darkness.— Diminish the semi-diameter of the shadow by the semidiameter of the moon, and from the center of the shadow describe a circle, with a radius equal to the remainder; a portion of the moon's path must come within this circle; that portion, measured or divided by the hourly motion, will give the time of total darkness.

When the moon's latitude is north, as in the present example, the southern limb of the moon is eclipsed — and conversely.

CHAPTER IV.

SOLAR ECLIPSES — GENERAL AND LOCAL.

CHAP. IV.

General directions to find the elements.

The elements for a solar eclipse are computed in the same manner as the elements of a lunar eclipse; all of which are found by the solar and lunar tables.

The approximate time of new moon is first computed, and for this time, compute the sun's longitude, declination, parallax, semidiameter, and hourly motion; and for the same time compute the moon's longitude, latitude, hourly motion in longitude and latitude, horizontal parallax, and semidiameter.

If the longitudes of both sun and moon are found to be the same, then the approximate time of conjunction; found by the lunation tables, is the same as the true time; if not, we proportion to the true time, as described in the last chapter.

The elements for a general solar eclipse are:

What elements are necessary.

1. The time of ☌ * at some known meridian. 2. Longitude of ☉ and ☾. 3. ☉'s declination. 4. ☾'s latitude. 5. ☉'s hourly motion. 6. ☾'s hourly motion in longitude. 7. ☾'s hourly motion in latitude. 8. The angle of the ☾'s visible path with the ecliptic. 9. ☾'s horizontal parallax. 10. ☾'s semidiameter. 11. ☉'s semidiameter. 12. ☉'s horizontal parallax.

For a local eclipse, the latitude of the particular locality must also be given, or considered as one of the elements.

A definite example proposed.

As we can best illustrate general principles by taking a *particular example*, we now propose to show *the general course of an eclipse of the sun, which will occur in May* 1854; *where it will first commence on the earth; in what latitude and longitude the sun will be centrally eclipsed at noon, and where; in what latitude and longitude the eclipse will finally leave the earth.*

Some general preliminary explanations.

We speak of an eclipse of the sun *being on the earth;* by this we mean *the moon's shadow on the earth.* If an observer is in the moon's shadow, of course, the sun would be in an eclipse to him; and, if a tangent line be drawn between the

* Sign of conjunction.

CHAP. IV.

sun and moon, and that line strike the eye of an observer on the earth, to that observer the limbs of the sun and moon would apparently meet, and all projections of eclipses are on the principle of lines drawn from some part of the sun to some part of the moon, and those lines striking the earth. When no such lines can strike the earth there can be no eclipse. For the sake of simplicity in explaining a projection of a solar eclipse, whether it be general or local, an observer is supposed to be at the moon, looking down on the earth, viewing the moon's shadow as it passes over the earth's disc; and, of course, the earth to him appears as a plane, equal to the *moon's horizontal parallax.*

Point of view.

The approximate time of new moon will be found computed on page 254, and, if very close results are not required, we may compute the sun's longitude, declination, hourly motion, and semidiameter for this time, and take out the moon's horizontal parallax, hourly motion, and semidiameter from Table IX; but we have computed the elements more accurately by the lunar tables, and find them as follows:

Accurate elements for the solar eclipse, which will take place May 26, 1854.

		d.	h.	m.	s.
1.	Greenwich mean time of ☌ 1854, May	26	8	45	39
2.	Lon. of ☉ and ☽ - - -		65°	14′	6″
3.	Declination of the ☉ - -		21	11	43 N.
4.	Latitude of the ☽ - - -			21	19 N.
5.	☉'s hourly motion in lon., - - -			2	24
6.	☽'s hourly motion in lon., - - -			30	3
7.	☽'s hourly motion in lat., tending north,			2	46

From 5, 6, and 7 we obtain 8, as explained in the last chapter.

		°	′	″
8.	Angle of the moon's visible path with the eclip., - - - -	5	42	50
9.	The ☽'s horizontal equatorial parallax,		54	30
10.	The ☽'s semidiameter, - -		14	51
11.	The ☉'s semidiameter, - -		15	48
12.	The ☉'s horizontal parallax, always taken at			9

Add together the ☉'s horizontal parallax, the ☽'s horizontal parallax, and the semidiameters of ☉ and ☽, and if the moon's latitude is less than this sum, there will be an eclipse, otherwise not; and it is by comparing this sum with the moon's latitude *that all doubtful cases are decided.*

TO CONSTRUCT A GENERAL ECLIPSE.

1. Make, or procure, a convenient scale of equal parts, and from any point as C (Fig. 56) with the radius CB, equal to the difference of the parallaxes of ☉ and ☽ (in the present example 54′ 21″, the minute is the unit), describe the semicircle $CBPH$, or the whole circle, when the case requires it. When the moon has small latitude (less than 20′) describe the whole circle; when the moon has large *north* latitude, describe the northern semicircle; when *south*, describe the southern semicircle.

Through C draw $VCDPL$ perpendicular to HB. This perpendicular will represent the plane of the earth's axis, as seen from the moon.

From P take PA, PF, each equal to the obliquity of the ecliptic 23° 27′ 30″, and draw the chord AF.

On AF, as a diameter, describe the semicircle ALF.

How to find the axis of the ecliptic.

2. Find the distance of the sun from the tropic, nearest to it, by taking the difference between the sun's longitude and 90° or 270°, as the case may be. In the present example we subtract 65° 14′ from 90°, the remainder is 24° 46′. Take LT, equal to 24° 46′, and draw TE parallel to LC. Draw CE the axis of the ecliptic.

The axis of the ecliptic variable in position.

By the revolution of the earth round the sun, the axis of the ecliptic appears to coincide with the axis of the equator, when the sun is at either tropic, and it appears to depart from that line by the whole amount of the obliquity of the ecliptic; and the time of this greatest departure is when the sun is on the equator. That is, CE runs out to CA at the vernal equinox, and runs out to CF at the autumnal equinox. As a general rule, CE, the axis of the ecliptic, is to the left of CP, the axis of the equator, from the 20th of December to the 20th of June, and to the right of that line the rest of the year. Draw CG the axis of the moon's orbit, so that the angle GCE shall be equal to the angle of the moon's visible path with the ecliptic, and CG is to the left of

How to find the axis of the lunar orbit.

x*

CHAP. IV. CE when the eclipse is about the ascending node, as in this example, but at the right when the eclipse is about the decending node.

For this projection to appear natural, the reader should face the north, so that H will appear to the west, and B on the east of the figure.

The shadow of the moon across the earth is from a western to an eastern direction, therefore, the moon is conceived to come in on the earth from the west side.

The equator. The point C is perpendicular to the sun's declination, and $C\,V$ is the sine of the declination, and the curved line $H\,V\,B$ is a representation of the equator as seen from the moon. When the sun has no declination, the equator draws up into a straight line.

How to draw the moon's path. 3. Take $C\,n$ from the scale of equal parts, making it equal to the moon's latitude, and through the point n, and at right angles to $C\,G$, draw the line $k\,l\,m\,n\,r\,p\,q$, which represents the center of the shadow, or the moon's path across the disc.

From C as a center, at the distance $C\,O$, describe the outer semicircle, equal to the sum of the moon's horizontal parallax, the sun's horizontal parallax, and the semidiameter of both sun and moon; then $O\,H$ is the semidiameter of the sun and moon.

When the eclipse first commences, the center of the moon is at k, and the center of the sun is on the circumference of the other circle, in a direct line to C, not represented in the figure, therefore, the two limbs must then just touch.

As C is the center of the earth, and H on the equator, therefore $C\,H\,O$ is a line in the plane of the equator, and the point k is a little below the equator; which shows that the eclipse first commences on the earth a little south of the equator.

How to determine the duration of a general eclipse. The time that the eclipse is on the earth is measured by the time required for the moon to pass from k to q with its true angular motion from the sun.

The length of this line, $k\,q$, can be found from the elements, and trigonometry, as in an eclipse of the moon, and the time of describing it is found in the same way.

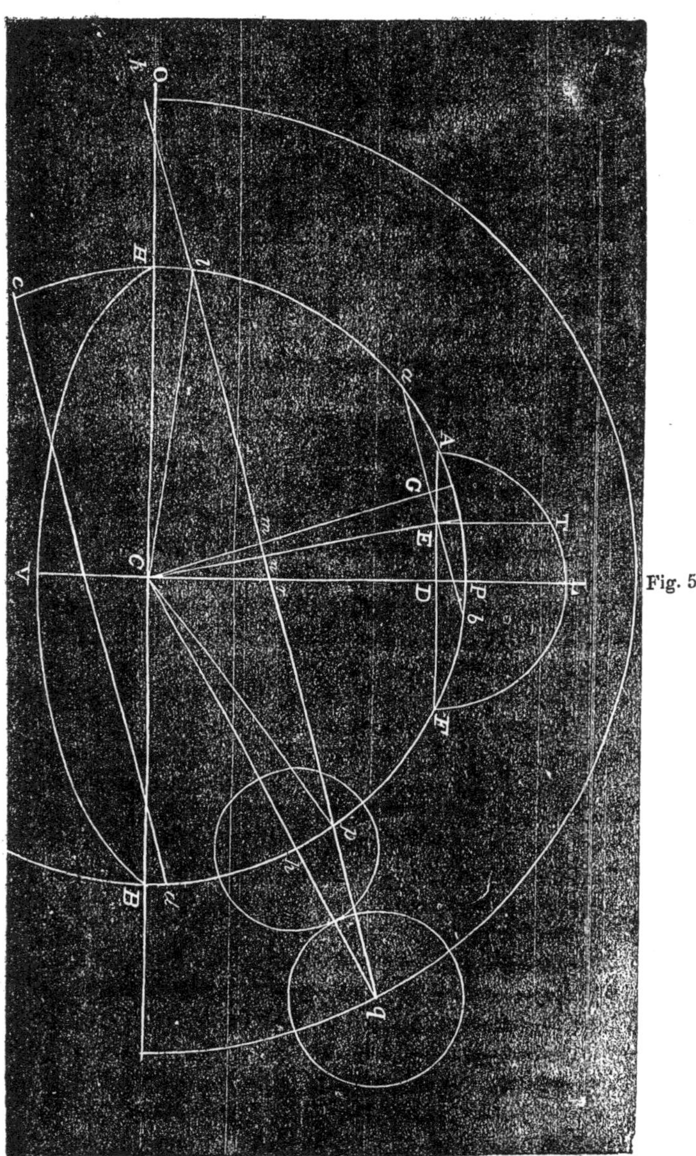

Fig. 56.

CHAP. IV.

How to determine in what latitudes the eclipse will enter, pass over, and pass off the earth.

When the moon's center comes to l, the central eclipse commences, and the arc Hl shows that it must be about in the latitude of 7° north. When the moon's center comes to r, the sun will be centrally eclipsed at apparent noon; and Cr is the sine of the number of degrees north of the sun's declination, which, in this case, is about 23°; hence to the sun's declination, 21° 12′, add 23°, making 44° 12′; showing, as near as a mere projection can show, that the sun will be centrally eclipsed at noon on *some meridian*, in latitude 44° 12′ north. The central eclipse will end, or pass off the earth, when the moon's center arrives at p and the arc Bp from the equator, shows that the latitude must be about 41° north. The eclipse will entirely leave the earth when the moon's center arrives at q, and for its limb to touch the sun, the sun's center must be at h, and the arc Bh shows that the latitude must be about 30° north.

The lines, cd and ab, parallel to the moon's path, and distant from it equal to the sum of the semidiameters of sun and moon, represent the lines of simple contacts across the earth, or limits of the eclipse; cd is the southern line of simple contact, and ab is the northern line of simple contact, and the latitudes at which these lines make their transits over the earth, are determined precisely as the latitudes on the central line.

We may make accurate computations by plane trigonometry.

But we need not stop at coarse approximations: we have all the data for correct mathematical results, on the same principles as we determined those in relation to a lunar eclipse.

In the triangle Cnr, we have the side Cn, the moon's latitude in seconds, which may be used as linear measure, as yards or feet and in proportion thereto, we may compute Cr and nr, *when we know the angle* nCr.

An equation for the position of the axis of the ecliptic.

But the following equation always gives the tangent of the angle ECD or nCr, calling the sun's distance from the solstice D, the obliquity of the ecliptic E, and the radius unity.

$$\tan. ECD = \tan. E \sin. D.^*$$

* The student who has acquired a little skill in analytical trigonometry can discover the preliminary steps to this equation; the principles are all visible in the construction of the figure.

To the angle ECD, add the angle GCE, the angle of the moon's visible path with the ecliptic, and we have the whole angle GCD, or mCr. Cmn is a right angle; and in the two triangles Cmn and Cmr, we have all the data, and can compute nr and rC.

When the moon arrives at m, it is in the line of conjunction in her orbit; when it arrives at n, it is in ecliptic conjunction; and when it arrives at r, it attains *conjunction in right ascension.*

Recent changes in the English Nautical Almanac

For the last six or eight years, the English Nautical Almanac has given the conjunctions and oppositions in right ascension, in place of conjunctions and oppositions in longitude, and has given the difference of declinations between the sun and moon, in place of giving the moon's latitude; that is, it has given the time that the moon arrives at r, in place of n, and given the line Cr in place of Cn.

All lunar tables give the ecliptic conjunction at n, and from this we can compute the time at r by means of the triangle Cnr.

Having explained the principle of finding the latitude on the earth, when a solar eclipse first commences, we are now ready to show another important principle—how to find the longitude; and with the latitude and longitude, we have the exact point on the earth.

The method of finding the longitude where the eclipse first strikes the earth.

Where an eclipse first commences on the earth, it commences with the rising sun, and finally leaves the earth with the setting sun. In this example, we have decided that the eclipse must commence very near the equator, not more than one degree south; but in that latitude the sun rises at 6 h. A. M. apparent time; therefore, at the place where the eclipse commences, it is six in the morning, apparent time.

From the scale of equal parts, take the moon's hourly motion from the sun in the dividers (27′ 39″), and apply it on the line kq: it will extend three times, and a little over, to the point n. This shows that three hours, and a little more (we say 3 h. 3 m.) must elapse from the first commencement of the eclipse to the change of the moon at n. Hence, by the local time at the place of the commencement of the eclipse,

CHAP. IV. the moon changes at 9 h. 3 m. in the morning, apparent time; but the apparent time of new moon at Greenwich is 8 h. 49 m. P. M., making a difference of 11 h. 46 m. for mere locality: the absolute instant is the same; the difference is only in meridians which correspond to a difference of longitude of 175° 30′; and it is west, because it is later in the day at Greenwich.

The method of finding where the central eclipse first strikes the earth.

The central eclipse also first comes on the earth at a place where the sun is rising. In this example it first strikes the earth at the point *l*, in latitude about 7° N.; but, in latitude 7° N., and declination 21° N., the sun rises at 5 h. 48 m., A. M. apparent time (Prob. II); and from that time to the change of the moon, namely, the time required for the moon to move from *l* to *n*, is (as near as we can estimate it by the construction), 1 h. 56 m.; therefore, the time of new moon, in the locality where the central eclipse first commences, is 7 h. 44 m. in the morning. From this to 8 h. 49 m. in the evening, the time at Greenwich, gives a difference of 13 h. 5 m., reckoned eastward from the locality, or 10 h. 55 m. reckoned westward; which corresponds to 196° 15′ *west* longitude from Greenwich, or 163° 45′ *east* longitude; the meridian is the same. If the longitude is called east, the day of the month must be one later; but, to avoid this, we had better call the longitude *west*.

To find the longitude where the sun will be centrally eclipsed at noon.

Where the sun is centrally eclipsed on the meridian, it is just 12, apparent time; the moon's center is then at *r*, and, by the construction, it must be about seven minutes after conjunction in that locality; hence, the conjunction is seven minutes before 12, and at Greenwich it is 8 h. 49 m. after 12, giving 8 h. 56 m. for difference of longitude, or 134° west longitude.

The central eclipse will leave the earth with the setting sun, when the center of the moon and sun are both at *p*; but the latitude of *p* we decided to be 40° north, and in this latitude, when the sun's declination is 21° 11′, as it now is, the sun sets at 7 h. 15 m. apparent time; but this is 1 h. 40 m. after conjunction, therefore the conjunction in that locality must be at 5 h. 35 m.; but, at Greenwich, it is

8 h. 49 m., giving, for difference of longitude, 3 h. 14 m., or 48° 30′ west. CHAP. IV.

The eclipse finally leaves the earth in latitude 46° north; but, in this latitude, the sun sets at 6 h. 51 m., and the conjunction will be 3 h. 0 m. sooner (the time required for the moon to pass from *n* to *q*), therefore the conjunction in this locality must be at 3 h. 51 m.; but, at Greenwich, it will be 8 h. 49 m., giving 4 h. 58 m. for difference of longitude, or 74° 30′ *west.*

To find the longitude where the eclipse will leave the earth.

Thus, by the mere geometrical construction, we have roughly determined the following important particulars:

	App. time Gr. h. m.	Lat. °	Longitude. ° ′
Eclipse commences, May 26,	5 46	1 S.	175 30 W.
Cen. eclipse commences,	6 53	7 N.	196 15 W.
Cen. eclipse at local noon,	8 56	46	134 00 W.
Cen. eclipse ends,	10 34	40	48 30 W.
End of eclipse,	11 46	30	73 30 W.

Results mechanically taken from the projection.

To find the latitude of the first commencement of simple contact on the southern line, all we have to do is to find the arc Hc; and for the latitude on the northern line, we find the arc Ha; the point c is in latitude about 27° south, and a in about 54° north.

The localities of the southern and northern lines of simple contact.

The southern line of simple contact leaves the earth at d, between the seventh and eighth degrees of north latitude, and the northern line passes off beyond the pole.

We have, thus far, taken the results but approximately from the projection, and the projection is sufficient to *teach us principles;* and it must be our guide, if we attempt to obtain more minute results; and with the elements and the figure we have the whole subject before us as minutely accurate as it is magnificent, and as simple as it is sublime.

To complete our illustration, we now go through the trigonometrical computation.

In the triangle Cnm, we have $Cn=21' \, 19''=1279$, the angle $mCn=5° \, 42' \, 50''$, and the angle m a right angle.

Whence $Cm=1273''$, and $mn=127''.3$.

CHAP. IV.

In these computations the moon's latitude and the distances from the center, C, to the circumferences are given lines.

tan. $ECD = nCr =$ tan. (23° 27′ 32″) sin. (24° 45′ 54″) (page 284).

Whence $ECD =$ 10° 18′ 8″,
Add $GCE =$ 5° 42′ 50″,

Sum is $GCD = mCr =$ 16° 0′ 58″.

In the triangle mCr, we have Cm (1273), the perpendicular, and the angle mCr as just determined; whence,

$$mr = 365''.3; \quad Cr = 1324''.3.$$

In the triangle Cmp, Cp is the horizontal parallax of moon and sun (54′ 30″)—9″, or 54′ 21″=3260″.

By the well-known property of the right-angled triangle,

$$Cm^2 + mp^2 = Cp^2.$$

Or $mp^2 = Cp^2 - Cm^2 = (Cp + Cm)(Cp - Cm)$,

That is, $mp = \sqrt{(4533)(1987)} = 3001''.7$.

Therefore, lp, the whole chord, is 6003″.4, which, divided by 1659′ (the moon's motion from the sun), gives 3.616 h. or 3 h. 37 m. 40 s. for the time that the central eclipse will be on the earth.

In the same manner the line mq is found.

That is, $mq = \sqrt{(Cq + Cm)(Cq - Cm)}$,

But, $Cq = 54' \, 21'' + 14' \, 51'' + 15' \, 48'' = 5100''$.

Or $mq = \sqrt{(6373)(3827)} = 4938''.3$.

Therefore, the whole chord, kq, is 9876.6, which, divided by 1659″, gives 5 h. 57 m. 20 s. for the entire duration of the general eclipse on the earth.

On the supposition that the moon's motion from the sun is uniform for the six hours that the eclipse will be on the earth, the several parts of the moon's path will be passed over by the moon, as follows:

Accurate results on the condition of invariable elements.

	h.	m.	s.	
From k to l in	1	9	54	
From l to m in	1	49	00	to ☌ in orbit.
From m to n in		4	36	to ☌ in ecliptic.
From n to r in		8	37	to ☌ in right ascension.
From r to p in	1	35	40	
From p to q in	1	9	54	

The apparent time of ecliptic conjunction, at Greenwich, as determined by the tables (and applying the equation of time), is at

time), is at	8 h.	49 m.	0 s.
Subtract from k to ecliptic ☌,	3	3	30
Eclipse commences, Greenwich app. time,	5	45	30
Central eclipse commences (add 1 9 54),	6	55	24
Sun centrally eclipsed *on some meridian*, or ☌ in right ascension, Greenwich time, at (add 2 2 36),	8	58	00
Central eclipse ends at (add 1 33 58),	10	33	48
End of eclipse at (add 1 9 54),	11	43	41

A careful projection more accurate, than is generally supposed.

By comparing these times with those obtained simply by the projection, we perceive that the projection is not far out of the way, notwithstanding the terms *rough* and *roughly* that we have been compelled to use concerning it. Indeed, a good draftsman, with a delicate scale and good dividers, can decide the times within two minutes, and the latitudes and longitudes within half a degree; but all mathematical minds, of course, prefer more accurate results; yet, however great the care, *absolute accuracy* cannot be attained; the nature of the case does not admit of it.*

To find whether the point k is north or south of the equa-

* The astronomer, by making use of his judgment, can be very accurate with very little trouble: he perceives, at a glance, what elements vary, and what the effects of such variation will be; but a learner, who is supposed not to be able to take a comprehensive view of the whole subject, must go through the tedious process of computing the elements for the times of the beginning and end of the eclipse, as well as the time of conjunction, if he aims at accuracy, but an astronomer can be at once brief and accurate. In computing the moon's longitude, in the present example, the astronomer would notice in particular the moon's anomaly, and, by it, he perceives whether the moon's hourly motion is on the increase or decrease, and at what rate.

It is on the decrease, and the first part of the chord $k\,m$ is passed over by the moon in about 7 seconds less time than our computation made it, and the last part requires about 7 seconds longer time; but the times of passing m and n should be considered accurate, and the times of beginning and end should be modified for the variation of the moon's motion, making the beginning and end 7 seconds later, and the beginning and end of the central eclipse about 4 seconds later.

CHAP. IV. tor, we conceive k and C joined, and if the angle $m\,Ck$ is greater than the angle $m\,CH$, the point k is south, otherwise north.

By trigonometry, $Ck : km :: \text{sine } 90° : \text{sine } mCk$;

		° ′ ″
Or,	5138 : 4900″.3 : sin. 90 : sin. m Ck=	75 31 20
To this add $G\,CD$,	- - - -	16 0 58
Sum is the angle $r\,Ck$	- - -	91 32 18

This angle shows that the eclipse will first touch the earth in latitude 1° 32′ 18″ south.

To find the arc Hl, conceive the points Cl joined, and the two triangles Clm, $m\,Cp$ are equal.

And $Cl : lm :: \text{sin. } 90° : m\,Cl$;

		° ′ ″
Or,	3261 : 3003.7 : : sin. 90 : sin. m Cl=	67 7 50
To this add $G\,CD$,	- - -	16 0 58
The sum is,	- - - -	83 8 48

Where the eclipse first strikes the earth

This angle shows the latitude of the point l to be 6° 51′ 12″ north. That is, the central eclipse first touches the earth in 6° 51′ 12″ of north latitude; differing very little from the point determined by construction.

To find the latitude of the point p, we have $m\,Cl = m\,Cp = 67° \; 7' \; 50''$; and subtracting 16° 0′ 58″, we have the polar distance, or co-latitude; the result is, that the central eclipse passes off at latitude 38° 53′ 8″ north, and the general eclipse entirely leaves the earth in latitude 30° 25′ 38″.

To find the latitude of the point r, we consider Cr to be a sine of an arc, and $C\,P$ the radius.

	° ′ ″
Therefore, 3261″ : 1324″.3 : : R : sin. x =	23 58 00
To this add the sun's declination, -	21 11 43
Sum is latitude where the sun will be centrally eclipsed on the meridian, -	45 9 43 N.

How to find the longitude of the place where the sun is centrally eclipsed on the meridian.

Wherever the sun is centrally eclipsed on the meridian, it is apparent noon at that place, but at Greenwich the apparent time is 8 h. 57 m. 37 s., P. M.; this difference, changed into longitude, gives 134° 25′ west, within a degree of the result determined from the projection; and it is not important to go over a trigonometrical computation for the longitudes, since we are sure of knowing how to do it; and we are also sure that the results will not differ much from those already determined. CHAP. IV.

In short, from the elements, the figure, and a knowledge of trigonometry, we can determine all the important points in each of the three lines *c d*, *k q*, and *a b*, for between them we have, or may have, a complete *net-work* of plane triangles. Sufficient data in the figure.

CHAPTER V.

LOCAL ECLIPSES, ETC.

WE now close the subject of eclipses by showing how to project and accurately compute every circumstance in relation to a local eclipse. CHAP. V.

For an example, we take the eclipse of May, 1854, and for the locality, we take Boston, Mass., because we anticipated a central eclipse at that place, but the result of computations shows that it will not be quite central even there. We use the same elements as for the general eclipse.

THE CONSTRUCTION.

Draw a line *CD*, and divide it into 65 equal parts, and consider each part or unit as corresponding to one minute of the moon's horizontal parallax. From *C*, as a center, at a distance equal to the diff. of parallax of the sun and moon (54 21), describe a semicircle north or south according to the latitude, or describe a whole circle if the latitude is near the equator. The scale

From *C* draw *C*♋, *the universal meridian*, at right angles to *CD*, and from ♋ take ♋ ♈ and ♋ ♎, each equal to the obliquity of the ecliptic (23° 27′) and draw the straight line ♈ ♎, ♎ on the right. Subtract the sun's longitude from 90° or 270° to find its distance from the nearest solstitial point, and note the difference (in this example 24° 46′).

From the point *a*, with *a* ♈ as radius, make *a G* equal to How to find the axis of ecliptic.

CHAP. V. the sine of 24° 46′,* and join $C\,G$, and produce it to E; $C\,E$ is the axis of the ecliptic: this line is variable, and is on the other side of the line C♋ between June 20 and December 21.

How to find the axis of the moon's orbit. From E take the arc $E\,L$ equal to the moon's visible path with the ecliptic, to the right of E when the moon is descending, but to the left when ascending as in the present example. Join $C\,L$, a line representing the axis of the moon's orbit.

To and from the reduced latitude of the place add and subtract the sun's declination:

Thus, Boston, reduced latitude,	-	42° 6′ 39″ N.
Sun's declination,	-	21 11 43 N.

Sum is 63° 19′ 22″, and difference is 20° 54′ 56″.

How to find the points in the ellipse marking the visible path of the place over the earth's disc. From C, make $C\,12$ equal to the sine of the difference of the two arcs (20° 54′ 56″); and $C\,d$ the sine of the sum (63° 19′ 22″).

Divide (12) d into two equal parts at the point g; and on g (12), as radius, mark the sine of 15°, 30°, 45°, 60°, 75°, 90°; the line 7, 5, runs through the first point; 8, 4, through the second, &c.

Subtract the latitude (42° 6′ 39″) from 90°, thus finding the co-latitude (47° 53′ 21″). On the semidiameter of the earth's disc, as radius, take the sine of the co-latitude (47° 53′), and set off that distance from g, both ways to 6; thus making a line, 6, 6, at right angles to the universal meridian, $C\,g$. On g (6) as radius, and from the point g as a center, find the sine of 15°, 30°, 45°, &c., and set off those distances each way from g and through the points thus found, draw lines parallel to $g\,C$; these lines, meeting the lines drawn parallel to $6\,g\,6$, will define the points 5, 6, 7, 8, &c. to 12, and 1, 2, 3, &c. to 7, the hours of the day on the elliptic curve.

Explanation of the hours round the ellipse. That is, our supposed observer at the moon would see Boston (or any other place in the same latitude as Boston), at the point 9 when it is 9 o'clock at the place, and at 12 when it is noon at the place, &c.

---

* The reader is supposed to understand how to draw a sine to any arc, corresponding to any radius, either with or without a sector

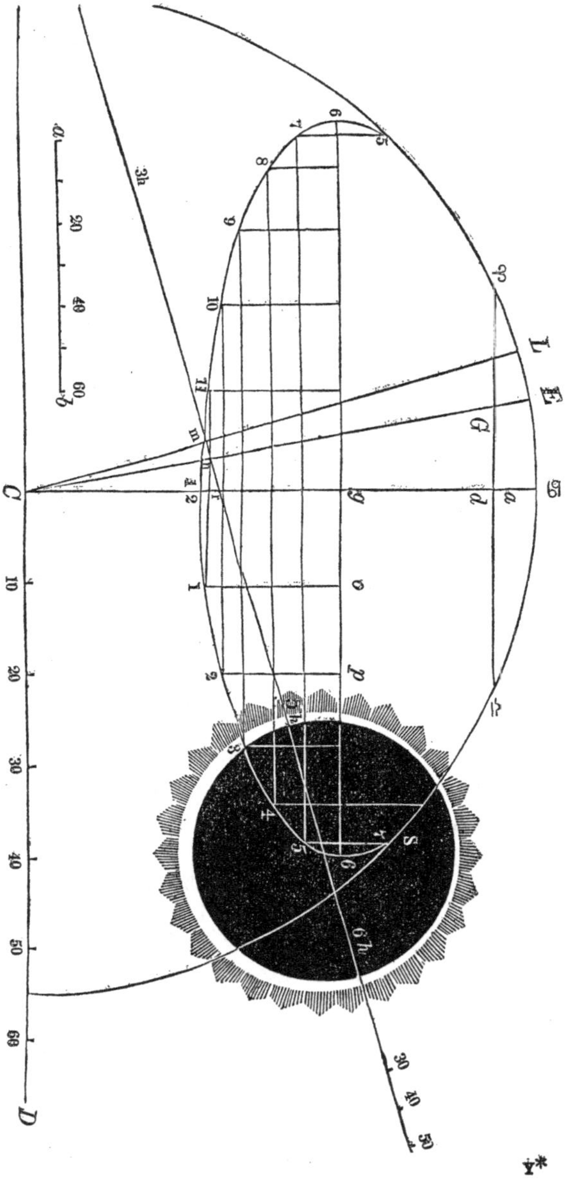

CHAP. V.

As this curve touches the disc before 5 and after 7, it shows that, in that latitude, on the day in question, the sun will rise before 5 in the morning, and set after 7 in the evening. If the declination of the sun had been as much south as now north, the point *d* would have been 12 at noon, and all the hours would have been on the upper part of the ellipse, which is not now represented.

From *C*, as in the general eclipse, set off the distance *Cn* equal to the moon's latitude, and, through the point *n*, draw the moon's path at right angles to *CL*.

As the ellipse represents the sun's path on the disc, and as the point (12) refers, of course, to apparent noon, *and not to mean noon*, therefore, we will mark off the time on the moon's path corresponding to apparent time.

How to mark time on the moon's path.

When the moon's center passes the point *n*, it is at ecliptic conjunction, apparent time, at Boston, or it must be considered the apparent time corresponding to any other meridian for which the projection may be intended.

The ecliptic ☌, apparent time, Greenwich, is	8 h.	49 m.	0 s.
For the longitude of Boston, subtract	4	44	16
Conjunction, apparent time, at Boston,	4	4	44

The moon's hourly motion from the sun is 27′ 39″: take this distance from the scale, in the dividers, and make the small scale *ab*, which divide into 60 equal parts; then each part corresponds with a minute of the moon's motion from the sun, and the distance *ab* will correspond with one hour of the moon's motion along its path. At 4 h. 4 m. 44 s. the moon's center will be at the point *n*; the sun's center, at the same time, will be just beyond the point 4 on the ellipse; and, as the distance between these two points is greater than the sum of the semidiameters of sun and moon, therefore the eclipse will not then have commenced; but the moon moves rapidly along its path, and, at 5 o'clock, the center of the moon will be at the point marked 5 on the moon's path, and the center of the sun will be at the point marked 5 on the ellipse; and these two points are manifestly so near each other, that the limb of the moon must cover a part of that of the sun, show-

In this case, the ellipse must commence between 4 and 5 o'clock.

ing that the eclipse must have commenced prior to that time. To find the time of commencement more exactly, let the hour on the moon's path be subdivided into 10 or 5-minute spaces, and take the sum of the semidiameter of the sun and moon in your dividers from the scale *CD*, and, with the dividers thus open, apply one foot on the moon's path and the other on the sun's path, and so adjust them that each foot will stand at the same hour and minute on each path as near as the eye can decide. The result in this case is 4 h. 28 m. The end of the eclipse is decided by the dividers in the same manner, and, as near as we can determine, must take place at 6 h. 44 m.

CHAP. V.

To find the more exact time.

To find the time of greatest obscuration, we must look along the moon's path, and discover, as near as possible, from what point a line drawn at right angles from that path will strike the sun's path at the same hour and minute; the time, thus marked on both paths, will be the time of greatest obscuration.

How to find the time of greatest obscuration.

In this case it appears to be 5 h. 40 m., and the two centers are very nearly together; so near, that we cannot decide on which side of the sun's center the moon's center will be, without a trigonometrical calculation.

To show a representation of an eclipse at any time during its continuance, we must take the semidiameter of the sun in the dividers from the scale; and, from the point of time on the sun's path, describe the sun; and, from the same point of time on the moon's path, describe a circle with the radius of the moon's semidiameter; the portion of the sun's diameter eclipsed, measured by the dividers, and compared with the whole diameter, will give the magnitude of the eclipse as near as it can be determined by projection.

How to find the magnitude of the eclipse.

The results of this projection are as follows:

	App. time.		Mean time.		
Beginning of the eclipse, P. M.,	4 h.	28 m.	4 h.	24 m.	39 s.
Greatest obscuration,	5	40	5	36	39
End of the eclipse,	6	44	6	40	39

Accuracy of the results.

From the projection the two centers are nearer together than the difference of the semidiameter of the sun and moon,

CHAP. V. and the moon's diameter being least, the eclipse will be *annular*, as represented in the projection.

The above results are, probably, to be relied upon to within *three minutes*.

We have now done with the projection, as far as the particular locality, Boston, is concerned; but, in consequence of the facility of solution, we cannot forbear to solve the following problem: *In the same parallel of latitude as Boston, find the longitude where the greatest obscuration will be exactly at* 2 P. M. *apparent time*.

A very easy and important problem.

How solved.

From the point 2, in the ellipse, draw a line at right angles to the moon's path, and that point must also be 2h. on the moon's path; running back to conjunction, we find it must take place at 1 h. 50 m.; but the conjunction for Greenwich time is 8 h. 49 m., the difference is 6 h. 59 m., corresponding to 104° 45′ west longitude; we further perceive that the sun would there be about 9 digits eclipsed on the sun's southern limb.

How to find more accurate results.

Now, admitting this construction to be on mathematical principles (as it really is, except the variability of the elements), we can determine the beginning and end of a local eclipse to great accuracy, by the application of ANALYTICAL GEOMETRY.

General equations to aid in computing all the circumstances of an eclipse as seen at any one place.

Let CD and C ♋ be two rectangular co-ordinates, then the distance of any point in the projection from the center can be determined by means of *equations*.

Let x and y be the co-ordinates of any point on the sun's path or elliptic curve, and X and Y the co-ordinates of any point on the moon's path, then we have the following equations:

$$\left.\begin{array}{ll}(1) & y=p\sin. L\cos. D\pm p\cos. L\sin. D\cos. t\\ (2) & x=p\cos. L\sin. t\end{array}\right\}\text{solar co-ordin.}$$

$$\left.\begin{array}{ll}(3) & Y=d\pm hi\sin. B\\ (4) & X=hi\cos. B\end{array}\right\}\text{lunar co-ordinates.}$$

In these remarkable equations, p is the semidiameter of projection, L the latitude, D the sun's declination, t the time from apparent noon, d the difference in declination between sun and moon at the instant of conjunction in right ascension, h the moon's hourly motion from the sun, i the interval of time from conjunction in right ascension—*minus*, if before conjunction—*plus*, if after; and B is the angle $L\ C$ ☊, or the angle which the moon's path makes with $C\ D$.

In the equations, x and X are horizontal distances. In equation (1), the *plus* sign is taken when the hours are on the upper side of the ellipse, as in winter; when on the lower side, take the *minus* sign.

In equation (3), the *plus* sign is taken when the motion of the moon is *northward*, and the *minus* sign when *southward*. The sin. t, or cos. t, means the sin. or cos. of an arc, corresponding to the time at the rate of 15° to one hour.

Explanation of the symbols.

The solar and lunar co-ordinates, or equations (1), (2), (3), and (4), are connected together by the following equations; the *minus* sign applies to forenoon, the *plus* sign to afternoon:

The symbol ☌ expresses the time of conjunction in right ascension.

$$☌ - i = -t;$$
$$☌ + i = t.$$

To apply these equations, and, of course, the former ones, i, the interval of time from conjunction must be *assumed*, and, as the time of conjunction is known, t thus becomes known; d, h, and B, are known by the elements; therefore, x, y, and X, Y, are all known. But the distance between any two points referred to co-ordinates, is always expressed by

$$\sqrt{(x \backsim X)^2 + (y \backsim Y)^2}.$$

When an eclipse first commences, or just as it ends, this expression must be just equal to the semidiameter of the sun and moon; and if, on computing the value of this expression, it is found to be less than that quantity, the sun is eclipsed; if greater, the sun is not eclipsed; and the result will show how much of the moon's limb is over the sun, or how far asunder the limbs are, and will, of course, indicate what change in the time must be made to correspond with a contact, or a particular phase of the eclipse.

For an eclipse absolutely central, and at the time of being central, the last expression must equal zero; and, in that

CHAP. V.

case, $x=X$, and $y=Y$. In cases of annular eclipses, to find the time of formation or rupture of the ring, the expression must be put equal to the difference of the semidiameters of sun and moon. In short, these expressions accurately, efficiently, and briefly cover the whole subject; and we now close by showing their application to the case before us.

Application of the preceding expressions.

By the projection we decided that the beginning of the eclipse would be at 4 h. 28 m., apparent time at Boston. Call this the assumed or approximate time, and for this instant we will compute the exact distance between the center of the sun and the center of the moon, and if that distance is equal to the sum of their semidiameter, then 4 h. 28 m. is, in fact, the time, otherwise it is not, &c.

An accurate computation for the beginning of the eclipse as seen from Boston.

	h. m. s.
Conjunc. in R. A., app. time, Boston,	4 13 21
Assume i equal to	15
Therefore, t is equal to	4 28 21 $=67^\circ\ 5'\ 15''$.

$p=54'\ 21''=3261$. Reduced lat., $L=42^\circ\ 6'\ 38''$.

$D=21^\circ\ 11'\ 43''$; $d=Cr=1324''.3$; $h=1659$ $i=\frac{1}{4}$;

$B=16^\circ\ 0'\ 58''$.

p 3261	- log. 3.513511	-	log. 3.513511
L 42° 6′ 38″	sin. 9.826437	-	cos. 9.870315
D 21 11 43	cos. 9.969583	-	sin. 9.558149
t 67 5 15			cos. 9.590288
2039.1	log. 3.309531	346.3	log. 2.532263
346.3			
$y=$1692.8			

p	3.513511
cos. L	9.870315
sin. t	9.964303
$x=$2228.5 log.	3.348129

For Y and X:

B 16° 0′ 58″	- sin. 9.440775	-	cos. 9.982804
hi 414″.75	- log. 2.617800	-	log. 2.617800
114.5	2.058575	398.6	2.600604
add 1324.3			
$Y=$ 1438.8		$X=$398.6	
$(Y \backsim y)=264$		$(x \backsim X)=1829.9$.	

Here are two sides of a right-angled triangle, and the hypothenuse of that triangle is 1857''.8, which is the distance between the center of the sun and moon at that instant; but the semidiameter of the sun and moon is only 1853''; therefore the eclipse has not yet commenced, and will not until the moon moves over 4''.8; which will require about 9 s., as we determined by proportion, because the apparent motion of the moon will be almost directly toward the sun.

The eclipse must be annular.

When the apparent motion of the moon is not so nearly in a line with the sun, as it is in this case, we cannot proportion directly to the result of the correction. In fact, the apparent motion of the moon is on one side of a plane right-angled triangle, and the distance between the center of sun and moon is the hypothenuse to that triangle, and the variation of the moon on its base varies the hypothenuse, and the computation must be made accordingly.

	h.	m.	s.
Hence, to the assumed time of beginning,	4 h.	28 m.	21 s.
Add - - - - - - -			19
Beginning, apparent time, - -	4	28	40
Mean time, - - - - -	4	25	19

By the application of the same expressions, we learn that the greatest obscuration will take place at 4 h. 41 m. mean time at Boston; and the apparent distance of the moon's center will be 18'' north of the sun's center; and, as the moon's semidiameter is 57'' less than that of the sun, a ring will be formed of between 10'' and 11'' wide at the narrowest point. End of the eclipse, 6 h. 46 m. 58 s. mean time.

The moon's center apparently 18'' N. of the sun's at apparent conjunction.

In computing for the end of the eclipse, we assumed i=1 h. 33 m., and as t is more than 6 h., the second part of y changes sign, as we see by the figure; the sun after 6, must be above the line 6 g 6.

Occultations of stars are computed on the same principles as an eclipse of the sun, the star having neither diameter nor parallax.

As problems, to give practice to the learner, we take the elements of two solar eclipses for 1846, from the Nautical Almanac, with their results as answers to the problems:

CHAP. V.

ELEMENTS OF THE ECLIPSES OF THE SUN.

Examples given for practice.

1846.	April 25.	October 19.
	h. m. s.	h. m. s.
Greenwich M. T. of ☌ in R. A.,	4 55 54 ·5	19 50 12 .2
☉ and ☾'s Right Ascension,	2 11 8 ·31	13 38 31 ·54
	° ′ ″	° ′ ″
☾'s declination,	N. 13 25 19 ·8	S. 10 23 43 ·0
☉'s declination,	N. 13 13 21 ·2	S. 10 15 3 ·9
☾'s hourly motion in R. A.,	33 55 ·1	30 42 ·2
☉'s hourly motion in R. A.,	2 21 ·3	2 21 ·5
☾'s hourly motion in dec.	N. 8 23 ·6	S. 8 37 ·0
☉'s hourly motion in dec.	N. 0 48 ·8	S. 0 54 ·1
☾'s equatorial hor. parallax,	57 53 ·8	55 33 ·4
☉'s equatorial hor. parallax,	8 ·5	8 ·6
☾'s true semidiameter,	15 46 ·5	15 8 ·4
☉'s true semidiameter,	15 54 ·5	16 5 ·6

THE APRIL ECLIPSE.

General results.

Begins on the earth generally April 25 d. 2 h. 2 m. 4 s., mean time at Greenwich, in longitude 119° 40′ W. of Greenwich, and latitude 6° 15′ S.

Central Eclipse begins generally April 25 d. 3 h. 3 m. 3 s. in longitude 135° 51′ W. of Greenwich, and lat. 2° 11′ S.

Central eclipse at noon, April 25 d. 4 h. 55 m. 9 s. in longitude 74° 31′ W. of Greenwich, and lat. 25° 21′ N.

Central eclipse ends generally April 25 d. 6 h. 37 m. 6 s. in longitude 3° 43′ W. of Greenwich, and lat. 24° 56′ N.

Ends on the earth generally April 25 d. 7 h. 38 m. 5 s. in longitude 20° 4′ W. of Greenwich, and lat. 20° 52′ N.

THE OCTOBER ECLIPSE.

Begins on the earth generally October 19 d. 16 h. 46 m. 7 s. mean time at Greenwich, in longitude 16° 21′ E. of Greenwich, and latitude 9° 50′ N.

Central eclipse begins generally October 19 d. 17 h. 52 m. 0 s. in longitude 0° 32′ W. of Greenwich, and lat. 6° 44′ N.

Central eclipse at noon, October 19 d. 19 h. 50 m. 2 s in longitude 58° 41′ E. of Greenwich, and lat. 19° 22′ S.

Central Eclipse ends generally October 19 d. 21 h. 38 m. 9 s. in longitude 126° 5′ E. of Greenwich, and lat. 23° 51′ S.
Ends on the earth generally October 19 d. 22 h. 44 m. 1 s. in longitude 109° 6′ E. of Greenwich, and lat. 20° 47′ S.

The following is a catalogue of the solar eclipses that will be visible in New England and New York, between the years 1850 and 1900; the dates are given in civil, not astronomical, time.

Statistics of eclipses from 1850 to 1900.

1851, July 28th. Digits eclipsed, 3¾, on sun's northern limb.
1854, May 26th. As computed in the work.
1858, March 15th. Sun rises eclipsed. Greatest obscuration, 5½ digits on sun's southern limb.
1859, July 29th. Digits eclipsed, 2½, on sun's northern limb.
1860, July 18th. Digits eclipsed, 6, on sun's northern limb.
1861, December 31st. Sun rises eclipsed. Digits eclipsed at greatest obscuration, 4⅓, on sun's southern limb.
1865, October 19th. Digits eclipsed, 8⅓, on sun's southern limb.
1866, October 8th. ½ digit eclipsed. South of New York no eclipse.
1869, August 7th. Digits eclipsed, 10, on sun's southern limb. This eclipse will be total in North Carolina.
1873, May 25th. Sun and moon in contact at sunrise, Boston.
1875, September 29th. Sun rises eclipsed. This eclipse will be annular in Boston, Maine, New Hampshire, and Vermont.
1876, March 25th. Digits eclipsed, 3½, on sun's northern limb.
1878, July 29th. Digits eclipsed, 7⅓, on sun's southern limb. This is the fourth return of the total eclipse of 1806.
1880, December 31st. Sun rises eclipsed. Digits eclipsed at greatest obscuration, 5½, on sun's northern limb.
1885, March 16th. Digits eclipsed, 6½, on sun's northern limb.

CHAP. V. Statistics of eclipses from 1850 to 1900.

1886, August 28th. North of Massachusetts no eclipse; south, sun eclipsed.

1892, October 20th. Digits eclipsed, 8, on sun's northern limb.

1897, July 29th. Digits eclipsed, 4½, on sun's southern limb.

1900, May 28th. Digits eclipsed, 11, on sun's southern limb. The sun will be totally eclipsed in the State of Virginia.

EXTRACTS FROM THE NAUTICAL ALMANAC FOR JANUARY, 1846.

Day of the Month.	THE SUN'S *Apparent*		Logar. of the Radius Vector of the Earth.	THE MOON'S			
	Longitude.	Latitude.		Longitude.	Latitude.	Semi-diam.	Hor. Paral.
	Noon.	Noon.	Noon.	Noon.	Noon.	Noon.	Noon.
	° ′ ″	″		° ′ ″	° ′ ″	′ ″	′ ″
1	280 46 15.3	N.0.49	9.99266	330 42 13.9	N.4 54 8.5	16 21.6	60 2.3
2	281 47 26.1	0.45	9.99266	345 7 12.0	4 24 8.7	16 8.3	59 13.5
3	282 48 36.5	0.37	9.99267	359 4 55.4	3 39 5.9	15 53.9	58 20.5
4	283 49 46.5	0.27	9.99267	12 35 34.7	2 43 1.9	15 39.8	57 28.7
5	284 50 56.1	0.16	9.99268	25 41 31.5	1 39 55.7	15 26.7	56 40.8
6	285 52 5.3	N.0.03	9.99268	38 26 25.0	N.0 33 28.3	15 15.2	55 58.7
7	286 53 13.9	S. 0.11	9.99270	50 54 23.2	S. 0 33 3.6	15 5.6	55 23.3
8	287 54 22.0	0.25	9.99271	63 9 30.1	1 36 46.8	14 57.6	54 54.1
9	288 55 29.7	0.38	9.99272	75 15 21.8	2 35 8.6	14 51.5	54 31.6
10	289 56 36.8	0.49	9.99274	87 14 56.3	3 25 55.4	14 46.9	54 14.6
11	290 57 43.4	0.58	9.99277	99 10 31.3	4 7 13.7	14 43.8	54 3.3
12	291 58 49.5	0.65	9.99279	111 3 50.8	4 37 30.7	14 42.1	53 57.0
13	292 59 55.3	0.70	9.99282	122 56 17.6	4 55 38.9	14 41.7	53 55.7
14	294 1 0.5	0.71	9.99285	134 49 7.9	5 0 56.4	14 42.8	53 59.8
15	295 2 5.4	0.69	9.99288	146 43 48.4	4 53 7.6	14 45.5	54 9.7
16	296 3 9.9	0.64	9.99292	158 42 11.3	4 32 23.1	14 50.0	54 26.0
17	297 4 14.0	0.57	9.99295	170 46 44.8	3 59 17.1	14 56.3	54 49.0
18	298 5 17.8	0.47	9.99299	183 0 38.7	3 14 47.1	15 4.6	55 19.7
19	299 6 21.2	0.35	9.99304	195 27 41.8	2 20 14.2	15 15.2	55 58.4
20	300 7 24.2	0.23	9.99308	208 12 10.4	1 17 27.8	15 27.7	56 44.4
21	301 8 26.7	S. 0.09	9.99313	221 18 27 5	S. 0 8 53.1	15 42.0	57 37.0
22	302 9 28.9	N.0.04	9.99318	234 50 26.7	N.1 2 20.5	15 57.3	58 32.9
23	303 10 30.4	0.15	9.99323	248 50 42.5	2 12 11.7	16 12.5	59 28.8
24	304 11 31.3	0.25	9.99328	263 19 30.4	3 15 50.9	16 26.2	60 19.0
25	305 12 31.5	0.33	9.99334	278 13 48.8	4 8 2.8	16 36.8	60 57.9
26	306 13 30.9	0.38	9.99339	293 26 49.2	4 43 49.4	16 42.9	61 20.2
27	307 14 29.3	0.40	9.99345	308 48 22.8	4 59 32.4	16 43.5	61 22.6
28	308 15 26.8	0.40	9.99351	324 6 34.0	4 53 45.4	16 38.7	61 4.9
29	309 16 23.3	0.37	9.99357	339 9 55.3	4 27 32.9	16 28.9	60 29.1
30	310 17 18.5	0.30	9.99363	353 49 32.0	3 44 8.2	16 15.6	59 40.2
31	311 18 12.6	0.21	9.99369	8 0 13.1	2 47 58.7	16 0.2	58 43.7
32	312 19 5.3	N.0.10	9.99375	21 40 34.3	N.1 43 50.6	15 44.2	57 45.1

TABLE I.

MEAN ASTRONOMICAL REFRACTIONS.

Barometer 30 in. Thermometer, Fah. 50°.

Ap. Alt.	Refr.	Ap. Alt.	Refr.	Ap. Alt.	Refr.	Alt.	Refr.
0° 0′	33′ 51″	4° 0′	11′ 52″	12° 0′	4′ 28.1″	42°	1 4.6′
5	32 53	10	11 30	10	4 24.4	43	1 2.4
10	31 58	20	11 10	20	4 20.8	44	1 0.3
15	31 5	30	10 50	30	4 17.3	45	0 58.1
20	30 13	40	10 32	40	4 13.9	46	56.1
25	29 24	50	10 15	50	4 10.7	47	54.2
30	28 37	5 0	9 58	13 0	4 7.5	48	52.3
35	27 51	10	9 42	10	4 4.4	49	50.5
40	27 6	20	9 27	20	4 1.4	50	48.8
45	26 24	30	9 11	30	3 58.4	51	47.1
50	25 43	40	8 58	40	3 55.5	52	45.4
55	25 3	50	8 45	50	3 52.6	53	43.8
1 0	24 25	6 0	8 32	14 0	3 49.9	54	42.2
5	23 48	10	8 20	10	3 47.1	55	40.8
10	23 13	20	8 9	20	3 44.4	56	39.3
15	22 40	30	7 58	30	3 41.8	57	37.8
20	22 8	40	7 47	40	3 39.2	58	36.4
25	21 37	50	7 37	50	3 36.7	59	35.0
30	21 7	7 0	7 27	15 0	3 34.3	60	33.6
35	20 38	10	7 17	15 30	3 27.3	61	32.3
40	20 10	20	7 8	16 0	3 20.6	62	31.0
45	19 43	30	6 59	16 30	3 14.4	63	29.7
50	19 17	40	6 51	17 0	3 8.5	64	28.4
55	18 52	50	6 43	17 30	3 2.9	65	27.2
2 0	18 29	8 0	6 35	18 0	2 57.6	66	25.9
5	18 5	10	6 28	19	2 47.7	67	24.7
10	17 43	20	6 21	20	2 38.7	68	23.5
15	17 21	30	6 14	21	2 30.5	69	22.4
20	17 0	40	6 7	22	2 23.2	70	21.2
25	16 40	50	6 0	23	2 16.5	71	19.9
30	16 21	9 0	5 54	24	2 10.1	72	18.8
35	16 2	10	5 47	25	2 4.2	73	17.7
40	15 43	20	5 41	26	1 58.8	74	16.6
45	15 25	30	5 36	27	1 53.8	75	15.5
50	15 8	40	5 30	28	1 49.1	76	14.4
55	14 51	50	5 25	29	1 44.7	77	13.4
3 0	14 35	10 0	5 20	30	1 40.5	78	12.3
5	14 19	10	5 15	31	1 36.6	79	11.2
10	14 4	20	5 10	32	1 33.0	80	10.2
15	13 50	30	5 5	33	1 29.5	81	9.2
20	13 35	40	5 0	34	1 26.1	82	8.2
25	13 21	50	4 56	35	1 23.0	83	7.1
30	13 7	11 0	4 51	36	1 20.0	84	6.1
35	12 53	10	4 47	37	1 17.1	85	5.1
40	12 41	20	4 43	38	1 14.4	86	4.1
45	12 28	30	4 39	39	1 11.8	87	3.1
50	12 16	40	4 35	40	1 9.3	88	2.0
55	12 3	50	4 31	41	1 6.9	89	1.0

TABLE C.

CORRECTION OF MEAN REFRACTION.

Hight of the Thermometer.

App. Alt. ° ′	24° ′+″	28° ′+″	32° ′+″	36° ′+″	40° ′+″	44° ′+″	52° ′—″	56° ′—″	60° ′—″	64° ′—″	68° ′—″	72° ′—″	76° ′—″	80° ′—″
0.00	2.18	1.55	1.33	1.11	51	31	10	29	48	1.07	1.25	1 43	2.01	2.19
0.10	2.12	1.49	1.28	1.08	48	29	9	27	45	1.04	1.21	1.38	1.54	2.12
0.20	2.05	1.44	1.24	1.04	46	28	9	26	44	1.01	1.17	1.33	1.49	2.05
0.30	1.59	1.39	1.20	1.01	44	26	8	25	41	58	1.13	1.28	1.43	1.59
0.40	1.53	1.34	1.16	58	42	25	8	24	39	55	1.10	1.24	1.38	1.53
0.50	1.48	1.29	1.12	55	40	24	8	23	37	52	1.06	1.20	1.34	1.48
1.00	1.43	1.25	1.09	53	38	23	7	21	36	50	1.03	1.17	1.30	1.43
1.10	1.38	1.21	1.06	50	36	22	7	20	34	48	1.00	1.13	1.26	1.38
1.20	1.33	1.17	1.03	48	34	21	6	19	32	45	57	1.09	1.21	1.33
1.30	1.29	1.14	1.00	46	32	20	6	18	31	43	54	1.06	1.18	1.29
1.40	1.25	1.11	57	44	31	18	6	18	30	41	52	1.04	1.15	1.25
1.50	1.21	1.08	55	42	30	17	6	17	28	39	50	1.01	1.11	1.21
2.00	1.18	1.05	53	39	29	17	5	16	27	37	48	58	1.08	1.18
2.20	1.11	1.00	48	37	26	16	5	15	25	35	44	54	1.03	1.11
2.40	1.06	55	44	34	24	14	5	14	23	32	41	50	58	1.06
3.00	1.01	51	41	32	22	13	4	13	21	30	38	46	54	1.01
3.20	57	47	38	29	21	13	4	12	20	28	35	43	50	.57
3.40	53	44	36	28	20	12	4	11	18	26	33	40	47	53
4.00	49	41	33	26	18	11	4	10	17	24	31	37	44	50
4.30	45	38	31	24	17	10	3	9	16	22	28	34	40	45
5.00	41	35	28	22	16	9	3	9	14	20	26	31	36	40
5.30	38	32	26	20	14	9	3	8	13	19	24	29	34	38
6.00	35	30	24	19	13	8	2	7	12	17	22	26	31	35
6.30	33	28	22	17	12	7	2	7	11	15	20	24	29	33
7.00	31	26	21	16	12	7	2	6	10	14	19	23	27	31
8	27	23	19	15	10	6	2	5	9	13	16	20	24	27
9	24	20	16	13	9	5	2	5	8	11	14	18	21	24
10	22	18	15	12	8	5	1	4	7	10	13	16	19	22
11	20	17	14	11	8	5	1	4	7	9	12	15	18	20
12	18	15	13	10	7	4	1	4	6	9	11	13	16	18
13	17	14	12	9	7	4	1	3	6	8	10	12	15	17
14	16	13	11	8	6	4	1	3	5	7	9	11	14	16
15	15	12	10	8	6	3	1	3	5	7	9	11	13	15
16	14	12	9	7	5	3	1	3	5	6	8	10	12	14
17	13	11	9	7	5	3	1	3	4	6	8	9	11	13
18	12	10	8	6	5	3	1	2	4	6	7	9	10	12
19	11	9	8	6	4	3	1	2	4	5	7	8	10	11
20	11	9	7	6	4	2	1	2	4	5	6	8	9	11
21	10	9	7	5	4	2	1	2	3	5	6	7	9	10
22	10	8	7	5	4	2	1	2	3	5	6	7	8	10
23	9	8	6	5	4	2	1	2	3	4	6	7	8	9
24	9	7	6	5	3	2	1	2	3	4	5	6	8	9
25	8	7	6	5	3	2	1	2	3	4	5	6	7	8
26	8	7	6	4	3	2	1	2	3	4	5	6	7	8
27	8	6	5	4	3	2	1	2	3	4	5	6	7	8
28	7	6	5	4	3	2	0	1	2	3	5	5	6	7
30	7	6	5	4	3	2	0	1	2	3	4	5	6	7
			—	—	—	—	+	+	+	+	+			
Hight of the Barometer.			28.26	28.56	28.85	29.15	29.75	30 05	30.35	30.64	30.93			

TABLE II.

MEAN PLACES FOR 100 PRINCIPAL FIXED STARS, FOR JAN. 1, 1846.

Star's Name.	Mag.	Right Ascen.	Annual Var.	Declination.	Ann. Var.
		h. m. s.	s.	deg. min. sec.	sec.
α Andromedæ, ...	1	0 0 26.257	+ 3.0720	N.28 14 25.40	+20.055
γ Pegasi (*Algenib*), ...	2.3	0 5 18.691	3.0784	N.14 19 37.80	20.050
β Hydri, ...	3	0 17 34.168	3.3054*	S. 78 7 24.40	19.997
α Cassiopeæ, ...	3	0 31 48.294	3.3418	N.55 41 31.08	19.862
β Ceti, ...	2.3	0 35 51.339	+ 2.9995	S. 18 49 59.01	+19.810
α Urs. Min. (*Polaris*), ...	2.3	1 3 52.226	17.1346*	N.88 29 17.88	19.279
θ¹ Ceti, ...	3	1 16 19.692	3.0015	S. 8 58 45.93	18.952
α Eridani (*Achernar*), ...	1	1 31 58.291	2.2339	S. 58 1 14.34	18.461
α Arietis, ...	3	1 58 30.193	+ 3.3475	N.22 43 53.86	+17.432
γ Ceti, ...	3	2 35 19.633	3.1085	N. 2 35 1.17	15.621
α Ceti, ...	2.3	2 54 14.072	3.1266	N. 3 28 55.70	14.532
α Persei, ...	2.3	3 13 21.403	4.2324	N.49 18 28.20	13.329
η Tauri, ...	3	3 38 20.382	+ 3.5473	N.23 37 27.73	+11.620
γ¹ Eridani, ...	2.3	3 50 50.760	2.7898	S. 13 57 1.50	10.711
α Tauri (*Aldebaran*), ...	1	4 27 5.404	3.4274	N.16 11 41.39	7.907
α Aurigæ (*Capella*), ...	1	5 5 19.317	4.4082	N.45 50 6.56	4.737
β Orionis (*Rigel*), ...	1	5 7 8.383	+ 2.8787	S. 8 23 3.33	+ 4.583
β Tauri	2	5 16 33.662	3.7827	N.28 28 17.49	3.776
δ Orionis, ...	2	5 24 8.428	3.0609	S 0 25 4.86	3.123
α Lepris, ...	3.4	5 25 56.406	2.6425	S. 17 56 12.77	2.968
ε Orionis, ...	2.3	5 28 24.062	+ 3.0404	S. 1 18 17.53	+ 2.754
α Columbæ, ...	2	5 34 4.531	2.1691	S. 34 9 36.95	2.262
α Orionis, ...	1	5 46 50.189	3.2433	N. 7 22 22.32	+ 1.149
μ Geminorum, ...	3	6 13 38.621	3.6257	N.22 35 13.16	— 1.196
α Argus (*Canopus*), ...	1	6 20 32.145	+ 1.3279	S. 52 36 49.17	— 1.796
51 (Hev.) Cephei, ...	6	6 26 30.287	30.7946	N.87 15 31.20	2.337
α Canis Maj. (*Sirius*), ...	1	6 38 21.883	2.6459*	S. 16 30 32.83	4.484*
ε Canis Majoris, ...	2.3	6 52 34.440	2.3558	S. 28 45 59.38	4.562
δ Geminorum, ...	3.4	7 10 55.298	+ 3.5918	N.22 15 37.47	— 6.110
α² Geminor. (*Castor*), ...	3	7 24 46.065	3.8561	N.32 13 12.93	7.253
α Can. Min. (*Procyon*), ..	1.2	7 31 14.237	3.1445*	N. 5 36 54.95	8.758*
β Geminor. (*Pollux*), ...	2	7 35 53.153	3.6829*	N.28 23 34.06	8.152
15 Argus, ...	3.4	8 0 59.232	+ 2.5596	S. 23 51 50.94	—10.104
ε Hydræ, ...	4	8 38 37.154	3.1966	N. 6 58 48.51	12.800
ι Ursæ Majoris, ...	3.4	8 48 38.088	4.1261*	N.48 38 32.35	13.464
ι Argus, ...	2	9 12 58.192	1.6100	S. 58 37 49.78	14.961
α Hydræ, ...	2	9 20 1.170	+ 2.9499	S. 7 59 39.05	—15.366
θ Ursæ Majoris, ...	3	9 22 31.453	4.0504*	N.52 22 31.09	16.108*
ε Leonis, ...	3	9 37 6.098	3.4258	N.24 28 49.46	16.283
α Leonis (*Regulus*), ...	1	10 0 10.062	+ 3.2211	N.12 43 2.96	—17.377

Star's Name.	Mag.	Right Ascen.	Annual Var.	Declination.	Ann. Var.
		h. m. s.	s.	deg. min. sec.	s.
η Argus,	2	10 39 6.223	+ 2.3051	S. 58 52 34.26	—18.82
α Ursæ Majoris,	1.2	10 54 10.737	3.8001	N.62 34 51.81	19.24
δ Leonis,	3	11 5 54.583	3.1928	N.21 21 59.86	19.50
δ Hydræ et Crateris,	3.4	11 11 38.718	3.0010	S. 13 56 46.85	19.61
β Leonis,	2.3	11 41 12.066	+ 3.0654*	N.15 25 58.12	—19.99
γ Ursæ Majoris,	2	11 45 42.219	3.1874	N.54 33 3.18	20.02
β Chamæleontis,	5	12 9 26.893	3.3409	S. 78 27 26.15	20.04
α1 Crucis,	1	12 18 4.916	3.2710	S. 62 14 39.74	19.99
β Corvi,	2.3	12 26 18.465	+ 3.1342	S. 22 32 39.93	—19.92
12 Canum Venaticorum,	2.3	12 48 49.007	2.8403	N.39 9 4.18	19.60
α Virginis (*Spica*),	1	13 17 5.233	3.1512	S. 10 21 20.80	18.94
η Ursæ Majoris,	2.3	13 41 27.894	2.3525*	N.50 5 1.45	18.12
η Bootis,	3	13 47 21.140	+ 2.8606	N.19 10 21.03	—17.89
β Centauri,	1	13 53 0.800	4.1508	S. 59 37 33.93	17.67
α Bootis, (*Arcturus*),	1	14 8 38.366	2.7336*	N.19 59 12.07	18.94*
α2 Centauri,	1	14 29 11.925	4.0165*	S. 60 11 37.00	15.12*
ε Bootis,	3	14 38 15.706	+ 2.6229	N.27 43 35.23	—15.46
α2 Libræ,	3	14 42 22.132	+ 3.3102	S. 15 23 53.52	15.23
β Ursæ Minoris,	3	14 51 13.199	— 0.2692	N.74 47 5.58	14.71
β Libræ,	2.3	15 8 43.595	+ 3.2226	S. 8 48 38.53	13.63
α Coronæ Borealis,	2	15 28 10.083	+ 2.5279	N.27 14 11.07	—12.33
α Serpentis,	2.3	15 36 41.077	+ 2.9391	N. 6 54 49.88	11.74
ζ Ursæ Minoris,	4	15 49 41.194	— 2.3520	N.78 15 55.43	10.80
β1 Scorpii,	2	15 56 29.397	+ 3.4742	S. 19 22 44.18	10.29
δ Ophiuchi,	3	16 6 16.830	+ 3.1382	S. 3 17 35.67	— 9.55
α Scorpii, (*Antares*),	1	16 19 58.461	3.6638	S. 26 5 4.58	8.48
η Draconis,	3	16 21 55.119	0.7960	N.61 51 50.58	8.32
α Trianguli Australis,	2	16 32 25.090	+ 6.2587	S. 68 44 4.75	7.48
ε Ursæ Minoris,	4	17 1 55.988	— 6.5328*	N.82 16 52.30	— 5.03
α Herculis,	3.4	17 7 37.617	+ 2.7320	N.14 34 12.67	4.54
σ Octantis,	6	17 22 55.004	106.8627	S. 89 16 10.25	3.14
β Draconis,	2	17 26 57.473	1.3513	N.52 25 3.28	2.88
α Ophiuchi,	2	17 27 47.219	+ 2.7727	N.12 40 37.11	— 2.81
γ Draconis,	2	17 53 1.955	1.3900	N.51 30 33.50	— 0.61
μ1 Sagittarii,	3.4	18 4 33.276	+ 3.5861	S. 21 5 36.14	+ 0.40
δ Ursæ Minoris,	3	18 22 0.703	—19.2683*	N.86 35 42.58	+ 1.91
α Lyræ (*Vega*),	1	18 31 43.386	+ 2.0118	N.38 38 35.33	+ 2.77
β Lyræ,	3	18 44 23.696	2.2124	N.33 11 14.80	3.86
ζ Aquilæ,	3	18 58 19.965	2.7566	N.13 38 20.49	5.05
δ Aquilæ.	3.4	19 17 43.889	+ 3.0086	N. 2 48 43.64	+ 6.67
γ Aquilæ,	3	19 38 56.278	+ 2.8511	N.10 14 31.50	+ 8.39
α Aquilæ, (*Altair*),	1.2	19 43 16.128	2.9254*	N. 8 27 54.32	8.74
β Aquilæ.	3.4	19 47 44.866	2.9446	N. 6 1 33.90	8.55*
α2 Capricorni,	3	20 9 30.316	3.3315	S. 13 1 4.19	10.74

Star's Name.	Mag.	Right Ascen.	Annual Var.	Declination.	Ann. Var
		h. m. s.	s.	deg. min. sec.	sec.
α Pavonis, ...	2	20 13 25.814	+ 4.8046	S. 57 13 19.50	+11.03
γ Ursæ Minoris, ...	5	20 16 31.309	—52.1273	N.88 50 53.54	11.22
α Cygni, ...	1	20 36 11.005	+ 2.0418	N.44 43 57.43	12.64
61 ı Cygni, ...	5.6	20 59 59.947	2.6908*	N.37 59 42.08	17.48*
Cygni, ...	3	21 6 23.073	+ 2.5486	N.29 35 53.03	+14.57
α Cephei, ...	3	21 14 53.940	1.4163	N.61 56 4.55	15.07
β Aquarii, ...	3	21 23 26.875	3.1628	S. 6 14 44.46	15.56
β Cephei, ...	3	21 26 39.120	0.8059	N.69 53 7.21	15.73
ε Pegasi, ...	2.3	21 36 37.346	+ 2.9441	N. 9 10 17.35	+16.26
α Aquarii, ...	3	21 57 52.326	3.0831	S. 1 3 56.72	17.28
α Gruis, ...	2	21 58 29.837	3.8134	S. 47 42 12.42	17.30
ζ Pegasi, ...	3	22 33 46.976	2.9837	N.10 1 44.67	18.65
α Pis. Aus. (*Fomalhaut*), .	1	22 49 7.531	+ 3.3095	S. 30 26 12.28	+19.11
α Pegasi (*Markab*), ...	2	22 57 5.584	2.9776	N.14 22 40.12	19.31
ι Piscium, ...	4.5	23 32 1.736	3.0569	N. 4 47 30.74	19.36*
γ Cephei, ...	3	23 33 4.581	+ 2.4042	N.76 46 22.01	+19.92

Those Annual Variations which include proper motion are distinguished by an Asterisk.

SUN'S RIGHT ASCENSION FOR 1846.

Day of Mo.	January.	February.	March.	April.	May.	June.
	h. m. s.	h. m. s.	h. m. s.	h. m. s.	h. m. s.	h. m. s.
1	18 46 52	20 59 11	22 48 17	0 41 52	2 23 6	4 35 48
5	19 4 30	21 15 22	23 3 12	0 56 26	2 48 25	4 52 12
10	19 26 21	21 35 18	23 21 40	1 14 43	3 7 47	5 12 50
15	19 47 57	21 54 54	23 40 0	1 33 6	3 27 24	5 33 34
20	20 9 17	22 14 12	23 58 14	1 51 38	3 47 15	5 54 22
25	20 30 19	22 33 14	0 16 25	2 10 22	4 7 20	6 15 10
30	20 51 0		0 34 36	2 29 17	4 27 38	6 35 55

Day of Mo.	July.	August.	September.	October.	November.	December.
	h. m. s.	h. m. s.	h. m. s.	h. m. s.	h. m. s.	h. m. s.
1	6 40 4	8 44 55	10 41 0	12 29 4	14 25 16	16 29 1
5	6 56 34	9 0 23	10 55 29	12 43 36	14 41 2	16 46 23
10	7 17 5	9 19 29	11 13 30	13 1 54	15 1 5	17 8 17
15	7 37 25	9 38 21	11 31 28	13 20 24	15 21 28	17 30 22
20	7 57 33	9 56 60	11 49 25	13 39 8	15 42 14	17 52 33
25	8 17 28	10 15 27	12 7 24	13 58 9	16 3 19	18 14 46
30	8 37 7	10 33 44	12 25 27	14 17 27	16 24 43	18 36 57

The R. A. in this table will answer for corresponding days in other years within four minutes; and for periods of four years, the difference is only about seven seconds for each period.

TABULAR VIEW OF THE SOLAR SYSTEM.

Names.	Mean diameter in miles.	Mean distance from the Sun in miles.	Mean dist.; the Earth's dist. unity.	Log. of mean distance.	Time of revolutions round the Sun.	Log. of times of revolution.
Sun	883000				DAYS.	
Mercury	3224	37 million	0.387098	9.587818	87.969258	1.944324
Venus	7687	68 "	0.723332	9.859306	224.700787	2.351610
The Earth	7912	95 "	1.000000	0.000000	365.256383	2.562598
Mars	4189	144 "	1.523692	0.182810	686.979646	2.836942
Vesta	238	224,340,000	2.36120	0.373100	1324.289	3.121991
Iris *	Unknown	226 million	2.37880	0.376384	1327.973	3.123190
Hebe *	Unknown	230 "	2.42190	0.384004	1375. nearly	3.138303
Flora *	Unknown	240 "	2.52630	0.402487	1469.76	3.167300
Astrea *	Unknown	246 "	2.5895	0.413211	1512. nearly	3.179547
Juno	1420	253,600,000	2.66514	0.425710	1594.721	3.202700
Ceres	Not well known. 160	263,236,000	2.76910	0.442334	1683.064	3.226086
Pallas	Not well known. 120	265 million	2.77125	0.442725	1685.162	3.226610
Jupiter	89170	490 "	5.202776	0.716212	4332.584821	3.636738
Saturn	79040	900 "	9.538786	0.979476	10759.219817	4.031718
Uranus	35000	1800 "	19.182390	1.282853	30686.8208	4.486953
Neptune	35000	2850 "	29.59	1.477121	60128.14	4.779076

TABLE III.

ELEMENTS OF ORBITS FOR THE EPOCH OF 1850, JANUARY 1, MEAN NOON AT GREENWICH.

Planets.	Inclination of orbits to ecliptic.	Variation in 100 years.	Long. of the ascending nodes.	Variation in 100 years.	Longitude of Perihelion.	Variation in 100 years.	Mean longitude at epoch.
	° ′ ″	″	° ′ ″	′	° ′ ″	′	° ′ ″
Mercury	7 0 18	+18.2	46 34 40		75 9 47	+93	327 17 9
Venus	3 23 26	— 4. 6	75 17 40	+51	129 22 53	+78	243 58 4
Earth					100 22 10	103	100 47 1
Mars	1 51 6	— 0. 2	48 20 24	+42	333 17 57	+110	182 9 30
Vesta	7 8 29	—12.	103 20 47	+26	254 4 34	157	113 28 12
Juno	13 2 53	....	170 53 0	....	54 18 32	....	165 17 38
Ceres	10 37 17	....	80 47 56	....	147 25 41	....	1 3 10
Pallas	34 37 44	....	172 42 38	....	121 30 13	....	327 31 24
Jupiter	1 18 42	—22.	98 55 19	+57	11 56 0	+ 95	160 21 50
Saturn	2 29 29	—15.	112 22 54	+51	90 7 0	+116	13 58 13
Uranus	0 46 27	3.	73 12 0	+24	168 14 47	+ 87	28 20 22

* It is with reluctance that we give these planets a place in the tables. The fact of their existence is as yet questionable, and their elements, at present, cannot be well known. We give the logarithms in the tables, that the data may be at hand to exercise the student on Kepler's third law.

TABLE III.

TABULAR VIEW OF THE SOLAR SYSTEM.

Names.	Mass.	Density.	Gravity.	Sidereal Rotation.	Light and Heat.
				h. m. s.	
Mercury..	$\frac{1}{2025810}$	3.244	1.22	24 5 28	6.680
Venus.....	$\frac{1}{401211}$	0.994	0.96	23 21 7	1.911
Earth.....	$\frac{1}{355000}$	1.000	1.00	24 0 0	1.000
Mars	$\frac{1}{2680337}$	0.973	0.50	24 39 21	.431
Jupiter	$\frac{1}{1048.7}$	0.232	2.70	9 55 50	.037
Saturn....	$\frac{1}{3500.2}$	0.132	1.25	10 29 17	.011
Uranus ...	$\frac{1}{17918}$	0.246	1.06	Unknown.	.003
Sun.......	1	0.256	28.19	25 12 0	
Moon.....	$\frac{1}{26620200}$	0.665	0.18	27 7 43	

TABLE III.

Planets.	Eccentricities of orbits.	Variation in 100 years.	Motion in mean long. in 1 year of 365 days.	Mean Daily Motion in longitude.
			° ′ ″	° ′ ″
Mercury....	0.20551494	+.000003868	53 43 3.6	4 5 32.6
Venus......	0.00686074	—.000062711	224 47 29.7	1 36 7.8
Earth	0.01678357	—.000041630	—0 14 19.5	0 59 8.3
Mars.......	0.09330700	+.000090176	191 17 9.1	0 31 26.7
Vesta.......	0.08856000	+.000004009	....	0 16 17.9
Juno.......	0.25556000	....	....	0 13 33.7
Ceres.......	0.07673780	—.000005830	....	0 12 49.4
Pallas	0.24199800	....	....	0 12 48.7
Jupiter......	0.04816210	+.000159350	30 20 31.9	0 4 59.3
Saturn......	0.05615050	—.000312402	12 13 36.1	0 0 0.6
Uranus.....	0.04661080	—.000025072	4 17 45.1	0 0 42.4

TABLE III.

LUNAR PERIODS.

Mean sidereal revolution, ...	27.321661418
Mean synodical revolution, . . .	29.530588715
Mean revolution of nodes (retrograde), .	6793.391080
Mean revolution of perigee (direct), ...	3232.575343
Mean inclination of orbit, ...	5° 8′ 48″
Mean distance in measure, of the equatorial radius of the earth, ...	29.98217
Mean distance in measure of the mean radius, . .	30.20000

SUN'S EPOCHS.

Years.	M. Long.				Long. Perigee.				I.	II.	III.	N.
	s.	°	′	″	s.	°	′	″				
1846	9	8	45	8	9	8	17	17	124	673	897	379
1847	9	8	30	48	9	8	18	19	484	588	623	433
1848 B.	9	9	15	37	9	8	19	20	878	505	151	487
1849	9	9	1	17	9	8	20	22	238	420	775	540
1850	9	8	46	58	9	8	21	23	598	336	400	594
1851	9	8	32	39	9	8	22	24	958	250	025	648
1852 B.	9	9	17	27	9	8	23	26	353	168	653	701
1853	9	9	3	8	9	8	24	27	713	083	277	755
1854	9	8	48	48	9	8	25	29	073	998	902	809
1855	9	8	34	29	9	8	26	30	433	913	527	863
1856 B.	9	9	19	18	9	8	27	32	827	832	153	916
1857	9	9	4	58	9	8	28	34	187	746	779	970
1858	9	8	50	39	9	8	29	35	547	661	404	024
1859	9	8	36	19	9	8	30	37	907	576	029	078
1860 B.	9	9	21	8	9	8	31	38	301	494	656	131
1861	9	9	6	49	9	8	32	39	661	409	281	185
1862	9	8	52	29	9	8	33	41	021	324	906	239
1863 B.	9	8	38	10	9	8	34	42	381	239	530	292
1864	9	9	22	58	9	8	35	44	775	157	157	346
1865	9	9	8	39	9	8	36	45	135	072	783	400
1866	9	8	54	20	9	8	37	47	495	985	408	453
1867	9	8	40	0	9	8	38	49	855	902	033	507
1868 B.	9	9	24	49	9	8	39	50	249	820	659	561
1869	9	9	10	30	9	8	40	52	609	734	285	615
1870	9	8	56	10	9	8	41	53	969	649	910	668
1882	9	9	1	41	9	8	54	10	391	638	416	313
1871	9	8	41	51	9	8	42	54	329	564	534	721
1872 B.	9	9	26	39	9	8	43	56	723	481	161	774
1873	9	9	12	20	9	8	45	58	083	396	785	828
1874	9	8	58	1	9	8	47	0	443	311	410	881
1875	9	8	43	41	9	8	48	2	803	226	034	935
1876 B.	9	9	28	30	9	8	49	4	297	143	661	989
1877	9	9	14	10	9	8	50	5	657	058	286	042
1878	9	8	59	51	9	8	51	6	017	974	912	096
1879	9	8	45	32	9	8	52	7	377	889	537	150
1880 B.	9	9	30	20	9	8	53	9	671	807	164	204
1881	9	9	16	1	9	8	54	10	031	722	790	257
1882	9	9	1	41	9	8	55	12	391	637	415	311
1883	9	8	47	22	9	8	56	13	751	552	040	364
1884 B.	9	9	32	10	9	8	57	15	145	469	666	418
1885	9	9	17	51	9	8	58	16	505	385	292	471
1886	9	9	3	32	9	8	59	17	865	300	918	525
1887	9	8	49	12	9	8	0	19	225	216	544	579
1888 B.	9	9	34	1	9	8	1	20	619	133	169	632

TABLE V.

SUN'S MOTIONS FOR MONTHS.

Months.	Longitude.				Per.	I.	II.	III.	N.
	s.	°	′	″	″				
Jan. Com.	0	0	0	0	0	0	0	0	0
Jan. Bis.	11	29	0	52	0	966	997	998	0
Feb. Com.	1	0	33	18	5	47	78	53	4
Feb. Bis.	0	29	34	10	5	13	75	51	4
March.........	1	28	9	11	10	993	148	01	9
April..........	2	28	42	30	15	42	226	154	13
May............	3	28	16	40	20	59	301	206	18
June...........	4	28	49	58	26	110	379	259	22
July...........	5	28	24	8	31	129	454	310	27
August.........	6	28	57	26	36	182	531	363	31
September......	7	29	30	44	41	233	609	416	36
October........	8	29	4	54	46	250	684	468	40
November......	9	29	38	12	52	300	762	521	45
December.......	10	29	12	22	57	313	837	572	49

TABLE VI.

SUN'S HOURLY MOTION.

ARGUMENT.—Sun's Mean Anomaly.

	0s	Is	IIs	IIIs	IVs	V	
°	′ ″	′ ″	′ ″	′ ″	′ ″	′ ″	°
0	2 33	2 32	2 30	2 28	2 25	2 24	30
10	2 33	2 32	2 29	2 27	2 25	2 23	20
20	2 33	2 31	2 29	2 26	2 24	2 23	10
30	2 32	2 30	2 28	2 25	2 24	2 23	0
	XIs	Xs	IXs	VIIIs	VIIs	VIs	

SUN'S SEMIDIAMETER.

ARGUMENT.—Sun's Mean Anomaly.

	0s	Is	IIs	IIIs	IVs	Vs	
°	′ ″	′ ″	′ ′	′ ″	′ ″	′ ″	°
0	16 18	16 15	16 9	16 1	15 53	15 48	30
10	16 18	16 14	16 7	15 58	15 51	15 46	20
20	16 17	16 12	16 4	15 56	15 49	15 46	10
30	16 15	16 9	16 1	15 53	15 48	15 45	0
	XIs	Xs	IXs	VIIIs	VIIs	VIs	

SUN'S MOTIONS FOR DAYS AND HOURS.

Days.	Logitude.			Per.	I.	II.	III.	N.	Hours.	Long.		I.
	°	′	″	″						′	″	
1	0	0	0	0	0	0	0	0	1	2	28	1
2	0	59	8	0	34	3	2	0	2	4	56	3
3	1	58	17	0	68	5	3	0	3	7	23	4
4	2	57	25	0	101	8	5	0	4	9	51	6
5	3	56	33	1	135	10	7	1	5	12	19	7
6	4	55	42	1	169	13	9	1	6	14	47	8
7	5	54	50	1	203	15	10	1	7	17	15	10
8	6	53	58	1	236	18	12	1	8	19	43	11
9	7	53	7	1	270	20	14	1	9	22	11	13
10	8	52	15	1	304	23	15	1	10	24	38	14
11	9	51	23	2	338	25	17	1	11	27	6	16
12	10	50	32	2	371	28	19	2	12	29	34	17
13	11	49	40	2	405	30	21	2	13	32	2	18
14	12	48	48	2	439	33	22	2	14	34	30	20
15	13	47	57	2	473	35	24	2	15	36	58	21
16	14	47	5	3	506	38	26	2	16	39	26	23
17	15	46	13	3	540	40	27	2	17	41	53	24
18	16	45	22	3	574	43	29	2	18	44	21	25
19	17	44	30	3	608	45	31	3	19	46	49	27
20	18	43	38	3	641	48	33	3	20	49	17	28
21	19	42	47	3	675	50	34	3	21	51	45	30
22	20	41	55	4	709	53	36	3	22	54	13	31
23	21	41	3	4	743	55	38	3	23	56	40	32
24	22	40	12	4	777	58	39	3	24	59	8	34
25	23	39	20	4	810	60	41	4				
26	24	38	28	4	844	63	43	4				
27	25	37	37	4	878	65	45	4				
28	26	36	45	5	912	68	46	4				
29	27	35	53	5	945	70	48	4				
30	28	35	2	5	979	73	50	4				
31	29	34	10	5	13	75	51	4				

SUN'S MOTIONS FOR MINUTES.

Min.	Longitude.		Min.	Longitude.	
	′	″		′	″
1	0	2	30	1	16
5	0	12	35	1	26
10	0	25	40	1	39
15	0	37	45	1	51
20	0	49	50	2	3
25	1	2	55	2	16
30	1	14	60	2	28

EQUATIONS OF THE SUN'S CENTER.

ARGUMENT.—Sun's Mean Anomaly.

	0s			Is			IIs			IIIs			IVs			Vs		
°	°	′	″	°	′	″	°	′	″	°	′	″	°	′	″	°	′	″
0	1	59	30	2	58	15	3	40	27	3	54	50	3	38	21	2	56	9
1	2	1	33	3	0	0	3	41	25	3	54	47	3	37	18	2	54	25
2	2	3	37	3	1	44	3	42	21	3	54	41	3	36	14	2	52	40
3	2	5	40	3	3	27	3	43	15	3	54	33	3	35	8	2	50	54
4	2	7	43	3	5	9	3	44	8	3	54	23	3	34	1	2	49	8
5	2	9	46	3	6	49	3	44	58	3	54	11	3	32	51	2	47	20
6	2	11	49	3	8	28	3	45	47	3	53	57	3	31	41	2	45	32
7	2	13	51	3	10	6	3	46	33	3	53	41	3	30	28	2	43	43
8	2	15	54	3	11	43	3	47	17	3	53	23	3	29	14	2	41	53
9	2	17	56	3	13	18	3	48	0	3	53	3	3	27	58	2	40	3
10	2	19	57	3	14	51	3	48	40	3	52	40	3	26	41	2	38	11
11	2	21	58	3	16	24	3	49	18	3	52	16	3	25	22	2	36	19
12	2	23	59	3	17	54	3	49	55	3	51	50	3	24	2	2	34	27
13	2	25	59	2	19	24	3	50	29	3	51	21	3	22	40	2	32	34
14	2	27	59	3	20	51	3	51	1	3	50	51	3	21	17	2	30	40
15	2	29	58	3	22	18	3	51	31	3	50	18	3	19	52	2	28	46
16	2	31	57	3	23	42	3	51	59	3	49	44	3	18	26	2	26	52
17	2	33	55	3	25	5	3	52	25	3	49	7	3	16	58	2	24	56
18	2	35	52	3	26	26	3	52	49	3	48	29	3	15	30	2	23	0
19	2	37	49	3	27	46	3	53	10	3	47	49	3	14	0	2	21	4
20	2	39	45	3	29	4	3	53	30	3	47	7	3	12	28	2	19	8
21	2	41	40	3	30	24	3	53	47	3	46	22	3	10	55	2	17	11
22	2	43	34	3	31	35	3	54	3	3	45	36	3	9	22	2	15	14
23	2	45	28	3	32	48	3	54	16	3	44	48	3	7	46	2	13	16
24	2	47	20	3	33	59	3	54	27	3	43	58	3	6	10	2	11	19
25	2	49	12	3	35	8	3	54	36	3	43	7	3	4	33	2	9	21
26	2	51	2	3	36	16	3	54	43	3	42	13	3	2	54	2	7	23
27	2	52	52	3	37	21	3	54	48	3	41	18	3	1	14	2	5	25
28	2	54	41	3	38	25	3	54	51	3	40	21	2	59	33	2	3	27
29	2	56	28	3	39	27	3	54	52	3	39	22	2	57	52	2	1	28
30	2	58	15	3	40	27	3	54	50	3	38	21	2	56	9	1	59	30

EQUATIONS OF THE SUN'S CENTER.

ARGUMENT.—Sun's Mean Anomaly.

	VIs			VIIs			VIIIs			IXs			Xs			XIs		
°	°	′	″	°	′	″	°	′	″	°	′	″	°	′	″	°	′	″
0	1	59	30	1	2	51	0	20	39	0	4	10	0	18	33	1	0	45
1	1	57	32	1	1	8	0	19	38	0	4	8	0	19	33	1	2	32
2	1	55	33	0	59	27	0	18	39	0	4	9	0	20	35	1	4	19
3	1	53	35	0	57	46	0	17	42	0	4	12	0	21	39	1	6	8
4	1	51	37	0	56	6	0	16	47	0	4	17	0	22	44	1	7	58
5	1	49	39	0	54	27	0	15	53	0	4	24	0	23	52	1	9	48
6	1	47	41	0	52	47	0	15	2	0	4	33	0	25	1	1	11	40
7	1	45	44	0	51	14	0	14	12	0	4	44	0	26	12	1	13	32
8	1	43	46	0	49	38	0	13	24	0	4	57	0	27	25	1	15	26
9	1	41	49	0	48	5	0	12	38	0	5	13	0	28	40	1	17	20
10	1	39	52	0	46	32	0	11	53	0	5	30	0	29	56	1	19	15
11	1	37	56	0	45	0	0	11	11	0	5	50	0	31	14	1	21	11
12	1	36	0	0	43	30	0	10	31	0	6	11	0	32	34	1	23	8
13	1	34	4	0	42	1	0	9	53	0	6	35	0	33	55	1	25	5
14	1	32	9	0	40	34	0	9	16	0	7	1	0	35	18	1	27	3
15	1	30	14	0	39	8	0	8	42	0	7	29	0	36	42	1	29	2
16	1	28	20	0	37	43	0	8	9	0	7	59	0	38	9	1	31	1
17	1	26	26	0	36	20	0	7	39	0	8	31	0	39	36	1	33	1
18	1	24	33	0	34	58	0	7	10	0	9	5	0	41	9	1	35	1
19	1	22	41	0	33	38	0	6	44	0	9	42	0	42	36	1	37	1
20	1	20	49	0	32	19	0	6	20	0	10	20	0	44	9	1	39	3
21	1	18	57	0	31	2	0	5	57	0	11	0	0	45	42	1	41	4
22	1	17	7	0	29	46	0	5	37	0	11	43	0	47	17	1	43	6
23	1	15	17	0	28	32	0	5	19	0	12	27	0	48	54	1	45	9
24	1	13	28	0	27	19	0	5	3	0	13	13	0	50	32	1	47	11
25	1	11	40	0	26	9	0	4	49	0	14	2	0	52	11	1	49	14
26	1	9	52	0	24	59	0	4	37	0	14	52	0	53	51	1	51	17
27	1	8	6	0	23	52	0	4	27	0	15	45	0	55	33	1	53	20
28	1	6	20	0	22	46	0	4	19	0	16	39	0	57	16	1	55	23
29	1	4	35	0	21	41	0	4	13	0	17	35	0	59	0	1	57	27
30	1	2	51	0	20	39	0	4	10	0	18	33	1	0	45	1	59	30

TABLE IX.

SMALL EQUATIONS OF THE SUN'S LONGITUDE.

Arg.	I	II.	III.	Arg.	I.	II.	III.
	″	″	″		″	″	″
0	10	10	10	500	10	10	10
10	10	11	9	510	10	10	9
20	11	11	9	520	9	10	8
30	11	12	8	530	9	10	7
40	11	13	8	540	9	10	7
40	12	14	7	550	8	10	6
60	12	14	7	560	8	9	5
70	12	15	7	570	8	9	4
80	13	15	7	580	7	9	3
90	13	16	7	590	7	9	3
100	13	16	7	600	7	9	2
110	14	17	7	610	6	8	1
120	14	17	7	620	6	8	1
130	14	18	8	630	6	8	1
140	15	18	8	640	5	7	0
150	15	18	9	650	5	7	0
160	15	18	9	660	5	6	0
170	15	18	10	670	5	6	1
180	15	18	10	680	5	6	1
190	16	18	11	690	4	5	2
200	16	18	11	700	4	5	2
210	16	18	12	710	4	4	3
220	16	18	12	720	4	4	3
230	16	18	13	730	4	4	4
240	16	17	14	740	4	3	5
250	16	17	14	750	4	3	6
260	16	17	15	760	4	3	6
270	16	16	16	770	4	2	7
480	16	16	17	780	4	2	8
290	16	16	17	790	4	2	8
300	16	15	18	800	4	2	9
310	16	15	18	810	4	2	9
320	15	14	19	820	5	2	10
330	15	14	19	830	5	2	10
340	15	14	20	840	5	2	11
350	15	13	20	850	5	2	11
360	15	13	20	860	5	2	12
370	14	12	19	870	6	2	12
380	14	12	19	880	6	3	13
390	14	12	19	890	6	3	13
400	13	11	18	900	7	4	13
410	13	11	17	910	7	4	13
420	13	11	17	920	7	5	13
430	12	11	16	930	8	5	13
440	12	11	15	940	8	6	13
450	12	10	14	950	8	6	13
460	11	10	13	960	9	7	12
470	11	10	13	970	9	8	12
480	11	10	12	980	9	9	11
490	10	10	11	990	10	9	11
500	10	10	10	1000	10	10	10

TABLE X.

NUTATIONS.

ARGUMENT.—Supplement of the Node, or N.

N.	Long.	R. Asc.	Obliq.	N.	Long.	R. Asc.	Obliq.
	"	"	"		"	"	"
0	+ 0	+ 0	+ 10	500	— 0	— 0	— 10
20	2	2	10	520	2	2	9
40	4	4	9	540	4	4	9
60	7	6	9	560	7	6	9
80	9	8	8	580	9	8	8
100	+ 11	+ 10	+ 8	600	— 11	— 10	— 8
120	12	11	7	620	12	11	7
140	14	13	6	640	14	13	6
160	15	14	5	660	15	14	5
180	16	15	4	680	16	15	4
200	+ 17	+ 16	+ 3	700	— 17	— 16	— 3
220	18	16	2	720	18	16	2
240	18	16	1	740	18	16	1
260	18	16	— 1	760	18	16	+ 1
280	18	16	2	780	18	16	2
300	+ 17	+ 16	— 3	800	— 17	— 16	+ 3
320	16	15	4	820	16	15	4
340	15	14	5	840	15	14	5
360	14	13	6	860	14	13	6
380	12	11	7	880	12	11	7
400	+ 11	+ 10	— 8	900	— 11	— 10	+ 8
420	9	8	8	920	9	8	8
440	7	6	9	940	7	6	9
460	4	4	9	960	4	4	9
480	2	2	10	980	2	2	10
500	+ 0	+ 0	— 10	1000	— 0	— 0	+ 10

TABLE XI.

EARTH'S RADIUS VECTOR.—ARGUMENT. Sun's Mean Anomaly.

	0s	Is	IIs	IIIs	IVs	Vs	
0°	0.98313	0.98545	0.99173	1.00018	1.00850	1.01450	30°
2	0.98314	0.98576	0.99225	1.00077	1.00899	1.01477	28
4	0.98317	0.98608	0.99278	1.00135	1.00947	1.01503	26
6	0.98322	0.98643	0.99331	1.10193	1.00994	1.01527	24
8	0.98330	0.98679	0.99386	1.00251	1.01040	1.01549	22
10	0.98339	0.98717	0.99441	1.00308	1.01084	1.01569	20
12	0.98350	0.98756	0.99497	1.00366	1.01128	1.01588	18
14	0.98364	0.98797	0.99554	1.00422	1.01170	1.01604	16
16	0.98380	0.98840	0.99611	1.00478	1.01210	1.01619	14
18	0.98397	0.98883	0.99668	1.00534	1.01249	1.01632	12
20	0.98417	0.98929	0.99726	1.00588	1.01286	1.01643	10
22	0.98439	0.98975	0.99784	1.00642	1.01322	1.01652	8
24	0.98462	0.99023	0.99843	1.00695	1.01357	1.01659	6
26	0.98486	0.99072	0.99901	1.00748	1.01389	1.01663	4
28	0.98515	0.99122	0.99960	1.00799	1.01420	1.01666	2
30	0.98545	0.99173	1.00018	1.00850	1.01450	1.01667	0
	XIs	Xs	IXs	VIIIs	VIIs	VIs	

TABLE XI.

MEAN NEW MOONS AND ARGUMENTS IN JANUARY.

	Mean New Moon in January.	I.	II.	III.	IV.	N.
A. D.	D. H. M.					
1836 B.	17 10 32	0469	1246	17	08	669
1837	5 19 20	0171	9852	00	97	692
1838	24 16 53	0681	9175	99	85	799
1839	14 1 42	0383	7780	82	74	822
1840 B.	3 10 30	0085	6386	65	63	844
1841	21 8 3	0595	5709	63	51	951
1842	10 16 51	0297	4314	46	40	974
1843	29 14 24	0807	3637	44	28	081
1844 B.	18 23 13	0509	2243	28	17	104
1845	7 8 1	0211	0848	11	06	126
1846	26 5 34	0721	0171	09	94	234
1847	15 14 22	0423	8777	92	84	256
1848 B.	4 23 11	0125	7382	75	73	278
1849	22 20 43	0635	6705	73	61	386
1850	12 5 32	0337	5311	56	50	408
1851	1 14 21	0038	3916	40	39	431
1852 B.	20 11 53	0549	3239	38	27	538
1853	8 20 42	0251	1845	21	16	560
1854	27 18 14	0761	1168	19	04	668
1855	17 3 3	0463	9773	02	93	690
1856 B.	6 11 51	0164	8379	85	82	713
1857	24 9 24	0675	7702	84	70	820
1858	13 18 13	0376	6307	67	59	843
1859	3 3 1	0078	4913	50	48	865
1860 B.	22 0 34	0588	4236	48	36	972
1861	10 9 22	0290	2840	31	25	995
1862	29 6 55	0800	2163	30	14	102
1863	18 15 44	0504	0769	13	03	125
1864 B.	8 0 32	0204	9374	96	92	147
1865	25 22 5	0714	8698	94	80	256
1866	15 6 53	0416	7303	77	69	277
1867	4 15 42	0118	5909	60	58	299
1868 B.	23 13 14	0628	5231	59	46	407
1869	11 22 3	0330	3837	42	35	429
1870	1 6 51	0032	2442	25	24	451
1871	20 4 24	0542	1765	23	12	559
1872 B.	8 13 13	0244	0371	05	01	581
1873	27 10 46	0754	9694	03	89	689
1874	[illegible] 19 35	0456	8300	86	78	711
1875	[illegible] 4 24	0158	6906	69	67	733
1876 B.	26 1 57	0668	6229	67	55	841
1877	14 10 45	0370	4835	50	44	863
1878	3 18 38	0072	3441	33	23	885
1879	22 6 11	0582	2764	31	21	993
1880 B.	11 15 0	0284	1370	14	10	015

TABLE XII.

MEAN LUNATIONS AND CHANGES OF THE ARGUMENTS.

Num.	Lunations.			I.	II.	III.	IV.	N.
	d.	h.	m.					
½	14	18	22	404	5359	58	50	43
1	29	12	44	808	717	15	99	85
2	59	1	28	1617	1434	31	98	170
3	88	14	12	2425	2151	46	97	256
4	118	2	56	3234	2869	61	96	341
5	147	15	40	4042	3586	76	95	425
6	177	4	24	4851	4303	92	95	511
7	206	17	8	5659	5020	7	94	596
8	236	5	52	6468	5737	22	93	682
9	265	18	36	7276	6454	37	92	767
10	295	7	20	8085	7117	53	91	852
11	324	20	5	8893	7889	68	90	937
12	354	8	49	9702	8606	83	89	22
13	383	21	33	510	9323	98	88	108

TABLE XIII.

NUMBER OF DAYS FROM THE COMMENCEMENT OF THE YEAR TO THE FIRST OF EACH MONTH.

Months.	Com.	Bis.
	Days.	Days.
January...	0	0
February..	31	31
March	59	60
April	90	91
May......	120	121
June......	151	152
July......	181	182
August....	212	213
September.	243	244
October...	273	274
November.	304	305
December.	334	335

TABLE XIV.

Arg. II.	☾ H. Par.		☾ S. D.		☾ H. Mo.		Arg. II.
	′	″			′	″	
0	60	29	16	29	36	48	10000
250	60	26	16	26	36	44	9750
500	60	17	16	25	36	19	9500
750	60	0	16	21	36	8	9250
1000	59	47	16	17	35	48	9000
1250	59	24	16	11	35	28	8750
1500	58	56	16	3	34	57	8500
1750	58	30	15	56	34	34	8250
2000	58	7	15	50	33	58	8000
2250	57	37	15	42	33	32	7750
2500	57	1	15	31	32	42	7500
2750	56	32	15	23	32	9	7250
3000	56	2	15	16	31	36	7000
3250	55	40	15	10	31	13	6750
3500	55	22	15	7	30	52	6500
3750	55	12	15	3	30	29	6250
4000	54	51	14	56	30	7	6000
4250	54	39	14	54	29	55	5750
4500	54	26	14	50	29	43	5500
4750	54	18	14	48	29	37	5250
5000	54	13	14	45	29	35	5000

EQUATIONS FOR NEW AND FULL MOON.

Arg.	I.	II.	Arg.	I.	II.	Arg.	III.	IV.	Arg.
	h. m.	h. m.		h. m.	h. m.		m.	m.	
0	4 20	10 10	5000	4 20	10 10	25	3	31	25
100	4 36	9 36	5100	4 5	10 50	26	3	31	24
200	4 52	9 2	5200	3 49	11 30	27	3	30	23
300	5 8	8 28	5300	3 34	12 9	28	3	30	22
400	5 24	7 55	5400	3 19	12 48	29	3	30	21
500	5 40	7 22	5500	3 4	13 26	30	3	30	20
600	5 55	6 49	5600	2 49	14 3	31	3	30	19
700	6 10	6 17	5700	2 35	14 39	32	4	20	18
800	6 24	5 46	5800	2 21	15 13	33	4	29	17
900	6 38	5 15	5900	2 8	15 46	34	4	29	16
1000	6 51	4 46	6000	1 55	16 18	35	4	29	15
1100	7 4	4 17	6100	1 42	16 48	36	5	28	14
1200	7 15	3 50	6200	1 31	17 16	37	5	28	13
1300	7 27	3 24	6300	1 19	17 42	38	5	27	12
1400	7 37	2 59	6400	1 9	18 6	39	5	27	11
1500	7 47	2 35	6500	0 59	18 28	40	6	26	10
1600	7 55	2 14	6600	0 50	18 48	41	6	26	9
1700	8 3	1 53	6700	0 42	19 6	42	7	25	8
1800	8 10	1 35	6800	0 34	19 21	43	7	25	7
1900	8 16	1 18	6900	0 28	19 33	44	7	24	6
2000	8 21	1 3	7000	0 22	19 44	45	8	23	5
2100	8 25	0 51	7100	0 17	19 52	46	8	23	4
2200	8 29	0 40	7200	0 14	19 57	47	9	22	3
2300	8 31	0 32	7300	0 11	20 0	48	9	21	2
2400	8 32	0 25	7400	0 9	20 1	49	10	21	1
2500	8 32	0 21	7500	0 8	19 59	50	10	20	0
2600	8 31	0 19	7600	0 8	19 55	51	10	19	99
2700	8 29	0 20	7700	0 9	19 48	52	11	19	98
2800	8 26	0 23	7800	0 11	19 40	53	11	18	97
2900	8 23	0 28	7900	0 15	19 29	54	12	17	96
3000	8 18	0 36	8000	0 19	19 17	55	12	17	95
3100	8 12	0 47	8100	0 24	19 2	56	13	16	94
3200	8 6	0 59	8200	0 30	18 45	57	13	15	93
3300	7 58	1 14	8300	0 37	18 27	58	13	15	92
3400	7 50	1 32	8400	0 45	18 6	59	14	14	91
3500	7 41	1 52	8500	0 53	17 45	60	14	14	90
3600	7 31	2 14	8600	1 3	17 21	61	15	13	89
3700	7 21	2 38	8700	1 13	16 56	62	15	13	88
3800	7 9	3 4	8800	1 25	16 30	63	15	12	87
3900	6 58	3 32	8900	1 36	16 3	64	15	12	86
4000	6 45	4 2	9000	1 49	15 34	65	16	11	85
4100	6 32	4 34	9100	2 2	15 5	66	16	11	84
4200	6 19	5 7	9200	2 16	14 34	67	16	11	83
4300	6 5	5 41	9300	2 30	14 3	68	16	10	82
4400	5 51	6 17	9400	2 45	13 31	69	17	10	81
4500	5 36	6 54	9500	3 0	12 58	70	17	10	80
4600	5 21	7 32	9600	3 16	12 25	71	17	10	79
4700	5 6	8 11	9700	3 32	11 52	72	17	10	78
4800	4 51	8 50	9800	3 48	11 18	73	17	10	77
4900	4 35	9 30	9900	4 4	10 44	74	17	9	76
5000	4 20	10 10	10000	4 20	10 10	75	17	9	75

TABLE E, SHOWING THE EQUATION OF TIME *

ARGUMENT.—Sun's true Longitude.

	0s	Is	IIs	IIIs	IVs	Vs	VIs	VIIs	VIIIs	IXs	Xs	XIs
	m. s.	m. s.	m. s.	m. s.	m. s.	m. s.	m. s.	m. s.	m. s.	m. s.	m. s.	m. s.
0	+7 31.2	—1 7.3	—3 43.5	+1 15.0	+6 8.2	+2 27.7	— 7 33.4	—15 33.3	—13 39.4	— 1 23.6	+11 19.0	+14 9.8
2	6 56.0	1 32.4	3 35.1	1 50.7	6 9.7	1 57.8	8 16.8	15 48.7	13 6.1	— 0 21.1	11 51.7	13 58.2
4	6 19.8	1 53.6	3 25.0	2 16.4	6 10.0	1 19.9	8 58.3	16 0.4	12 29.3	+ 0 35.9	12 22.3	13 42.5
6	5 43.8	2 16.8	3 12.6	2 42.2	6 8.0	0 44.0	9 38.1	16 9.0	11 51.0	1 34.3	12 49.0	13 25.1
8	5 04.4	2 35.0	2 57.1	3 10.0	6 3.5	+0 6.6	10 18.0	16 14.3	11 9.1	2 31.2	13 12.8	13 4.8
10	4 25.6	2 55.5	2 38.0	3 23.7	5 58.1	—0 30.7	10 55.7	16 16.5	10 24.4	3 30.0	13 34.0	12 45.0
12	3 49.4	3 9.5	2 20.4	3 56.8	5 46.2	1 12.4	11 32·4	16 15.3	9 37.0	4 25.8	13 50.3	12 19.1
14	3 13.2	3 22.0	2 0.0	4 18.3	5 33.2	1 42.3	12 7.0	16 11.4	8 48.0	5 20.8	14 6.6	11 52.8
16	2 39.4	3 32.8	1 37.7	4 38.1	5 13.3	2 32.0	12 40.6	16 3.8	7 57.0	6 11.9	14 17.5	11 22.8
18	2 2.4	3 44.1	1 14.8	4 57.9	5 0.2	3 15.9	13 12.1	15 52.9	7 2.9	7 2.6	14 25.0	10 52.4
20	1 25.6	3 50.6	0 49.0	5 15.7	4 41.2	4 04.0	13 42.2	15 40.7	6 9.4	7 50.3	14 30.6	10 23.7
22	0 53.0	3 54.2	0 24.1	5 30.4	4 17.1	4 42.9	14 8.9	15 11.0	5 11.1	8 37.5	14 32.2	9 52.2
24	+0 21.4	3 54.9	—0 22.0	5 42.7	3 53.0	5 25.8	14 34.1	15 0.6	4 15.3	9 21.5	14 30.9	9 18.1
26	—0 8.6	3 53.3	+0 28.6	5 54.1	3 26.6	6 9.8	14 56.4	14 36.2	3 17.9	10 3.2	14 27.4	8 42.4
28	0 38.7	3 49.4	0 55.9	6 1.1	2 58.3	6 50.5	15 20.7	14 9.1	2 20.2	10 43.1	14 20.5	8 7.7
30	1 7.3	3 43.5	1 15.0	6 8.2	2 27.7	7 33.4	15 33.3	13 39.4	1 23.6	11 19.0	14 9.8	7 31.2

* The Sun gives apparent time; to convert this into mean time apply the minutes and seconds, as found in this Table; the MINUS sign indicates subtraction, the plus sign, addition.

TABLE XVI.

MOON'S EPOCHS.

Years.	1	2	3	4	5	6	7	8	9
1846	0013	2475	3275	1688	0773	4880	3179	0800	9542
1847	0006	9683	2941	6432	3245	0678	4239	3257	8406
1848 B.	0026	7542	3646	1463	6052	6847	5358	6106	7295
1849	0019	4750	3312	6207	8524	2644	6418	8563	6158
1850	0012	1958	2978	0951	0995	8442	7479	1020	5022
1851	0005	9167	2644	5695	3467	4239	8539	3477	3885
1852 B.	0025	7025	3350	0726	6274	0408	9658	6326	2774
1853	0018	4233	3016	5469	8746	6206	0718	8782	1637
1854	0011	1442	2681	0213	1217	2003	1778	1240	0501
1855	0004	8650	2347	4957	3689	7801	2839	3697	9365
1856 B.	0024	6509	3053	9988	6496	3970	3957	6446	8254
1857	0017	3717	2719	4732	8968	9767	5018	9002	7117
1858	0010	0925	2385	9476	1439	5565	6078	1460	5981
1859	0003	8134	2051	4220	3911	1362	7139	3917	4845
1860 B.	0023	5992	2756	9551	6718	7531	8257	6765	3734
1861	0016	3200	2423	3995	9190	3329	9317	9222	2597
1862	0009	0409	2088	8739	1661	9126	0378	1679	1461
1863	0002	7617	1754	3483	4133	4923	1438	4137	0324
1864 B.	0022	5476	2460	8514	6941	1093	2557	6984	9212
1865	0015	2684	2126	3257	9412	6890	3617	9442	8076
1866	0008	9893	1792	8001	1883	2687	4678	1899	6940
1867	0001	7101	1457	2745	4355	8485	5738	4357	5804
1868 B.	0021	4959	2163	7776	7163	4654	6857	7204	4692
1869	0014	2168	1829	2520	9634	0452	7917	9662	3556
1870	0007	9376	1495	7264	2105	6249	8978	2119	2420
1871	0000	6584	1161	2008	4576	2046	0039	4576	1284
1872 B.	0020	4432	1867	7039	7383	8215	1158	7423	0172
1873	0013	1640	1533	1783	9854	4012	2239	9880	9036
1874	0006	8848	1199	6527	2325	9809	3300	2337	7900
1875	9999	6056	0865	1271	4796	5606	4361	4794	6764
1876 B.	0019	3914	1571	6292	7603	1775	5480	7641	5652
1877	0012	1122	1247	1036	0074	7572	6541	0098	4516
1878	0005	8330	0913	5780	2545	3369	7602	2555	3380
1879	9998	5538	0579	0524	5016	9166	8663	5012	2244
1880 B.	0018	3396	1285	5545	7823	5335	9782	7859	1132
1881	0011	0604	0951	0289	0294	1132	0843	0316	9996
1882	0004	7812	0617	5033	2765	6929	1904	2873	8860
1883	9997	5020	0283	9777	5236	2726	2965	5330	7724
1884 B.	0017	2878	0989	4798	8043	8895	4084	8177	6612
1885	0010	0086	0655	9542	0514	4692	5145	0634	5476
1886	0003	7294	0321	4286	2985	0489	6206	3091	4340
1887	9996	4502	9987	9030	5456	6286	7267	5548	3204
1888 B.	0016	2360	0693	4051	8263	2455	8386	8395	2092
1889	0009	9568	0359	8795	0734	8252	9447	0852	0956
1890	0002	6776	0025	3539	3205	4049	0508	3309	9820

MOON'S EPOCHS.

Years.	10	11	12	13	14	15	16	17	18	19	20
1846	203	123	250	171	419	760	126	396	167	379	204
1847	810	484	970	644	613	901	486	749	643	433	371
1848 B.	486	876	759	151	905	072	881	143	144	487	539
1849	093	237	479	624	099	212	241	496	619	540	705
1850	700	597	199	097	293	352	600	848	094	594	871
1851	306	958	918	570	487	493	960	201	569	648	038
1852 B.	983	350	707	077	780	664	355	595	070	701	206
1853	589	711	427	550	974	804	715	948	545	755	372
1854	196	072	147	023	168	944	074	300	020	809	539
1855	802	432	866	496	361	085	434	653	495	863	705
1856 B.	479	824	656	003	654	256	829	047	996	916	873
1857	086	185	375	476	848	396	189	400	471	970	039
1858	692	546	095	949	042	537	548	752	947	024	206
1859	299	907	814	422	236	677	908	105	422	078	372
1860 B.	975	298	604	929	529	848	303	499	923	131	540
1861	581	659	323	402	723	988	662	852	398	185	706
1862	187	020	042	875	916	129	021	204	873	239	873
1863	794	381	761	348	110	269	381	557	348	292	039
1864 B.	470	773	551	855	403	440	777	951	849	346	207
1865	077	134	271	328	597	580	136	304	324	400	373
1866	684	494	990	801	791	721	495	657	799	453	540
1867	290	855	710	274	985	861	855	009	274	507	707
1868 B.	967	247	500	781	277	032	251	404	775	561	874
1869	573	608	219	254	471	172	610	756	251	615	040
1870	180	968	939	737	665	313	969	109	726	668	207
1871	787	328	659	200	859	554	328	562	201	721	374
1872 B.	464	720	549	707	151	725	724	957	702	785	531
1873	071	080	269	180	345	966	083	410	177	838	698
1874	678	440	989	653	539	205	442	863	642	891	865
1875	285	800	709	126	733	446	801	316	117	944	032
1876 B.	962	192	599	633	025	617	197	711	618	008	199
1877	569	552	319	106	219	858	556	164	093	061	366
1878	176	912	039	579	413	099	915	617	568	114	533
1879	783	272	759	052	607	340	274	070	043	167	700
1880 B.	460	664	649	559	899	511	670	465	544	231	867
1881	067	024	369	032	093	752	029	918	019	284	034
1882	674	384	089	505	287	993	388	371	494	337	201
1883	281	744	809	978	481	234	747	824	969	390	368
1884 B.	958	136	699	485	773	405	143	219	470	454	535
1885	565	496	419	958	967	646	502	672	945	507	702
1886	172	856	139	431	161	887	861	125	420	560	869
1887	779	216	859	904	355	128	320	578	895	613	036
1888 B.	456	608	749	411	647	299	716	973	396	677	203
1889	063	968	469	884	841	540	075	426	871	730	370
1890	670	328	189	357	035	781	434	879	346	783	537

TABLE XVII.

MOON'S MOTIONS FOR MONTHS.

Months.	1	2	3	4	5	6	7	8	9
Jan. Com.	0000	0000	0000	0000	0000	0000	0000	0000	0000
Jan. Bis.	9973	9350	8960	9713	9664	9628	9942	9610	9976
Feb. Com.	849	146	2246	8896	402	1533	1789	2099	753
Feb. Bis.	821	9497	1205	8609	66	1161	1731	1709	729
March......	1615	8343	1371	6931	9797	1951	3404	3027	1433
April......	2464	8490	3616	5827	199	3484	5193	5126	2186
May	3285	7986	4822	4436	265	4646	6924	6835	2914
June	4134	8133	7067	3332	666	6179	8713	8934	3667
July.......	4955	7629	8273	1942	732	7341	444	643	4396
August.....	5804	7776	518	838	1134	8874	2233	2742	5148
September .	6653	7922	2764	9734	1536	408	4021	4842	5901
October.....	7474	7419	3969	8343	1602	1569	5752	6550	6630
November..	8323	7565	6215	7239	2004	3102	7541	8649	7382
December..	9144	7062	7420	5848	2070	4264	9272	358	8111

TABLE XVIII.

MOON'S MOTIONS FOR DAYS.

Days.	1	2	3	4	5	6	7	8	9
1	0000	0000	0000	0000	0000	0000	0000	0000	0000
2	27	650	1040	287	336	372	58	390	24
3	55	1300	2080	574	671	744	115	781	49
4	82	1950	3121	861	1007	1116	173	1171	73
5	109	2600	4161	1148	1342	1488	231	1561	97
6	137	3249	5201	1435	1678	1860	289	1952	121
7	164	3899	6241	1722	2013	2232	346	2342	146
8	192	4549	7281	2009	2349	2604	404	2732	170
9	219	5199	8321	2296	2684	2976	462	3122	194
10	246	5849	9362	2583	3020	3348	519	3513	219
11	274	6499	402	2870	3355	3720	577	3903	243
12	301	7149	1442	3157	3691	4093	635	4293	267
13	328	7799	2482	3444	4026	4465	692	4684	291
14	356	8449	3522	3731	4362	4837	750	5074	316
15	383	9098	4563	4018	4698	5209	808	5464	340
16	411	9748	5603	4305	5033	5581	866	5854	364
17	438	398	6643	4592	5369	5953	923	6245	389
18	465	1048	7683	4878	5704	6325	981	6635	413
19	493	1698	8723	5165	6040	6697	1039	7025	437
20	520	2348	9763	5452	6375	7069	1096	7416	461
21	548	2998	804	5739	6711	7441	1154	7806	486
22	575	3648	1844	6026	7046	7813	1212	8196	510
23	602	4298	2884	6313	7382	8185	1269	8586	534
24	630	4947	3924	6600	7717	8557	1327	8977	559
25	657	5597	4964	6887	8053	8929	1385	9367	583
26	684	6247	6005	7174	8389	9301	1443	9757	607
27	712	6897	7045	7461	8724	9673	1500	148	631
28	739	7547	8085	7748	9060	45	1558	538	656
29	767	8197	9125	8035	9395	417	1616	928	680
30	794	8847	165	8322	9731	789	1673	1319	704
31	821	9497	1205	8609	66	1161	1731	1709	729

TABLE XVII.

MOON'S MOTIONS FOR MONTHS.

Months.	10	11	12	13	14	15	16	17	18	19	20
Jan. Com.	000	000	000	000	000	000	000	000	000	000	000
Jan. Bis.	930	969	930	966	901	969	963	958	974	000	000
Feb. Com.	175	965	184	59	74	946	135	304	805	5	14
Feb. Bis.	105	934	114	25	975	916	98	262	779	5	14
March.....	139	836	157	16	851	801	159	482	532	9	27
April......	314	801	342	76	925	747	294	786	336	13	41
May	419	735	556	101	899	663	392	47	115	18	55
June	593	700	640	160	973	609	527	351	920	22	69
July.......	698	634	754	185	948	525	625	613	699	27	83
August....	873	599	938	245	22	471	759	917	503	31	97
September .	48	563	123	304	96	417	894	221	308	36	111
October....	152	497	237	329	71	333	992	483	87	40	125
November..	327	462	421	388	145	279	127	787	892	45	139
December..	432	396	535	414	120	194	225	49	670	49	153

TABLE XVIII.

MOON'S MOTIONS FOR DAYS.

Days.	10	11	12	13	14	15	16	17	18	19	20
1	000	000	000	000	000	000	000	000	000	000	000
2	70	31	70	34	99	31	37	42	26	0	0
3	140	62	141	68	198	61	73	84	52	0	1
4	210	93	211	103	297	92	110	126	78	0	1
5	281	125	282	137	397	122	146	168	104	1	2
6	351	156	352	171	496	153	183	210	130	1	2
7	421	187	423	205	595	183	220	252	156	1	3
8	491	218	493	239	694	214	256	294	182	1	3
9	561	249	564	273	793	244	293	336	208	1	4
10	631	280	634	308	892	275	329	379	234	1	4
11	702	311	705	342	992	305	366	421	260	1	5
12	772	342	775	376	91	336	403	463	286	2	5
13	842	374	845	410	190	366	439	505	312	2	5
14	912	405	916	444	289	397	476	547	337	2	6
15	982	436	986	478	388	427	512	589	368	2	6
16	52	467	57	513	487	458	549	631	389	2	7
17	122	498	127	547	587	488	586	673	415	2	7
18	193	529	198	581	686	519	622	715	441	2	8
19	263	560	268	615	785	549	659	757	467	3	8
20	333	591	339	649	884	580	695	799	493	3	9
21	403	623	409	683	983	611	722	841	517	3	9
22	473	654	480	718	82	641	769	883	545	3	10
23	543	685	550	752	182	672	805	925	571	3	10
24	614	716	621	786	281	702	842	967	597	3	11
25	684	747	691	820	380	733	878	9	623	4	11
26	754	778	762	854	479	763	915	52	649	4	11
27	824	809	832	888	578	794	952	94	675	4	12
28	894	840	903	923	677	824	988	136	701	4	12
29	964	872	973	957	777	855	25	178	727	4	13
30	34	903	43	991	876	885	61	220	753	4	13
31	105	934	114	25	975	916	98	262	779	4	14

TABLE XIX.

MOON'S MOTIONS FOR HOURS.

Hours.	1	2	3	4	5	6	7	8	9
1	1	27	43	12	14	16	2	16	1
2	2	54	87	24	28	31	5	33	2
3	3	81	130	36	42	47	7	49	3
4	5	108	173	48	56	62	10	65	4
5	6	135	217	60	70	78	12	81	5
6	7	162	260	72	84	93	14	98	6
7	8	190	303	84	98	109	17	114	7
8	9	217	347	96	112	124	19	130	8
9	10	244	390	108	126	140	22	146	9
10	11	271	433	120	140	155	24	163	10
11	12	298	477	131	154	171	26	179	11
12	14	325	520	143	168	186	29	195	12
13	15	352	563	155	182	202	31	211	13
14	16	379	607	167	196	217	34	228	14
15	17	406	650	179	210	233	36	244	15
16	18	433	693	191	224	248	38	260	16
17	19	460	737	203	238	264	41	276	17
18	20	487	780	215	252	279	43	293	18
19	22	515	823	227	266	295	46	309	19
20	23	542	867	239	280	310	48	325	20
21	24	569	910	251	294	326	50	341	21
22	25	596	953	263	308	341	53	358	22
33	26	623	997	275	322	357	55	374	23
24	27	650	1040	287	336	372	58	390	24

TABLE XIX.

MOON'S MOTIONS FOR MINUTES.

Min.	1	2	3	4	5	6	7	8	9	10	11	12	13	14
1	0	0	1	0	0	0	0	0	0	0	0	0	0	0
5	0	2	4	1	1	1	0	1	0	0	0	0	0	0
10	0	5	7	2	2	3	0	3	0	0	0	0	0	1
15	0	7	11	3	3	4	1	4	0	1	0	1	0	1
20	0	9	14	4	5	5	1	5	0	1	0	1	0	1
25	0	11	18	5	6	6	1	7	0	1	1	1	1	2
30	1	14	22	6	7	8	1	8	0	1	1	1	1	2
35	1	16	25	7	8	9	1	10	1	2	1	2	1	2
40	1	18	29	8	9	10	2	11	1	2	1	2	1	3
45	1	20	32	9	10	12	2	12	1	2	1	2	1	3
50	1	23	36	10	11	13	2	13	1	2	1	2	1	3
55	1	25	40	11	13	14	2	15	1	3	1	3	1	4
60	1	27	43	12	14	15	2	16	1	3	1	3	1	4

HELIOCENTRIC LONGITUDES, ETC. OF THE PLANET VENUS, AT THE TIMES OF THE NEXT TWO TRANSITS OVER THE SUN'S DISC.

The subject matter of this table is connected with Chapter IX, page 119.

Times.	Hel. Long. from true Equinox.	Hel. Lat.	Rad. Vec.
1874, Dec. 8th, at 12h.	76° 41′ 36.6″	4′ 6.3″ N.	0.7203632
16h.	76 57 44.1	5 3.5	0.7203449
20h.	77 13 51.5	6 1.0	0.7203268
1882, Dec. 6th, at noon.	74 12 55.7	4 58.1 S.	0.7205502
4h.	74 29 2.5	4 0.8	0.7205315
8h.	74 45 9.7	3 3.5	0.7205127

DIP OF THE HORIZON.

For the principle of computing the dip of the horizon see text-note, page 54.

Hight in feet.	Dip.		Hight in feet.	Dip.	
1	1′	1″	16	4′	3″
2	1	26	17	4	11
3	1	45	18	4	18
4	2	2	19	4	25
5	2	16	20	4	32
6	2	29	21	4	39
7	2	41	22	4	45
8	2	52	23	4	52
9	3	2	24	4	58
10	3	12	25	5	4
11	3	22	26	5	10
12	3	31	28	5	22
13	3	39	30	5	33
14	3	48	35	6	1
15	3	56	40	6	25

SUN'S SEMIDIAMETER FOR EVERY TENTH DAY OF THE YEAR.

Days.	Jan.		July.		Days.	April.		Oct.	
	′	″	′	″		′	″	′	″
1	16	18	15	46	1	16	1	16	1
11	16	17	15	46	11	15	58	16	3
21	16	17	15	46	21	15	55	16	7
	Feb.		August.			May.		Nov.	
1	16	15	15	47	1	15	53	16	9
11	16	13	15	49	11	15	51	16	12
21	16	11	15	51	21	15	49	16	14
	March.		Sept.			June.		Dec.	
1	16	10	15	53	1	15	48	16	16
11	16	7	15	56	11	15	46	16	17
21	16	4	15	58	21	15	46	16	18

TABLE XX.

MOON'S EPOCHS.

Years.	Evection.				Anomaly.				Variation.				Longitude.			
	s	°	′	″	s	°	′	″	s	°	′	″	s	°	′	″
1846	2	0	45	6	0	26	21	2	1	5	48	4	10	15	48	23
1847	7	21	16	35	3	25	4	23	5	15	25	29	2	25	11	28
1848 B.	1	23	7	5	7	6	51	37	10	7	14	21	7	17	45	8
1849	7	13	38	35	10	5	34	57	2	16	51	46	11	27	8	14
1850	1	4	10	4	1	4	18	18	6	26	29	11	4	6	31	20
1851	6	24	41	35	4	3	1	38	11	6	6	36	8	15	54	25
1852 B.	0	26	32	5	7	14	48	53	3	27	55	29	1	8	28	6
1853	6	17	3	34	10	13	32	13	8	7	32	53	5	17	51	11
1854	0	7	35	4	1	12	15	34	0	17	10	19	9	27	14	17
1855	5	28	6	33	4	10	58	54	4	26	47	43	2	6	37	22
1856 B.	11	29	57	3	7	22	46	9	9	18	36	36	6	29	11	3
1857	5	20	28	33	10	21	29	29	1	28	14	1	11	8	34	9
1858	11	11	0	2	1	20	12	50	6	7	51	26	3	17	57	14
1859	5	1	31	33	4	18	56	10	10	17	28	52	7	27	20	20
1860 B.	11	3	22	3	8	0	43	25	3	9	17	44	0	19	54	0
1861	4	23	53	33	10	29	26	45	7	18	55	9	4	29	17	6
1862	10	14	25	3	1	28	10	6	11	28	32	34	9	8	40	12
1863	4	4	56	33	4	26	53	27	4	8	10	0	1	18	3	18
1864 B.	10	6	47	2	8	8	40	41	8	29	58	51	6	10	36	58
1865	3	27	18	32	11	7	24	2	1	9	36	17	10	20	0	4
1866	9	17	50	2	2	6	7	23	5	19	13	42	2	29	23	10
1867	3	8	21	32	5	4	50	43	9	28	51	8	7	8	46	15
1868 B.	9	10	12	2	8	16	37	58	2	20	40	0	0	1	19	56
1869	3	0	43	33	11	15	21	19	7	0	17	25	4	10	43	2
1870	8	21	15	2	2	14	4	40	11	9	54	50	8	20	6	8
1871	2	11	45	31	5	12	47	1	3	19	31	16	0	29	28	13.7
1872 B.	8	2	17	0	8	11	30	21.7	7	29	8	41	5	8	51	19.4
1873	2	4	7	31	11	23	17	36.6	0	20	57	36	10	1	25	0.3
1874	7	24	39	0	2	22	0	57.3	5	0	35	0	2	10	48	6
1875	1	15	10	29	5	20	44	18	9	10	12	24	6	20	11	11.7
1876 B.	7	5	41	59	8	19	27	38.7	1	19	49	50	10	29	34	17.4
1877	1	7	32	30	0	1	14	53.6	6	11	38	40	3	22	7	58.3
1878	6	28	3	59	2	29	58	14.3	10	21	16	5	8	1	31	4
1879	0	18	35	28	5	28	41	35	3	0	53	30	0	10	54	9.7
1880 B.	6	9	6	58	8	27	24	55.7	7	10	30	55	4	20	17	15.4
1881	0	10	57	29	0	9	12	10.6	0	2	19	47	9	12	50	56.3
1882	6	1	28	58	3	7	55	31.3	4	11	57	12	1	22	14	2.0
1883	11	22	0	27	6	6	38	52.0	8	21	34	37	6	1	37	7.7
1884 B.	5	12	31	56	9	5	22	12.7	1	1	12	2	10	11	0	13.4
1885	11	14	22	28	0	17	9	27.6	5	23	0	54	3	3	33	54.3
1886	5	4	53	57	3	15	52	48.3	10	2	38	19	7	12	57	0.0
1887	10	25	25	26	6	14	36	9.0	2	12	15	44	11	22	20	5.7
1888 B.	4	15	56	57	9	13	19	29.7	6	21	53	9	4	1	43	11.0
1889	10	17	47	28	0	25	6	44.6	11	13	42	1	8	24	16	51.9
1890	4	8	18	57	3	23	50	5.3	3	23	19	26	1	3	39	57.6

TABLE X

MOON'S EPOCHS.

Years.	Supp. of Node.				II.			V.	VI.	VII.	VIII.	IX.	X.
	s	°	'	"	s	°	'						
1846	4	16	35	9	11	7	56	254	258	937	941	847	113
1847	5	5	54	52	2	28	38	668	670	245	247	927	053
1848 B.	2	25	17	45	7	0	9	116	122	582	587	042	997
1849	6	14	37	27	10	20	41	531	535	889	893	122	937
1850	7	3	57	9	2	11	13	944	947	196	200	202	876
1851	7	23	16	51	6	1	45	358	359	504	506	282	816
1852 B.	8	12	39	44	10	3	27	806	811	841	846	398	760
1853	9	1	59	26	1	23	59	220	223	148	152	477	700
1854	9	21	19	9	5	14	31	634	636	456	459	557	639
1855	10	10	38	51	9	5	3	047	048	763	765	637	579
1856 B.	11	0	1	44	1	6	44	495	500	100	105	753	523
1857	11	19	21	26	4	27	16	909	912	407	411	832	463
1858	0	8	41	8	8	17	48	323	325	715	718	912	402
1859	0	28	0	51	0	8	20	736	737	023	024	992	342
1860 B.	1	17	23	43	4	10	1	184	189	359	364	108	286
1861	2	6	43	27	8	0	33	598	601	666	670	187	226
1862	2	26	3	9	11	21	5	012	014	974	977	267	165
1863	3	15	23	11	3	11	37	426	426	282	283	347	105
1864 B.	4	4	45	44	7	13	18	873	878	618	623	463	049
1865	4	24	5	46	11	3	50	287	291	926	929	542	989
1866	5	13	25	28	2	24	22	701	703	233	236	622	928
1867	6	2	45	10	6	14	54	115	115	544	542	702	868
1868 B.	6	22	7	43	10	16	36	563	567	877	882	818	812
1869	7	11	27	46	2	7	8	977	980	185	188	897	752
1870	8	0	47	28	5	27	40	390	392	493	495	977	691
1871	8	20	6	49	9	18	11	803	804	800	802	057	630
1872 B.	9	9	26	31	1	8	43	216	216	108	110	137	569
1873	9	28	49	24	5	10	25	664	668	444	450	252	514
1874	10	18	9	6	9	0	57	077	080	752	758	332	453
1875	11	7	28	48	0	21	29	490	492	054	064	412	392
1876 B.	11	26	48	31	4	12	1	904	905	364	370	492	331
1877	0	16	11	24	8	13	42	352	357	700	710	607	276
1878	1	5	31	6	0	4	14	765	769	008	018	687	215
1879	1	24	50	48	3	24	46	178	181	316	326	767	154
1880 B.	2	14	10	30	7	15	18	593	593	624	630	847	093
1881	3	3	33	23	11	16	59	041	045	960	970	962	038
1882	3	22	53	5	3	7	31	454	457	268	278	042	977
1883	4	12	12	47	6	28	3	867	869	576	586	122	916
1884 B.	5	1	32	29	10	18	35	280	281	884	894	202	855
1885	5	20	55	22	2	20	16	728	733	220	234	317	800
1886	6	10	15	4	6	10	48	141	145	528	542	397	739
1887	6	29	34	46	10	1	20	554	557	836	850	477	678
1888 B.	7	18	54	28	1	21	52	967	969	144	158	557	617
1889	8	8	17	21	5	23	33	415	421	480	498	672	562
1890	8	27	36	3	9	14	5	828	833	788	806	752	501

TABLE XX.

MOON'S MOTIONS FOR MONTHS.

Months.	Evection.				Anomaly.				Variation.				M. Longitude.			
	s	°	′	″	s	°	′	″	s	°	′	″	s	°		″
Jan. Com..	0	0	0	0	0	0	0	0	0	0	0	0	0	0	0	0
Jan. Bis. .	11	18	41	1	11	16	56	6	11	17	48	33	11	16	49	25
Feb. Com..	11	20	48	42	1	15	0	53	0	17	54	48	1	18	28	6
Feb. Bis. .	11	9	29	43	1	1	56	59	0	5	43	21	1	5	17	31
March......	10	7	40	26	1	20	50	4	11	29	15	15	1	27	24	27
April.......	9	28	29	8	3	5	50	57	0	17	10	3	3	15	52	32
May........	9	7	58	51	4	7	47	56	0	22	53	24	4	21	10	3
June	8	28	47	33	5	22	48	49	1	10	48	11	6	9	38	9
July........	8	8	17	16	6	24	45	48	1	16	31	32	7	14	55	40
August	7	29	5	59	8	9	46	42	2	4	26	20	9	3	23	46
September...	7	19	54	41	9	24	47	35	2	22	21	7	10	21	51	52
October.....	6	29	24	24	10	26	44	34	2	28	4	28	11	27	9	22
November...	6	20	13	6	0	11	45	27	3	15	59	16	1	15	37	28
December ...	5	29	42	49	1	13	42	26	3	21	42	37	2	20	54	59

TABLE XX.

MOON'S MOTIONS FOR DAYS.

Days.	Evection.				Anomaly.				Variation.				Mean Longitude.			
1	0s	0°	0′	0″	0s	0°	0′	0″	0s	0°	0′	0″	0s	0°	0′	0″
2	0	11	18	59	0	13	3	54	0	12	11	27	0	13	10	35
3	0	22	37	59	0	26	7	48	0	24	22	53	0	26	21	10
4	1	3	56	58	1	9	11	42	1	6	34	20	1	9	31	45
5	1	15	15	58	1	22	15	36	1	18	45	47	1	22	42	20
6	1	26	34	57	2	5	19	30	2	0	57	13	2	5	52	55
7	2	7	53	57	2	18	23	24	2	13	8	40	2	19	3	30
8	2	19	12	56	3	1	27	18	2	25	20	7	3	2	14	5
9	3	0	31	55	3	14	31	12	3	7	31	34	3	15	24	40
10	3	11	50	55	3	27	35	6	3	19	43	0	3	28	35	15
11	3	23	9	54	4	10	39	0	4	1	54	27	4	11	45	50
12	4	4	28	54	4	23	42	54	4	14	5	54	4	24	56	25
13	4	15	47	53	5	6	46	48	4	26	17	20	5	8	7	0
14	4	27	6	53	5	19	50	42	5	8	28	47	5	21	17	35
15	5	8	25	52	6	2	54	36	5	20	40	14	6	4	28	10
16	5	19	44	51	6	15	58	29	6	2	51	40	6	17	38	45
17	6	1	3	51	6		2	23	6	15	3	7	7	0	49	20
18	6	12	22	50	7	12	6	17	6	27	14	34	7	13	59	55
19	6	23	41	50	7	25	10	11	7	9	26	1	7	27	10	30
20	7	5	0	49	8	8	14	5	7	21	37	27	8	10	21	5
21	7	16	19	49	8	21	17	59	8	3	48	54	8	23	31	40
22	7	27	38	48	9	4	21	53	8	16	0	21	9	6	42	16
23	8	8	57	47	9	17	25	47	8	28	11	47	9	19	52	51
24	8	20	16	47	10	0	29	41	9	10	23	14	10	3	3	26
25	9	1	35	46	10	13	33	35	9	22	34	41	10	16	14	1
26	9	12	54	46	10	26	37	29	10	4	46	7	10	29	24	36
27	9	24	13	45	11	9	41	23	10	16	57	34	11	12	35	11
28	10	5	32	45	11	22	45	17	10	29	9	1	11	25	45	46
29	10	16	51	44	0	5	49	11	11	11	20	28	0	8	56	21
30	10	28	10	43	0	18	53	5	11	23	31	54	0	22	6	56
31	11	9	29	43	1	1	56	59	0	5	43	21	1	5	17	31

TABLE XX.

MOON'S MOTIONS FOR MONTHS.

Months.	Supp. of Node.				II.			V.	VI.	VII.	VIII.	IX.	X.
	s	°	′	″	s	°	′						
Jan. Com..	0	0	0	0	0	0	0	000	000	000	000	000	000
Jan. Bis. .	11	29	56	49	11	18	51	966	961	972	966	964	995
Feb. Com..	0	1	38	30	11	15	43	54	224	875	45	111	165
Feb. Bis. .	0	1	35	19	11	4	34	20	185	847	11	75	159
March......	0	3	7	27	9	27	59	7	330	666	989	114	313
April.......	0	4	45	57	9	13	42	61	554	542	34	225	478
May........	0	6	21	16	8	18	15	81	738	389	46	300	638
June.......	0	7	59	46	8	3	58	136	962	264	91	411	802
July........	0	9	35	5	7	8	32	156	147	112	103	486	962
August.....	0	11	13	35	6	24	15	210	371	987	147	497	126
September...	0	12	52	5	6	9	58	265	595	862	193	708	291
October.....	0	14	27	24	5	14	32	285	780	710	204	783	451
November...	0	16	5	53	5	0	15	339	4	585	250	894	615
December...	0	17	41	13	4	4	49	359	188	432	261	969	775

TABLE XX.

MOON'S MOTIONS FOR DAYS.

Days.	Supp. of Node.			II.			V.	VI.	VII.	VIII.	IX.	X.
1	0°	0′	0″	0s	0°	0′	000	000	000	000	000	000
2	0	3	11		11	9	34	39	28	34	36	5
3	0	6	21		22	18	68	79	56	67	72	11
4	0	9	32	1	3	27	102	118	85	101	108	16
5	0	12	52	1	14	37	136	158	113	135	143	21
6	0	15	53	1	25	46	170	197	141	169	179	27
8	0	19	4	2	6	55	204	237	169	202	215	32
8	0	22	14	2	18	4	238	276	198	236	251	37
9	0	25	25	2	29	13	272	316	226	270	287	43
10	0	28	36	3	10	22	306	355	254	303	323	48
11	0	31	46	3	21	31	340	395	282	337	358	53
12	0	34	57	4	2	40	374	434	311	371	394	58
13	0	38	7	4	13	50	408	474	339	405	430	64
14	0	41	18	4	24	59	442	513	367	438	466	69
15	0	44	29	5	6	8	476	553	395	472	502	74
16	0	47	39	5	17	17	510	592	424	506	538	80
17	0	50	50	5	28	26	544	632	452	539	573	85
18	0	54	1	6	9	35	578	671	480	573	609	90
19	0	57	11	6	20	44	612	711	508	607	645	96
20	1	0	22	7	1	53	646	750	537	641	681	101
21	1	3	33	7	13	3	680	790	565	674	717	106
22	1	6	43	7	24	12	714	829	593	708	753	112
23	1	9	54	8	5	21	748	869	621	742	788	117
24	1	13	5	8	16	30	782	908	650	775	824	122
25	1	16	15	8	27	39	816	948	678	809	860	128
26	1	19	26	9	8	48	850	987	706	843	896	133
27	1	22	36	9	19	57	884	027	734	877	932	138
28	1	25	47	10	1	6	918	066	762	910	968	143
29	1	28	58	10	12	16	952	106	791	944	003	149
30	1	32	8	10	23	25	986	145	819	978	039	154
31	1	35	19	11	4	34	020	185	847	011	075	15[illegible]

TABLE XX.

MOON'S MOTIONS FOR HOURS.

Hours.	Evection.			Anomaly.			Variation.			Longitude.		
	°	′	″	°	′	″	°	′	″	°	′	″
1	0	28	17	0	32	40	0	30	29	0	32	56
2	0	56	35	1	5	19	1	0	57	1	5	53
3	1	24	52	1	37	59	1	31	26	1	38	49
4	1	53	10	2	10	39	2	1	54	2	11	46
5	2	21	27	2	43	19	2	32	23	2	44	42
6	2	49	45	3	15	58	3	2	52	3	17	39
7	3	18	2	3	48	38	3	33	20	3	50	35
8	3	46	20	4	21	18	4	3	49	4	23	32
9	4	14	37	4	53	58	4	34	17	4	56	28
10	4	42	55	5	26	37	5	4	46	5	29	25
11	5	11	12	5	59	17	5	35	15	6	2	21
12	5	39	30	6	31	57	6	5	43	6	35	17
13	6	7	47	7	4	37	6	36	12	7	8	14
14	6	36	5	7	37	16	7	6	40	7	41	10
15	7	4	22	8	9	56	7	37	9	8	14	7
16	7	32	40	8	42	36	8	7	38	8	47	3
17	8	0	57	8	15	16	8	38	6	9	20	0
18	8	29	15	9	47	55	9	8	35	9	52	56
19	8	57	32	10	20	35	9	39	3	10	25	53
20	9	25	50	10	53	15	10	9	32	10	58	49
21	9	54	7	11	25	55	10	40	1	11	31	46
22	10	22	24	11	58	34	11	10	29	12	4	42
23	10	50	42	12	31	14	11	40	58	12	37	39
24	11	18	59	13	3	54	12	11	27	13	10	35

TABLE XXI.

MOON'S MOTIONS FOR MINUTES.

Min.	Evec.		Anomaly.		Variations.		Longitude.		Sup. Node.	II.
	′	″	′	″	′	″	′	″	″	″
1	0	28	0	33	0	30	0	33	0	0
5	2	21	2	43	2	32	2	45	1	2
10	4	43	5	27	5	5	5	29	1	5
15	7	4	8	10	7	37	8	14	2	7
20	9	26	10	53	10	10	10	59	3	9
25	11	47	13	37	12	42	13	43	3	12
30	14	9	16	20	15	14	16	28	4	14
35	16	30	19	3	17	47	19	13	5	16
40	18	52	21	46	20	19	21	58	5	19
45	21	13	24	30	22	52	24	42	6	21
50	23	34	27	13	25	24	27	27	7	23
55	25	56	29	56	27	56	30	12	7	26
60	28	17	32	40	30	29	32	56	8	28

TABLE XX.

MOON'S MOTIONS FOR HOURS.

Hours.	Supp. of Node.	II.	V.	VI.	VII.	VIII.	IX.	X.
	′ ″	° ′						
1	0 8	0 28	1	2	1	1	1	0
2	0 16	0 56	3	3	2	3	3	0
3	0 24	1 24	4	5	4	4	4	1
4	0 32	1 52	6	7	5	6	6	1
5	0 40	2 19	7	8	6	7	7	1
6	0 48	2 47	9	10	7	9	9	1
7	0 56	3 15	10	12	8	10	10	2
8	1 4	3 43	11	13	9	11	12	2
9	1 11	4 11	13	15	11	13	13	2
10	1 19	4 39	14	16	12	14	15	2
11	1 27	5 7	16	18	13	15	16	2
12	1 35	5 35	17	20	14	17	18	3
13	1 43	6 2	18	21	15	18	19	3
14	1 51	6 30	20	23	16	19	21	3
15	1 59	6 58	21	25	18	21	22	3
16	2 7	7 26	23	26	19	22	24	4
17	2 15	7 54	24	28	20	24	25	4
18	2 23	8 22	26	29	21	25	27	4
19	2 31	8 50	27	31	22	27	28	4
20	2 39	9 18	28	32	24	28	30	4
21	2 47	9 45	30	34	25	29	31	5
22	2 55	10 13	31	36	26	31	33	5
23	3 3	10 41	33	38	27	32	34	5
24	3 11	11 9	34	39	28	34	36	5

TABLE A.*

PERTURBATIONS OF EARTH'S RADIUS VECTOR.

Arg.	I.	II.	III.	Arg.	I.	II.	III.
0	8	4	3	500	2	0	4
50	8	4	3	550	2	1	4
100	7	4	2	600	3	1	3
150	7	4	1	650	3	2	2
200	6	4	0	700	4	3	1
250	5	4	0	750	5	4	0
300	4	3	1	800	6	4	0
350	3	2	2	850	7	4	1
400	3	1	3	900	7	4	2
450	2	1	4	950	8	4	3
500	2	0	4	1000	8	4	3

TABLE B.

☽'s APPROX. LAT.—ARG. N.

N. A.	N. D.	S. D.	S. A.	☽'s Lat.
				′ ″
0	500	500	1000	0 0
5	495	505	995	9 41
10	490	510	990	19 22
15	485	515	985	29 3
20	480	520	980	38 40
25	475	525	975	48 18
30	470	530	970	58 40
35	465	535	965	67 28
40	460	540	960	76 45
45	455	545	955	86 21
50	450	550	950	95 26
55	445	555	945	04 56

* Tables A. and B. are put in this place on account of the convenience in the page.

FIRST EQUATION OF MOON'S LONGITUDE.—ARGUMENT I.

Arg.	1 ′	1 ″	Diff. ″	Arg	1 ′	1 ″	Diff. ″
0	12	40	42	5000	12	40	40
100	11	58	42	5100	13	20	41
200	11	16	42	5200	14	1	40
300	10	34	41	5300	14	41	39
400	9	53	41	5400	15	20	40
500	9	12	40	5500	16	0	38
600	8	32	38	5600	16	38	37
700	7	54	38	5700	17	15	37
800	7	16	36	5800	17	52	35
900	6	40	34	5900	18	27	34
1000	6	6	33	6000	19	1	32
1100	5	33	31	6100	19	33	31
1200	5	2	30	6200	20	4	29
1300	4	32	27	6300	20	33	28
1400	4	5	25	6400	21	1	26
1500	3	40	23	6500	21	27	23
1600	3	17	21	6600	21	50	22
1700	2	56	18	6700	22	12	19
1800	2	38	16	6800	22	31	17
1900	2	22	13	6900	22	48	15
2000	2	9	11	7000	23	3	12
2100	1	58	8	7100	23	15	10
2200	1	50	6	7200	23	25	7
2300	1	44	3	7300	23	32	5
2400	1	41	0	7400	23	37	2
2500	1	41	2	7500	23	39	0
2600	1	43	5	7600	23	39	3
2700	1	48	7	7700	23	36	6
2800	1	55	10	7800	23	30	8
2900	2	5	12	7900	23	22	11
3000	2	17	15	8000	23	11	13
3100	2	32	17	8100	22	58	16
3200	2	49	19	8200	22	42	18
3300	3	8	22	8300	22	24	21
3400	3	30	23	8400	22	3	23
3500	3	53	26	8500	21	40	25
3600	4	19	27	8600	21	15	27
3700	4	46	30	8700	20	48	30
3800	5	16	31	8800	20	18	31
3900	5	47	32	8900	19	47	33
4000	6	19	34	9000	19	14	34
4100	6	53	35	9100	18	40	36
4200	7	28	37	9200	18	4	38
4300	8	5	37	9300	17	26	38
4400	8	42	38	9400	16	48	40
4500	9	20	39	9500	16	8	41
4600	9	59	40	9600	15	27	41
4700	10	39	40	9700	14	46	42
4800	11	19	40	9800	14	4	42
4900	11	59	41	9900	13	22	42
5000	12	40		10000	12	40	

EQUATIONS 2 TO 7 OF MOON'S LONGITUDE.—ARGUMENTS 2 TO 7.

Arg.	2		3		4		5		6		7		Arg.
	′	″	′	″	′	″	′	″	′	″	′	″	
2500	4	57	0	2	6	30	3	39	0	6	0	1	2500
2600	4	57	0	2	6	30	3	39	0	6	0	1	2400
2700	4	56	0	3	6	29	3	38	0	7	0	1	2300
2800	4	55	0	3	6	27	3	37	0	8	0	2	2200
2900	4	53	0	4	6	24	3	36	0	9	0	3	2100
3000	4	50	0	5	6	21	3	34	0	10	0	4	2000
3100	4	47	0	6	6	17	3	32	0	12	0	5	1900
3200	4	43	0	8	6	12	3	29	0	14	0	6	1800
3300	4	39	0	9	6	7	3	26	0	17	0	8	1700
3400	4	34	0	11	6	1	3	22	0	19	0	10	1600
3500	4	29	0	13	5	54	3	18	0	22	0	12	1500
3600	4	23	0	15	5	47	3	14	0	25	0	14	1400
3700	4	17	0	18	5	39	3	10	0	29	0	17	1300
3800	4	11	0	20	5	30	3	5	0	33	0	19	1200
3900	4	4	0	23	5	21	3	0	0	37	0	22	1100
4000	3	57	0	26	5	12	2	54	0	41	0	25	1000
4100	3	49	0	29	5	2	2	49	0	45	0	28	900
4200	3	41	0	32	4	52	2	43	0	50	0	31	800
4300	3	33	0	35	4	41	2	37	0	54	0	35	700
4400	3	24	0	39	4	30	2	30	0	59	0	38	600
4500	3	15	0	42	4	19	2	24	1	4	0	42	500
4600	3	7	0	46	4	7	2	17	1	9	0	45	400
4700	2	58	0	49	3	56	2	10	1	14	0	49	300
4800	2	48	0	53	3	44	2	4	1	19	0	53	200
4900	2	39	0	56	3	32	1	57	1	25	0	56	100
5000	2	30	1	0	3	20	1	50	1	30	1	0	0000
5100	2	21	1	4	3	8	1	43	1	35	1	4	9900
5200	2	11	1	7	2	56	1	36	1	40	1	7	9800
5300	2	2	1	11	2	44	1	29	1	46	1	11	9700
5400	1	53	1	14	2	33	1	23	1	51	1	15	9600
5500	1	44	1	18	2	21	1	16	1	56	1	18	9500
5600	1	36	1	21	2	10	1	10	2	1	1	22	9400
5700	1	27	1	25	1	59	1	3	2	6	1	25	9300
5800	1	19	1	28	1	48	0	57	2	10	1	28	9200
5900	1	11	1	31	1	38	0	51	2	15	1	32	9100
6000	1	3	1	34	1	28	0	46	2	19	1	35	9000
6100	0	56	1	37	1	19	0	40	2	23	1	38	8900
6200	0	49	1	39	1	10	0	35	2	27	1	40	8800
6300	0	33	1	42	1	1	0	30	2	31	1	43	8700
6400	0	36	1	44	0	53	0	26	2	35	1	46	8600
6500	0	31	1	47	0	46	0	21	2	38	1	48	8500
6600	0	26	1	49	0	39	0	18	2	41	1	50	8400
6700	0	21	1	51	0	33	0	14	2	43	1	52	8300
6800	0	17	1	52	0	28	0	11	2	46	1	54	8200
6900	0	13	1	54	0	23	0	8	2	48	1	55	8100
7000	0	10	1	55	0	19	0	6	2	50	1	56	8000
7100	0	7	1	56	0	16	0	4	2	51	1	57	7900
7200	0	5	1	57	0	13	0	2	2	52	1	58	7800
7300	0	4	1	57	0	11	0	1	2	53	1	59	7700
7400	0	3	1	58	0	10	0	1	2	54	1	59	7600
7500	0	3	1	58	0	18	0	1	2	54	1	59	7500

TABLE XXIII.

EQUATIONS 8 TO 9 OF MOON'S LONGITUDE.—ARGUMENTS 8 TO 9.

Arg.	8		9		Arg.	8		9	
	′	″	′	″		′	″	′	″
0	1	20	1	20	5000	1	20	1	20
100	1	15	1	29	5100	1	24	1	26
200	1	11	1	37	5200	1	29	1	31
300	1	7	1	46	5300	1	33	1	37
400	1	2	1	54	5400	1	37	1	42
500	0	58	2	1	5500	1	42	1	47
600	0	54	2	8	5600	1	46	1	51
700	0	50	2	15	5700	1	50	1	55
800	0	46	2	20	5800	1	54	1	58
900	0	42	2	25	5900	1	58	2	0
1000	0	38	2	29	6000	2	1	2	1
1100	0	35	2	32	6100	2	5	2	2
1200	0	31	2	34	6200	2	8	2	2
1300	0	28	2	35	6300	2	11	2	1
1400	0	25	2	35	6400	2	14	1	59
1500	0	33	2	34	6500	2	17	1	56
1600	0	20	2	32	6600	2	19	1	52
1700	0	18	2	29	6700	2	22	1	48
1800	0	16	2	26	6800	2	24	1	43
1900	0	14	2	21	6900	2	25	1	38
2000	0	13	2	16	7000	2	27	1	32
2100	0	11	2	11	7100	2	28	1	25
2200	0	10	2	4	7200	2	29	1	18
2300	0	10	1	58	7300	2	30	1	11
2400	0	9	1	51	7400	2	30	1	4
2500	0	9	1	43	7500	2	31	0	56
2600	0	10	1	36	7600	2	30	0	49
2700	0	10	1	29	7700	2	30	0	42
2800	0	11	1	22	7800	2	29	0	36
2900	0	12	1	15	7900	2	28	0	29
3000	0	13	1	8	8000	2	27	0	24
3100	0	15	1	2	8100	2	26	0	18
3200	0	16	0	57	8200	2	24	0	14
3300	0	18	0	52	8300	2	22	0	10
3400	0	21	0	47	8400	2	20	0	8
3500	0	23	0	44	8500	2	17	0	6
3600	0	26	0	41	8600	2	15	0	5
3700	0	29	0	39	8700	2	12	0	5
3800	0	32	0	38	8800	2	9	0	6
3900	0	35	0	38	8900	2	5	0	8
4000	0	39	0	39	9000	2	2	0	11
4100	0	42	0	40	9100	1	58	0	15
4200	0	46	0	42	9200	1	54	0	20
4300	0	50	0	45	9300	1	50	0	25
4400	0	54	0	49	9400	1	46	0	32
4500	0	58	0	53	9500	1	42	0	39
4600	1	3	0	58	9600	1	38	0	46
4700	1	7	1	3	9700	1	33	0	54
4800	1	11	1	9	9800	1	29	1	3
4900	1	16	1	14	9900	1	24	1	11
5000	1	20	1	20	10000	1	20	1	20

EQUATIONS 10 AND 11.

Arg.	10	11	Arg.	10	11
	"	"		"	"
0	10	10	500	10	10
10	9	11	510	10	11
20	9	12	520	9	11
30	8	13	530	9	12
40	7	14	540	8	13
50	7	15	550	8	14
60	6	16	560	8	14
70	6	17	570	8	15
80	5	17	580	7	15
90	5	18	590	7	15
100	5	18	600	7	16
110	4	19	610	7	16
120	4	19	620	7	16
130	4	19	630	7	16
140	4	19	640	7	15
150	4	19	650	8	15
160	4	19	660	8	15
170	4	18	670	8	14
180	5	18	680	9	13
190	5	17	690	9	13
200	5	16	700	10	12
210	6	16	710	10	11
220	6	15	720	11	10
230	7	14	730	11	9
240	7	13	740	12	9
250	8	12	750	12	8
260	8	11	760	13	7
270	9	10	770	13	6
280	9	10	780	14	5
290	10	9	790	14	4
300	10	8	800	15	3
310	11	7	810	15	3
320	11	6	820	15	2
330	12	6	830	16	2
340	12	5	840	16	1
350	12	5	850	16	1
360	12	5	860	16	1
370	13	4	870	16	1
380	13	4	880	16	1
390	13	4	890	16	1
400	13	4	900	15	2
410	13	5	910	15	2
420	12	5	920	15	3
430	12	5	930	14	3
440	12	6	940	14	4
450	12	6	950	13	5
460	11	7	960	13	6
470	11	8	970	12	7
480	11	8	980	11	8
490	10	9	990	11	9
500	10	10	1000	10	10

EQUATIONS 12 TO 19.

Arg.	12	13	14	15	16	17	18	19	Arg.
	"	"		"		"	"	"	
250	2	2	8	0	34	3	17	3	250
260	2	2	8	0	34	3	17	3	240
270	2	2	8	0	34	3	17	3	230
280	3	2	8	0	33	3	17	3	220
290	3	2	8	0	33	4	16	3	210
300	3	2	8	0	33	4	16	3	200
310	3	3	9	1	33	4	16	3	190
320	4	3	9	1	32	4	16	4	180
330	4	4	9	1	32	4	16	4	170
340	5	4	10	2	32	4	16	4	160
350	6	5	10	2	31	5	15	4	150
360	6	6	11	2	31	5	15	5	140
370	7	7	11	3	30	5	15	5	130
380	8	7	12	3	29	5	15	5	120
390	9	8	12	4	29	6	14	6	110
400	10	9	13	4	28	6	14	6	100
410	10	10	13	5	27	6	14	6	90
420	11	11	14	5	27	7	13	7	80
430	12	12	15	6	26	7	13	7	70
440	13	13	15	6	25	8	12	7	60
450	14	14	16	7	24	8	12	8	50
460	16	15	17	7	23	8	12	8	40
470	17	16	18	8	23	9	11	9	30
480	18	18	18	9	22	9	11	9	20
490	19	19	19	9	21	10	10	10	10
500	20	20	20	10	20	10	10	10	000
510	21	21	21	11	19	10	10	10	990
520	22	22	21	11	18	11	9	11	980
530	23	23	22	12	17	11	9	11	970
540	24	25	23	12	17	12	8	12	960
550	25	26	24	13	16	12	8	12	950
560	26	27	24	14	15	12	7	13	940
570	27	28	25	14	14	13	7	13	930
580	28	29	26	15	13	13	7	13	920
590	29	30	26	15	13	13	6	14	910
600	30	31	27	16	12	14	6	14	900
610	31	32	28	16	11	14	6	14	890
620	32	33	28	17	11	14	5	15	880
630	33	33	29	17	10	15	5	15	870
640	34	34	29	18	9	15	5	15	860
650	34	35	30	18	9	15	5	16	850
660	35	36	30	18	8	16	4	16	840
670	35	36	31	19	8	16	4	16	830
680	36	37	31	19	8	16	4	16	820
690	36	37	31	19	7	16	4	17	810
700	37	37	32	19	7	16	4	17	800
710	37	38	32	20	7	16	3	17	790
720	37	38	32	20	6	16	3	17	780
730	38	38	32	20	6	16	3	17	770
740	38	38	32	20	6	17	3	17	760
750	38	38	32	20	6	17	3	17	750

EQUATION 20.

Arg.	20	Arg.
	"	
0	10	500
10	11	510
20	12	520
30	13	530
40	13	540
50	14	550
60	15	560
70	16	570
80	16	580
90	17	590
100	17	600
110	17	610
120	17	620
130	17	630
140	17	640
150	17	650
160	17	660
170	16	670
180	16	680
190	15	690
200	14	700
210	13	710
220	13	720
230	12	730
240	11	740
250	10	750
260	9	760
270	8	770
280	7	780
290	6	790
300	6	800
310	5	810
320	4	820
330	4	830
340	3	840
350	3	850
360	3	860
370	3	870
380	3	880
390	3	890
400	3	900
410	3	910
420	4	920
430	4	930
440	5	940
450	6	950
460	6	960
470	7	970
480	8	980
490	9	990
500	10	1000

TABLE XXIV.

EVECTION. *Argument.*—Evection Corrected.

	0s	Is	IIs	IIIs	IVs	Vs
0°	1° 30′ 0″	2° 10′ 43″	2° 40 10″	2° 50′ 25″	2° 39′ 8″	2° 9′ 42″
1	1 31 25	2 11 57	2 40 51	2 50 23	2 38 25	2 8 29
2	1 32 51	2 13 9	2 41 30	2 50 20	2 37 40	2 7 16
3	1 34 16	2 14 21	2 42 8	2 50 15	2 36 55	2 6 2
4	1 35 42	2 15 31	2 42 45	2 50 9	2 36 8	2 4 47
5	1 37 7	2 16 41	2 43 21	2 50 1	2 35 19	2 3 32
6	1 38 32	2 17 50	2 43 55	2 49 52	2 34 30	2 2 16
7	1 39 57	2 18 58	2 44 27	2 49 41	2 33 40	2 1 0
8	1 41 21	2 20 5	2 44 59	2 49 29	2 32 48	1 59 43
9	1 42 46	2 21 11	2 45 29	2 49 15	2 31 55	1 58 26
10	1 44 10	2 22 17	2 45 57	2 49 0	2 31 2	1 57 8
11	1 45 34	2 23 21	2 46 24	2 48 43	2 30 7	1 55 49
12	1 46 58	2 24 24	2 46 50	2 48 26	2 29 11	1 54 30
13	1 48 21	2 25 26	2 47 14	2 48 6	2 28 14	1 53 11
14	1 49 44	2 26 28	2 47 37	2 47 45	2 27 16	1 51 51
15	1 51 7	2 27 28	2 47 59	2 47 23	2 26 17	1 50 31
16	1 52 29	2 28 27	2 48 19	2 47 0	2 25 17	1 49 11
17	1 53 51	2 29 25	2 48 37	2 46 35	2 24 16	1 47 50
18	1 55 12	2 30 21	2 48 54	2 46 8	2 23 14	1 46 29
19	1 56 33	2 31 17	2 49 10	2 45 41	2 22 11	1 45 7
20	1 57 53	2 32 11	2 49 24	2 45 12	2 21 7	1 43 46
21	1 59 13	2 33 5	2 49 37	2 44 41	2 20 2	1 42 24
22	2 0 32	2 33 57	2 49 48	2 44 9	2 18 56	1 41 2
23	2 1 51	2 34 48	2 49 58	2 43 36	2 17 50	1 39 39
24	2 3 9	2 35 38	2 50 6	2 43 2	2 16 43	1 38 17
25	2 4 26	2 36 26	2 50 13	2 42 26	2 15 34	1 36 54
26	2 5 43	2 37 13	2 50 19	2 41 49	2 14 25	1 35 32
27	2 6 59	2 37 59	2 50 23	2 41 11	2 13 16	1 34 9
28	2 8 15	2 38 44	2 50 25	2 40 31	2 12 5	1 32 46
29	2 9 30	2 39 28	2 50 26	2 39 50	2 10 54	1 31 23
30	2 10 43	2 40 10	2 50 25	2 39 8	2 9 42	1 30 0

TABLE XXV.

MOON'S EQUATORIAL PARALLAX. *Argument.*—Arg. of the Evection.

	0s	Is	IIs	IIIs	IVs	Vs	
0°	1′ 28″	1′ 23″	1′ 9″	0′ 50″	0′ 32″	0′ 18″	30°
2	1 28	1 22	1 8	0 49	0 30	0 18	28
4	1 28	1 22	1 7	0 47	0 29	0 17	26
6	1 28	1 21	1 5	0 46	0 28	0 17	24
8	1 28	1 20	1 4	0 45	0 27	0 16	22
10	1 28	1 19	1 3	0 44	0 26	0 16	20
12	1 27	1 18	1 2	0 42	0 25	0 15	18
14	1 27	1 17	1 0	0 41	0 24	0 15	16
16	1 27	1 16	0 59	0 40	0 24	0 15	14
18	1 26	1 15	0 58	0 39	0 23	0 14	12
20	1 26	1 14	0 57	0 37	0 22	0 14	10
22	1 25	1 13	0 55	0 36	0 21	0 14	8
24	1 25	1 12	0 54	0 35	0 20	0 14	6
26	1 24	1 11	0 53	0 34	0 20	0 14	4
28	1 24	1 10	0 51	0 33	0 19	0 13	2
30	1 23	1 9	0 50	0 32	0 18	0 13	0
	XIs	Xs	IXs	VIIIs	VIIs	VIs	

TABLE XXIV.

EVECTION. *Argument.*—Evection Corrected.

	VIs	VIIs	VIIIs	IXs	Xs	XIs
0°	1° 30′ 0″	0° 50′ 18″	0° 20′ 52″	0° 9′ 34″	0° 19′ 50″	0° 49′ 16″
1	1 28 37	0 49 6	0 20 10	0 9 34	0 20 32	0 50 30
2	1 27 14	0 47 55	0 19 29	0 9 35	0 21 16	0 51 45
3	1 25 51	0 46 44	0 18 49	0 9 37	0 22 1	0 53 1
4	1 24 28	0 45 34	0 18 11	0 9 41	0 22 47	0 54 17
5	1 23 6	0 44 26	0 17 34	0 9 47	0 23 34	0 55 33
6	1 21 43	0 43 17	0 16 58	0 9 54	0 24 22	0 56 51
7	1 20 20	0 42 10	0 16 24	0 10 2	0 25 12	0 58 9
8	1 18 58	0 41 4	0 15 50	0 10 12	0 26 3	0 59 28
9	1 17 36	0 39 58	0 15 19	0 10 23	0 26 55	1 0 47
10	1 16 14	0 38 53	0 14 48	0 10 36	0 27 48	1 2 7
11	1 14 52	0 37 49	0 14 19	0 10 50	0 28 43	1 3 27
12	1 13 31	0 36 46	0 13 51	0 11 5	0 29 39	1 4 48
13	1 12 10	0 35 44	0 13 25	0 11 23	0 30 35	1 6 9
14	1 10 49	0 34 43	0 13 0	0 11 41	0 31 33	1 7 31
15	1 9 29	0 33 43	0 12 37	0 12 1	0 32 32	1 8 53
16	1 8 09	0 32 44	0 12 14	0 12 23	0 33 32	1 10 16
17	1 6 49	0 31 46	0 11 54	0 12 45	0 34 34	1 11 39
18	1 5 30	0 30 49	0 11 34	0 13 10	0 35 36	1 13 2
19	1 4 11	0 29 53	0 11 16	0 13 35	0 36 39	1 14 26
20	1 2 52	0 28 58	0 11 0	0 14 3	0 37 43	1 15 50
21	1 1 34	0 28 5	0 10 45	0 14 31	0 38 48	1 17 14
22	1 0 17	0 27 12	0 10 31	0 15 1	0 39 55	1 18 39
23	0 59 0	0 26 20	0 10 19	0 15 33	0 41 2	1 20 3
24	0 57 44	0 25 30	0 10 8	0 16 5	0 42 10	1 21 28
25	0 56 28	0 24 40	0 9 59	0 16 39	0 43 19	1 22 53
26	0 55 13	0 23 52	0 9 51	0 17 15	0 44 29	1 24 18
27	0 53 58	0 23 5	0 9 45	0 17 52	0 45 39	1 25 44
28	0 52 44	0 22 20	0 9 40	0 18 30	0 46 51	1 27 9
29	0 51 31	0 21 35	0 9 36	0 19 9	0 48 3	1 28 34
30	0 50 18	0 20 52	0 9 34	0 19 50	0 49 16	1 30 0

TABLE P.

MOON'S EQUATORIAL PARALLAX. *Argument.*—Arg. of the Variation.

	0s	Is	IIs	IIIs	IVs	Vs	
0°	56″	42″	16″	4″	18″	44″	30°
2	55	41	14	4	19	46	28
4	55	39	13	4	21	47	26
6	55	37	12	4	23	48	24
8	55	35	10	5	24	50	22
10	54	34	9	6	26	51	20
12	53	32	8	6	28	52	18
14	52	30	7	7	30	53	16
16	51	28	6	8	32	54	14
18	50	26	6	9	34	55	12
20	49	24	5	10	35	55	10
22	48	23	4	12	37	56	8
24	47	21	4	13	39	56	6
26	45	19	4	14	41	57	4
28	44	18	4	16	42	57	2
30	42	16	4	18	44	57	0
	XIs	Xs	IXs	VIIIs	VIIs	VIs	

TABLE XXV.

EQUATION OF MOON'S CENTER. *Argument.*—Anomaly corrected.

	0s	Is	IIs	IIIs	IVs	Vs
0°	7° 0′ 0″	10° 20′ 58″	12° 38′ 44″	13° 17′ 35″	12° 16′ 21″	9° 58 29″
1	7 7 5	10 26 52	12 41 43	13 17 5	12 12 48	9 52 58
2	7 14 10	10 32 42	12 44 35	13 16 28	12 9 11	9 47 24
3	7 21 15	10 38 27	12 47 20	13 15 44	12 5 29	9 41 48
4	7 28 19	10 44 8	12 49 59	13 14 53	12 1 41	9 36 10
5	7 35 23	10 49 43	12 52 30	13 13 56	11 57 49	9 30 29
6	7 42 26	10 55 14	12 54 55	13 12 52	11 53 52	9 24 46
7	7 49 28	11 0 39	12 57 12	13 11 41	11 49 50	9 19 1
8	7 56 28	11 6 0	12 59 23	13 10 24	11 45 44	9 13 13
9	8 3 28	11 11 15	13 1 26	13 9 1	11 41 33	9 7 24
10	8 10 26	11 16 24	13 3 23	13 7 31	11 37 17	9 1 32
11	8 17 22	11 21 29	13 5 12	13 5 54	11 32 57	8 55 39
12	8 24 17	11 26 27	13 6 55	13 4 12	11 28 33	8 49 44
13	8 31 10	11 31 20	13 8 30	13 2 23	11 24 5	8 43 47
14	8 38 1	11 36 8	13 9 59	13 0 27	11 19 32	8 37 49
15	8 44 50	11 40 49	13 11 20	12 58 26	11 14 55	8 31 49
16	8 51 36	11 45 25	13 12 34	12 56 18	11 10 14	8 25 48
17	8 58 20	11 49 54	13 13 41	12 54 5	11 5 30	8 19 46
18	9 5 1	11 54 18	13 14 41	12 51 45	11 0 41	8 13 42
19	9 11 39	11 58 35	13 15 34	12 49 19	10 55 49	8 7 38
20	9 18 15	12 2 47	13 16 20	12 46 47	10 50 53	8 1 32
21	9 24 47	12 6 52	13 16 59	12 44 10	10 45 53	7 55 26
22	9 31 16	12 10 50	13 17 31	12 41 27	10 40 50	7 49 18
23	9 37 42	12 14 42	13 17 56	12 38 38	10 35 43	7 43 10
24	9 44 4	12 18 28	13 18 14	12 35 43	10 30 33	7 37 1
25	9 50 23	12 22 7	13 18 24	12 32 43	10 25 20	7 30 52
26	9 56 38	12 25 40	13 18 28	12 29 37	10 20 4	7 24 42
27	10 2 49	12 29 6	13 18 25	12 26 26	10 14 45	7 18 32
28	10 8 56	12 32 25	13 18 16	12 23 10	10 9 22	7 12 21
29	10 14 59	12 35 38	13 17 59	12 19 48	10 3 57	7 6 11
30	10 20 58	12 38 44	13 17 35	12 16 21	9 58 29	7 0 0

TABLE XXVI.

MOON'S EQUATORIAL PARALLAX. *Argument.*—Corrected Anomaly.

	0s	Is	IIs	IIIs	IVs	Vs	
0°	58′ 58″	58′ 27″	57′ 8″	55′ 30″	54′ 2″	53′ 3″	30°
2	58 58	58 23	57 2	55 23	53 57	53 0	28
4	58 57	58 19	56 55	55 17	53 52	52 58	26
6	58 56	58 14	56 49	55 11	53 47	52 56	24
8	58 55	58 10	56 42	55 4	53 43	52 54	22
10	58 54	58 5	56 36	54 58	53 38	52 52	20
12	58 53	58 0	56 29	54 52	53 34	52 50	18
14	58 51	57 55	56 22	54 46	53 30	52 49	16
16	58 49	57 49	56 16	54 40	53 26	52 47	14
18	58 46	57 44	56 9	54 34	53 22	52 46	12
20	58 44	57 38	56 3	54 29	53 19	52 45	10
22	58 41	57 32	55 56	54 23	53 15	52 44	8
24	58 38	57 26	55 49	54 18	53 12	52 43	6
26	58 34	57 20	55 43	54 12	53 9	52 43	4
28	58 31	57 14	55 36	54 7	53 6	52 43	2
30	58 27	57 8	55 30	54 2	53 3	52 43	0
	XIs	Xs	IXs	VIIIs	VIIs	VIs	

EQUATION OF MOON'S CENTER. *Argument.*—Anomaly corrected.

	VIs	VIIs	VIIIs	IXs	Xs	XIs
0°	7° 0′ 0″	4° 1′ 31″	1° 43′ 39′	0° 42′ 25″	1° 21′ 16″	3° 39′ 2″
1	6 53 49	3 56 3	1 40 12	0 42 1	1 24 22	3 45 1
2	6 47 39	3 50 38	1 36 50	0 41 44	1 27 35	3 51 4
3	6 41 28	3 45 15	1 33 34	0 41 35	1 30 54	3 57 11
4	6 35 18	3 39 56	1 30 23	0 41 32	1 34 20	4 3 22
5	6 29 8	3 34 40	1 27 17	0 41 36	1 37 53	4 9 37
6	6 22 59	3 29 26	1 24 17	0 41 46	1 41 32	4 15 55
7	6 16 50	3 24 17	1 21 22	0 42 4	1 45 18	4 22 18
8	6 10 42	3 19 10	1 18 33	0 42 29	1 49 10	4 28 44
9	6 4 34	3 14 7	1 15 50	0 43 1	1 53 8	4 35 13
10	5 58 28	3 9 7	1 13 12	0 43 40	1 57 13	4 41 45
11	5 52 22	3 4 11	1 10 41	0 44 26	2 1 24	4 48 21
12	5 46 17	2 59 19	1 8 15	0 45 19	2 5 42	4 54 59
13	5 40 14	2 54 30	1 5 55	0 46 19	2 10 5	5 1 40
14	5 34 12	2 49 46	1 3 42	0 47 26	2 14 35	5 8 24
15	5 28 11	2 45 5	1 1 34	0 48 40	2 19 11	5 15 10
16	5 22 11	2 40 28	0 59 33	0 50 1	2 23 52	5 21 59
17	5 16 13	2 35 55	0 57 37	0 51 30	2 28 39	5 28 50
18	5 10 16	2 31 27	0 55 48	0 53 5	2 33 32	5 35 43
19	5 4 21	2 27 3	0 54 6	0 54 47	2 38 31	5 42 37
20	4 58 28	2 22 43	0 52 29	0 56 37	2 43 35	5 49 34
21	4 52 36	2 18 27	0 50 59	0 58 33	2 48 45	5 56 32
22	4 46 47	2 14 16	0 49 36	1 0 37	2 54 0	6 3 31
23	4 40 59	2 10 10	0 48 19	1 2 48	2 59 21	6 10 32
24	4 36 14	2 6 8	0 47 8	1 5 5	3 4 46	6 17 34
25	4 29 31	2 2 11	0 46 4	1 7 30	3 10 17	6 24 37
26	4 23 50	1 58 19	0 45 7	1 10 1	3 15 52	6 31 41
27	4 18 11	1 54 31	0 44 16	1 12 40	3 21 33	6 38 45
28	4 12 35	1 50 49	0 43 32	1 15 25	3 27 18	6 45 50
29	4 7 2	1 47 11	0 42 55	1 18 17	3 33 8	6 52 55
30	4 1 31	1 43 39	0 42 25	1 21 16	3 39 2	7 0 0

VARIATION.

ARGUMENT.—Variation, corrected.

	0s			Is			IIs			IIIs			IVs			Vs		
°	°	′	″	°	′	″	°	′	″	°	′	″	°	′	″	°	′	″
0	0	38	0	1	8	1	1	6	58	0	35	54	0	5	29	0	6	2
2	0	40	26	1	9	7	1	5	36	0	33	27	0	4	21	0	7	24
4	0	42	52	1	10	3	1	4	5	0	31	0	0	3	22	0	8	55
6	0	45	16	1	10	50	1	2	27	0	28	34	0	2	33	0	10	34
8	0	47	38	1	11	26	1	0	42	0	26	11	0	1	54	0	12	22
10	0	49	57	1	11	53	0	58	49	0	23	51	0	1	24	0	14	17
12	0	52	13	1	12	9	0	56	50	0	21	34	0	1	5	0	16	19
14	0	54	24	1	12	15	0	54	45	0	19	22	0	0	57	0	18	27
16	0	56	30	1	12	10	0	52	35	0	17	15	0	0	59	0	20	41
18	0	58	30	1	11	55	0	50	21	0	15	13	0	1	11	0	23	0
20	1	0	24	1	11	30	0	48	2	0	13	17	0	1	34	0	25	23
22	1	2	11	1	10	55	0	45	40	0	11	28	0	2	8	0	27	50
24	1	3	51	1	10	10	0	43	16	0	9	47	0	2	51	0	30	20
26	1	5	23	1	9	15	0	40	50	0	8	13	0	3	45	0	32	52
28	1	6	47	1	8	11	0	38	22	0	6	47	0	4	48	0	35	26
30	1	8	1	1	6	58	0	35	54	0	5	26	0	6	2	0	38	0

	VIs			VIIs			VIIIs			IXs			Xs			XIs		
°	°	′	″	°	′	″	°	′	″	°	′	″	°	′	″	°	′	″
0	0	38	0	1	9	58	1	10	30	0	40	6	0	9	2	0	7	58
2	0	40	34	1	11	11	1	9	13	0	37	38	0	7	49	0	9	13
4	0	43	8	1	12	15	1	7	47	0	35	10	0	6	45	0	10	37
6	0	45	40	1	13	9	1	6	13	0	32	44	0	5	50	0	12	9
8	0	48	10	1	13	52	1	4	31	0	30	19	0	5	5	0	13	49
10	0	50	37	1	14	26	1	2	42	0	27	58	0	4	29	0	15	36
12	0	53	0	1	14	48	1	0	47	0	25	39	0	4	4	0	17	30
14	0	55	19	1	15	1	0	58	45	0	23	25	0	3	50	0	19	30
16	0	57	33	1	15	3	0	56	38	0	21	15	0	3	45	0	21	36
18	0	58	41	1	14	54	0	54	25	0	19	10	0	3	51	0	23	47
20	1	1	43	1	14	35	0	52	9	0	17	11	0	4	7	0	26	3
22	1	3	38	1	14	6	0	49	49	0	15	18	0	4	34	0	28	22
24	1	5	25	1	13	27	0	47	26	0	13	33	0	5	10	0	30	44
26	1	7	5	1	12	38	0	45	0	0	11	54	0	5	57	0	33	8
28	1	8	36	1	11	39	0	42	33	0	10	24	0	6	53	0	35	33
30	1	9	58	1	10	30	0	40	6	0	9	2	0	7	58	0	38	0

TABLE XXVIII.

MOON'S DISTANCE FROM THE NORTH POLE OF THE ECLIPTIC.

ARGUMENT. Supplement of Node+Moon's Orbit Longitude.

	IIIs	IVs	Vs	VIs	VIIs	VIIIs	
0°	84° 39′ 16″	85° 20′ 43″	87° 13′ 47″	89° 48′ 0″	92° 22′ 13″	94° 15′ 17″	30°
2	84 39 27	85 26 16	87 23 12	89 58 46	92 31 27	94 20 31	28
4	84 40 1	85 32 9	87 32 48	90 9 31	92 40 30	94 25 25	26
6	84 40 58	85 38 20	87 42 33	90 20 14	92 49 19	94 29 59	24
8	84 42 17	85 44 50	88 52 28	90 30 55	92 57 56	94 34 12	22
10	84 43 58	85 51 37	88 2 31	90 41 33	93 6 18	94 38 4	20
12	84 46 2	85 58 42	88 12 42	90 52 7	93 14 27	94 41 35	18
14	84 48 27	86 6 3	88 23 0	91 2 36	93 22 20	94 44 45	16
16	84 51 15	86 13 40	88 33 24	91 13 0	93 29 57	94 47 32	14
18	84 54 25	86 21 33	88 43 53	91 23 18	93 37 18	94 49 58	12
20	84 57 56	86 29 42	88 54 27	91 33 29	93 44 23	94 52 2	10
22	85 1 48	86 38 4	89 5 5	91 43 32	93 51 10	94 53 43	8
24	88 6 1	86 46 41	89 15 46	91 53 27	93 57 40	94 55 2	6
26	85 10 35	86 55 30	89 26 29	92 3 12	94 3 51	94 55 59	4
28	85 15 29	87 4 32	89 37 14	92 12 48	94 9 44	94 56 33	2
30	85 20 48	87 13 47	89 48 0	92 22 13	94 15 17	94 56 44	0
	IIs	Is	0s	XIs	Xs	IXs	

TABLE XXIX.

EQUATION II OF THE MOON'S POLAR DISTANCE.

ARGUMENT II, corrected.

	IIIs	IVs	Vs	VIs	VIIs	VIIIs	
0°	0′ 14″	1′ 24″	4′ 37″	9′ 0″	13′ 23″	16′ 36″	30°
2	0 14	1 34	4 53	9 18	13 39	16 45	28
4	0 15	1 44	5 9	9 37	13 54	16 53	26
6	0 17	1 54	5 26	9 55	14 9	17 1	24
8	0 19	2 5	5 43	10 13	14 24	17 [illegible]	22
10	0 22	2 17	6 0	10 31	14 38	17 14	20
12	0 25	2 29	6 17	10 49	14 52	17 20	18
14	0 29	2 41	6 35	11 7	15 5	17 26	16
16	0 34	2 54	6 53	11 25	15 18	17 31	14
18	0 40	3 8	7 11	11 43	15 31	17 35	12
20	0 45	3 22	7 29	12 0	15 43	17 38	10
22	0 52	3 36	7 47	12 17	15 55	17 41	8
24	0 59	3 51	8 5	12 34	16 6	17 43	6
26	1 7	4 6	8 23	12 51	16 16	17 45	4
28	1 15	4 21	8 42	13 7	16 26	17 46	2
30	1 24	4 37	9 0	13 23	16 36	17 46	0
	IIs	Is	0s	XIs	Xs	IXs	

TABLE XXX.

EQUATION III OF THE POLAR DISTANCE.

ARGUMENT. Moon's True Longitude.

	IIIs	IVs	Vs	VIs	VIIs	VIIIs	
0°	16″	15″	12″	8″	4″	1″	30°
6	16	14	11	7	3	1	24
12	16	14	10	6	3	0	18
18	16	13	10	5	2	0	12
24	15	13	9	5	1	0	6
30	15	12	8	4	1	0	0
	IIs	Is	0s	XIs	Xs	IXs	

TABLE XXXI.

EQUATIONS OF POLAR DISTANCE.

ARGUMENTS.—20 of Longitude; V to IX, corrected; and X, not corrected.

Arg.	20	V.	VI.	VII.	VIII.	IX.	X	Arg.
260	0″	56″	6″	3″	25″	3″	11″	240
280	1	55	6	3	25	3	11	220
300	1	55	7	4	25	4	11	200
320	2	53	8	5	24	6	12	180
340	3	52	10	6	23	7	13	160
360	4	50	12	8	23	9	14	140
380	5	48	14	10	22	11	16	120
400	6	45	16	12	21	14	17	100
420	8	42	18	14	20	17	19	80
440	10	39	21	17	19	20	21	60
460	11	36	24	19	17	23	23	40
480	13	33	27	22	16	27	25	20
500	15	30	30	25	15	30	27	000
520	17	27	33	28	14	33	29	980
540	19	24	36	31	12	37	31	960
560	20	20	39	33	11	40	33	940
580	22	17	41	36	10	43	35	920
600	24	15	44	38	9	46	37	900
620	25	12	46	40	8	48	38	880
640	26	10	48	42	7	51	40	860
660	27	8	50	44	6	53	41	840
680	28	7	52	45	6	54	42	820
700	29	5	53	46	5	56	42	800
720	29	5	53	47	5	56	43	780
740	30	4	54	47	5	57	43	760

TABLE XXXII.

REDUCTION

ARGUMENT.—Supplement of Node + Moon's Orbit Longitude.

	0s VIs		Is VIIs		IIs VIIIs		IIIs IXs		IVs Xs		Vs XIs	
0°	7′	0″	1′	3″	1′	3″	7′	0″	13′	57′	12′	57″
2	6	31	0	49	1	18	7	29	13	10	12	42
4	6	3	0	38	1	35	7	57	13	22	12	25
6	5	34	0	28	1	54	8	26	13	32	12	6
8	5	6	0	20	2	14	8	54	13	40	11	46
10	4	39	0	14	2	35	9	21	13	46	11	25
12	4	12	0	10	2	58	9	48	13	50	11	2
14	3	46	0	8	3	22	10	13	13	52	10	38
16	3	22	0	8	3	46	10	38	13	52	10	13
18	2	58	0	10	4	12	11	2	13	50	9	48
20	2	35	0	14	4	39	11	25	13	46	9	21
22	2	14	0	20	5	6	11	46	13	40	8	54
24	1	54	0	28	5	34	12	6	13	32	8	26
26	1	35	0	38	6	3	12	25	13	22	7	57
28	1	18	0	49	6	31	12	42	13	10	7	29
30	1	3	1	3	7	0	12	57	12	57	7	0

TABLE XXXIV.

MOON'S SEMIDIAMETER.

ARGUMENT. Equatorial Parallax.

Eq. Parallax.	Semidiam.	Eq. Parallax.	Semidiam.	Eq. Parallax.	Semidiam.
53′ 0″	14′ 27″	56′ 0″	15′ 16″	59′ 0″	16′ 5″
53 20	14 32	56 20	15 21	59 20	16 10
53 40	14 37	56 40	15 26	59 40	16 16
54 0	14 43	57 0	15 32	60 0	16 21
54 20	14 48	57 20	15 37	60 20	16 26
54 40	14 54	57 40	15 43	60 40	16 32
55 0	14 59	58 0	15 48	61 0	16 37
55 20	15 5	58 20	15 54	61 20	16 43
55 40	15 10	58 40	15 59	61 40	16 48
56 0	15 16	59 0	16 5	62 0	16 54

TABLE XXXV.

AUGMENTATION OF MOON'S SEMIDIAMETER.

ARGUMENT. Apparent Altitude.

Ap. Alt.	Augm.
6°	2″
12	3
18	5
24	6
30	8
36	9
42	11
48	12
54	13
60	14
66	15
72	15
78	16
84	16
90	16

TABLE XXXVI.

MOON'S HOURLY MOTION IN LONGITUDE.

ARGUMENTS. 2, 3, 4, and 5 of Longitude.

Arg.	2	3	4	5	Arg.
0	6″	1″	3′	3	100
5	5	2	3	3	95
10	5	2	3	3	90
15	4	2	3	3	85
20	4	3	2	2	80
25	3	3	2	2	75
30	2	3	2	2	70
35	2	4	1	1	65
40	1	4	1	1	60
45	1	4	1	1	55
50	0	5	1	1	50

TABLE XXXVII.

MOON'S HOURLY MOTION IN LONGITUDE.

ARGUMENT. Argument of the Evection.

	0s	Is	IIs	IIIs	IVs	Vs	
0°	1′ 20″	1′ 15″	1′ 0″	0′ 39″	0′ 20″	0′ 6″	30°
2	1 20	1 14	0 58	0 38	0 19	0 5	28
4	1 20	1 13	0 57	0 37	0 18	0 5	26
6	1 20	1 12	0 56	0 35	0 16	0 4	24
8	1 20	1 11	0 54	0 34	0 15	0 4	22
10	1 20	1 11	0 53	0 33	0 14	0 3	20
12	1 19	. 10	0 52	0 31	0 13	0 3	18
14	1 19	1 9	0 50	0 30	0 12	0 2	16
16	1 19	1 8	0 49	0 29	0 11	0 2	14
18	1 18	1 7	0 48	0 27	0 11	0 2	12
20	1 18	1 5	0 46	0 26	0 10	0 1	10
22	1 17	1 4	0 45	0 25	0 9	0 1	8
24	1 17	1 3	0 44	0 23	0 8	0 1	6
26	1 16	1 2	0 42	0 22	0 7	0 1	4
28	1 15	1 1	0 41	0 21	0 7	0 1	2
30	1 15	1 0	0 39	0 20	0 6	0 1	0
	XIs	Xs	IXs	VIIIs	VIIs	VIs	

TABLE XXXVIII.

MOON'S HOURLY MOTION IN LONGITUDE.

ARGUMENTS. Sum of preceding equations, and Anomaly, corrected.

		0″	20″	40″	60″	80″	100″		
0s	0°	4″	6″	9″	11″	14″	16″	XIIs	0°
	10	4	7	9	11	13	16		20
	20	5	7	9	11	13	15		10
Is	0	5	7	9	11	13	15	XIs	0
	10	6	7	9	11	13	14		20
	20	7	8	9	11	12	13		10
IIs	0	7	8	9	11	12	13	Xs	0
	10	8	9	10	10	11	12		20
	20	9	10	10	10	10	11		10
IIIs	0	10	10	10	10	10	10	IXs	0
	10	11	11	10	10	9	9		20
	20	12	11	10	10	9	8		10
IVs	0	13	12	11	9	8	7	VIIIs	0
	10	14	12	11	9	8	6		20
	20	14	12	11	9	8	6		10
Vs	0	15	13	11	9	7	5	VIIs	0
	10	15	13	11	9	7	5		20
	20	15	13	11	9	7	5		10
VIs	0	15	13	11	9	7	5	VIs	0
		0″	20″	40″	60″	80″	100″		

TABLE XXXIX.

MOON'S HOURLY MOTION IN LONGITUDE.

ARGUMENT. Anomaly, corrected.

	0s	Is	IIs	IIIs	IVs	Vs	
0°	34′ 51″	34′ 14″	32′ 39″	30′ 45″	29′ 6″	28′ 1″	30°
2	34 51	34 9	32 32	30 38	29 0	27 58	28
4	34 51	34 4	32 24	30 31	28 55	27 55	26
6	34 50	33 59	32 17	30 23	28 50	27 53	24
8	34 49	33 53	32 9	30 16	28 45	27 50	22
10	34 47	33 47	32 2	30 9	28 40	27 48	20
12	34 45	33 41	31 54	30 2	28 35	27 46	18
14	34 43	33 35	31 46	29 56	28 30	27 45	16
16	34 41	33 28	31 38	29 49	28 26	27 43	14
18	34 38	33 22	31 31	29 42	28 22	27 42	12
20	34 34	33 15	31 23	29 36	28 18	27 41	10
22	34 31	33 8	31 15	29 30	28 14	27 40	8
24	34 27	33 1	31 8	29 23	28 10	27 39	6
26	34 23	32 54	31 0	29 17	28 7	27 39	4
28	34 19	32 47	30 53	29 12	28 4	27 38	2
30	34 14	32 39	30 45	29 6	28 1	27 38	0
	XIs	Xs	IXs	VIIIs	VIIs	VIs	

TABLE XL.

MOON'S HOURLY MOTION IN LONGITUDE.

ARGUMENTS. Sum of preceding equations, and Argument of Variation.

		27′	29′	31′	33′	35′	37′		
0s	0°	0″	2″	5″	7″	10″	12″	XIIs	0°
	10	0	3	5	7	9	12		20
	20	1	3	5	7	9	11		10
Is	0	3	4	5	7	8	9	XIs	0
	10	5	5	6	6	7	7		20
	20	7	7	6	6	5	5		10
IIs	0	9	8	7	5	4	3	Xs	0
	10	11	9	7	5	3	1		20
	20	12	10	7	5	2	0		10
IIIs	0	12	10	7	5	2	0	IXs	0
	10	12	10	7	5	2	0		20
	20	11	9	7	5	3	1		10
IVs	0	9	8	7	5	4	3	VIIIs	0
	10	7	7	6	6	5	5		20
	20	5	5	6	6	7	7		10
Vs	0	3	4	5	7	8	9	VIIs	0
	10	1	3	5	7	9	11		20
	20	0	2	5	7	10	12		10
VIs	0	0	2	5	7	10	12	VIs	0
		27′	29′	31′	33′	35′	37′		

TABLE XLI.

MOON'S HOURLY MOTION IN LONGITUDE.

ARGUMENT. Argument of the Variation.

	0s	Is	IIs	IIIs	IVs	Vs	
0°	1′ 17″	0′ 58″	0′ 20″	0′ 2″	0′ 22″	1′ 0″	30°
2	1 17	0 55	0 18	0 3	0 24	1 2	28
4	1 17	0 53	0 16	0 3	0 26	1 4	26
6	1 16	0 51	0 14	0 3	0 29	1 6	24
8	1 16	0 48	0 12	0 4	0 31	1 8	22
10	1 15	0 45	0 11	0 5	0 34	1 10	20
12	1 14	0 43	0 9	0 6	0 37	1 12	18
14	1 13	0 40	0 8	0 7	0 39	1 13	16
16	1 11	0 38	0 6	0 8	0 42	1 15	14
18	1 10	0 35	0 5	0 10	0 44	1 16	12
20	1 8	0 32	0 4	0 11	0 47	1 17	10
22	1 6	0 30	0 4	0 13	0 50	1 18	8
24	1 4	0 27	0 3	0 15	0 52	1 18	6
26	1 2	0 25	0 3	0 17	0 55	1 19	4
28	1 0	0 23	0 2	0 19	0 57	1 19	2
30	0 58	0 20	0 2	0 22	1 0	1 19	0
	XIs	Xs	IXs	VIIIs	VIIs	VIs	

TABLE XLII.

MOON'S HOURLY MOTION IN LONGITUDE.

ARGUMENT. Argument of the Reduction.

	0s	Is	IIs	IIIs	IVs	Vs	
0°	2″	6″	14″	18″	14″	6″	30°
2	2	7	14	18	13	6	28
4	2	7	15	18	13	5	26
6	2	8	15	18	12	5	24
8	2	8	16	18	12	4	22
10	3	9	16	17	11	4	20
12	3	9	16	17	11	4	18
14	3	10	17	17	10	3	16
16	3	10	17	17	10	3	14
18	4	11	17	16	9	3	12
20	4	11	17	16	9	3	10
22	4	12	18	16	8	2	8
24	5	12	18	15	8	2	6
26	5	13	18	15	7	2	4
28	6	13	18	14	7	2	2
30	6	14	18	14	6	2	0
	XIs	Xs	IXs	VIIIs	VIIs	VIs	

TABLE XLIII.

MOON'S HOURLY MOTION IN LATITUDE.

ARGUMENT. Argument I, of Latitude.

	0s+	Is+	IIs+	IIIs—	IVs—	Vs—	
0°	2′ 58″	2′ 34″	1′ 29″	0′ 0″	1′ 29″	2′ 34″	30°
2	2 58	2 31	1 24	0 6	1 35	2 37	28
4	2 58	2 28	1 18	0 12	1 40	2 40	26
6	2 57	2 24	1 13	0 19	1 45	2 43	24
8	2 56	2 20	1 7	0 25	1 50	2 45	22
10	2 55	2 17	1 1	0 31	1 55	2 47	20
12	2 54	2 12	0 55	0 37	1 59	2 49	18
14	2 53	2 8	0 49	0 43	2 4	2 51	16
16	2 51	2 4	0 43	0 49	2 8	2 53	14
18	2 49	1 59	0 37	0 55	2 12	2 54	12
20	2 47	1 55	0 31	1 1	2 17	2 55	10
22	2 45	1 50	0 25	1 7	2 20	2 56	8
24	2 43	1 45	0 19	1 13	2 24	2 57	6
26	2 40	1 40	0 12	1 18	2 28	2 58	4
28	2 37	1 35	0 6	1 24	2 31	2 58	2
30	2 34	1 29	0 0	1 29	2 34	2 58	0
	XIs+	Xs+	IXs+	VIIIs—	VIIs—	VIs—	

TABLE XLIV.

MOON'S HOURLY MOTION IN LATITUDE.

ARGUMENT. Argument II, of Latitude.

	0s+	Is+	IIs+	IIIs—	IVs—	Vs—	
0°	4″	4″	2″	0″	2″	4″	30°
6	4	3	2	0	3	4	24
12	4	3	1	1	3	4	18
18	4	3	1	1	3	4	12
24	4	3	0	2	3	4	6
30	4	2	0	2	4	4	0
	XIs+	Xs+	IXs+	VIIIs—	VIIs—	VIs—	

	0′	1′	2′	3′	4′	5′	6′	7′
0″	0 0000	1 7782	1 4771	1 3010	1 1761	1 0792	1 0000	9331
1	3 5563	1 7710	1 4735	1 2986	1 1743	1 0777	9988	9320
2	3 2553	1 7639	1 4699	1 2962	1 1725	1 0763	9976	9310
3	3 0792	1 7570	1 4664	1 2939	1 1707	1 0749	9964	9300
4	2 9542	1 7501	1 4629	1 2915	1 1689	1 0734	9952	9289
5	2 8573	1 7434	1 4594	1 2891	1 1671	1 0720	9940	9279
6	2 7782	1 7368	1 4559	1 2868	1 1654	1 0706	9928	9269
7	2 7112	1 7302	1 4525	1 2845	1 1636	1 0692	9916	9259
8	2 6532	1 7238	1 4491	1 2821	1 1619	1 0678	9905	9249
9	2 6021	1 7175	1 4457	1 2798	1 1601	1 0663	9893	9238
10	2 5563	1 7112	1 4424	1 2775	1 1584	1 0649	9881	9228
11	2 5149	1 7050	1 4390	1 2753	1 1566	1 0635	9869	9218
12	2 4771	1 6990	1 4357	1 2730	1 1549	1 0621	9858	9208
13	2 4424	1 6930	1 4325	1 2707	1 1532	1 0608	9846	9198
14	2 4102	1 6871	1 4292	1 2685	1 1515	1 0594	9834	9188
15	2 3802	1 6812	1 4260	1 2663	1 1498	1 0580	9823	9178
16	2 3522	1 6755	1 4228	1 2640	1 1481	1 0566	9811	9168
17	2 3259	1 6698	1 4196	1 2618	1 1464	1 0552	9800	9158
18	2 3010	1 6642	1 4165	1 2596	1 1457	1 0539	9788	9148
19	2 2775	1 6587	1 4133	1 2574	1 1430	1 0525	9777	9138
20	2 2553	1 6532	1 4102	1 2553	1 1413	1 0512	9765	9128
21	2 2341	1 6478	1 4071	1 2531	1 1397	1 0498	9754	9119
22	2 2139	1 6425	1 4040	1 2510	1 1380	1 0484	9742	9109
23	2 1946	1 6372	1 4010	1 2488	1 1363	1 0471	9731	9099
24	2 1761	1 6320	1 3979	1 2467	1 1347	1 0458	9720	9089
25	2 1584	1 6269	1 3949	1 2445	1 1331	1 0444	9708	9079
26	2 1413	1 6218	1 3919	1 2424	1 1314	1 0431	9697	9070
27	2 1249	1 6168	1 3890	1 2403	1 1298	1 0418	9686	9060
28	2 1091	1 6118	1 3860	1 2382	1 1282	1 0404	9675	9050
29	2 0939	1 6069	1 3831	1 2362	1 1266	1 0391	9664	9041
30	2 0792	1 6021	1 3802	1 2341	1 1249	1 0378	9652	9031
31	2 0649	1 5973	1 3773	1 2320	1 1233	1 0365	9641	9021
32	2 0512	1 5925	1 3745	1 2300	1 1217	1 0352	9630	9012
33	2 0378	1 5878	1 3716	1 2279	1 1201	1 0339	9619	9002
34	2 0248	1 5832	1 3688	1 2259	1 1186	1 0326	9608	8992
35	2 0122	1 5786	1 3660	1 2239	1 1170	1 0313	9597	8983
36	2 0000	1 5740	1 3632	1 2218	1 1154	1 0300	9586	8973
37	1 9881	1 5695	1 3604	1 2198	1 1138	1 0287	9575	8964
38	1 9765	1 5651	1 3576	1 2178	1 1123	1 0274	9564	8954
39	1 9652	1 5607	1 3549	1 2159	1 1107	1 0261	9553	8945
40	1 9542	1 5563	1 3522	1 2139	1 1091	1 0248	9542	8935
41	1 9435	1 5520	1 3495	1 2119	1 1076	1 0235	9532	8926
42	1 9331	1 5477	1 3468	1 2099	1 1061	1 0223	9521	8917
43	1 9228	1 5435	1 3441	1 2080	1 1045	1 0210	9510	8907
44	1 9128	1 5393	1 3415	1 2061	1 1030	1 0197	9499	8898
45	1 9031	1 5351	1 3388	1 2041	1 1015	1 0185	9488	8888
46	1 8935	1 5310	1 3362	1 2022	1 0999	1 0172	9478	8879
47	1 8842	1 5269	1 3336	1 2003	1 0984	1 0160	9467	8870
48	1 8751	1 5229	1 3310	1 1984	1 0969	1 0147	9456	8861
49	1 8661	1 5189	1 3284	1 1965	1 0954	1 0135	9446	8851
50	1 8573	1 5149	1 3259	1 1946	1 0939	1 0122	9435	8842
51	1 8487	1 5110	1 3233	1 1927	1 0924	1 0110	9425	8833
52	1 8403	1 5071	1 3208	1 1908	1 0909	1 0098	9414	8824
53	1 8320	1 5032	1 3183	1 1889	1 0894	1 0085	9404	8814
54	1 8239	1 4994	1 3158	1 1871	1 0880	1 0073	9393	8805
55	1 8[illegible]59	1 4956	1 3133	1 1852	1 0865	1 0061	9383	8796
56	1 8081	1 4918	1 3108	1 1834	1 0850	1 0049	9372	8787
57	1 8004	1 4881	1 3083	1 1816	1 0835	1 0036	9362	8778
58	1 7929	1 4844	1 3059	1 1797	1 0821	1 0024	9351	8769
59	1 7855	1 4808	1 3034	1 1779	1 0806	1 0012	9341	8760
60	1 7782	1 4771	1 3010	1 1761	1 0792	1 0000	9331	8751

	8′	9′	10′	11′	12′	13′	14′	15′	16′
0″	8751	8239	7782	7368	6990	6642	6320	6021	5740
1	8742	8231	7774	7361	6984	6637	6315	6016	5736
2	8733	8223	7767	7354	6978	6631	6310	6011	5731
3	8724	8215	7760	7348	6972	6625	6305	6006	5727
4	8715	8207	7753	7341	6966	6620	6300	6001	5722
5	8706	8199	7745	7335	6960	6614	6294	5997	5718
6	8697	8191	7738	7328	6954	6609	6289	5992	5713
7	8688	8183	7731	7322	6948	6603	6284	5987	5709
8	8679	8175	7724	7315	6942	6598	6279	5982	5704
9	8670	8167	7717	7309	6936	6592	6274	5977	5700
10	8661	8159	7710	7302	6930	6587	6269	5973	5695
11	8652	8152	7703	7296	6924	6581	6264	5968	5691
12	8643	8144	7696	7289	6918	6576	6259	5963	5686
13	8635	8136	7688	7283	6912	6570	6254	5958	5682
14	8626	8128	7681	7276	6906	6565	6248	5954	5677
15	8617	8120	7674	7270	6900	6559	6243	5949	5673
16	8608	8112	7667	7264	6894	6554	6238	5944	5669
17	8599	8104	7660	7257	6888	6548	6233	5939	5664
18	8591	8097	7653	7251	6882	6543	6228	5935	5660
19	8582	8089	7646	7244	6877	6538	6223	5930	5655
20	8573	8081	7639	7238	6871	6532	6218	5925	5651
21	8565	8073	7632	7232	6865	6527	6213	5920	5646
22	8556	8066	7625	7225	6859	6521	6208	5916	5642
23	8547	8058	7618	7219	6853	6516	6203	5911	5637
24	8539	8050	7611	7212	6847	6510	6198	5906	5633
25	8530	8043	7604	7206	6841	6505	6193	5902	5629
26	8522	8035	7597	7200	6836	6500	6188	5897	5624
27	8513	8027	7590	7193	6830	6494	6183	5892	5620
28	8504	8020	7583	7187	6824	6489	6178	5888	5615
29	8496	8012	7577	7181	6818	6484	6173	5883	5611
30	8487	8004	7570	7175	6812	6478	6168	5878	5607
31	8479	7997	7563	7168	6807	6473	6163	5874	5602
32	8470	7989	7556	7162	6801	6467	6158	5869	5598
33	8462	7981	7549	7156	6795	6462	6153	5864	5594
34	8453	7974	7542	7149	6789	6457	6148	5860	5589
35	8445	7966	7535	7143	6784	6451	6143	5855	5585
36	8437	7959	7528	7137	6778	6446	6138	5850	5580
37	8428	7951	7522	7131	6772	6441	6133	5846	5576
38	8420	7944	7515	7124	6766	6435	6128	5841	5572
39	8411	7936	7508	7118	6761	6430	6123	5836	5567
40	8403	7929	7501	7112	6755	6425	6118	5832	5563
41	8395	7921	7494	7106	6749	6420	6113	5827	5559
42	8386	7914	7488	7100	6743	6414	6108	5823	5554
43	8378	7906	7481	7093	6738	6409	6103	5818	5550
44	8370	7899	7474	7087	6732	6404	6099	5813	5546
45	8361	7891	7467	7081	6726	6398	6094	5809	5541
46	8353	7884	7461	7075	6721	6393	6089	5804	5537
47	8345	7877	7454	7069	6715	6388	6084	5800	5533
48	8337	7869	7447	7063	6709	6383	6079	5795	5528
49	8328	7862	7441	7057	6704	6377	6074	5790	5524
50	8320	7855	7434	7050	6698	6372	6069	5786	5520
51	8312	7847	7427	7044	6692	6367	6064	5781	5516
52	8304	7840	7421	7038	6687	6362	6059	5777	5511
53	8296	7832	7414	7032	6681	6357	6055	5772	5507
54	8288	7825	7407	7026	6676	6351	6050	5768	5503
55	8279	7818	7401	7020	6670	6346	6045	5763	5498
56	8271	7811	7394	7014	6664	6341	6040	5758	5494
57	8263	7803	7387	7008	6659	6336	6035	5754	5490
58	8255	7796	7381	7002	6653	6331	6030	5749	5486
59	8247	7789	7374	6996	6648	6325	6025	5745	5481
60	8239	7782	7368	6990	6642	6320	6021	5740	5477

	17′	18	19′	20′	21′	22′	23′	24′	25′
0″	5477	5229	4994	4771	4559	4357	4164	3979	3802
1	5473	5225	4990	4768	4556	4354	4161	3976	3799
2	5469	5221	4986	4764	4552	4351	4158	3973	3796
3	5464	5217	4983	4760	4549	4347	4155	3970	3793
4	5460	5213	4979	4757	4546	4344	4152	3967	3791
5	5456	5209	4975	4753	4542	4341	4149	3964	3788
6	5452	5205	4971	4750	4539	4338	4145	3961	3785
7	5447	5201	4967	4746	4535	4334	4142	3958	3782
8	5443	5197	4964	4742	4532	4331	4139	3955	3779
9	5439	5193	4960	4739	4528	4328	4136	3952	3776
10	5435	5189	4956	4735	4525	4325	4133	3949	3773
11	5430	5185	4952	4732	4522	4321	4130	3946	3770
12	5426	5181	4949	4728	4518	4318	4127	3943	3768
13	5422	5177	4945	4724	4515	4315	4124	3940	3765
14	5418	5173	4941	4721	4511	4311	4120	3937	3762
15	5414	5169	4937	4717	4508	4308	4117	3934	3759
16	5409	5165	4933	4714	4505	4305	4114	3931	3756
17	5405	5161	4930	4710	4501	4302	4111	3928	3753
18	5401	5157	4926	4707	4498	4298	4108	3925	3750
19	5397	5153	4922	4703	4494	4295	4105	3922	3747
20	5393	5149	4918	4699	4491	4292	4102	3919	3745
21	5389	5145	4915	4696	4488	4289	4099	3917	3742
22	5384	5141	4911	4692	4484	4285	4096	3914	3739
23	5380	5137	4907	4689	4481	4282	4092	3911	3736
24	5376	5133	4903	4685	4477	4279	4089	3908	3733
25	5372	5129	4900	4682	4474	4276	4086	3905	3730
26	5368	5125	4896	4678	4471	4273	4083	3902	3727
27	5364	5122	4892	4675	4467	4269	4080	3899	3725
28	5359	5118	4889	4671	4464	4266	4077	3896	3722
29	5355	5114	4885	4668	4460	4263	4074	3893	3719
30	5351	5110	4881	4664	4457	4260	4071	3890	3716
31	5347	5106	4877	4660	4454	4256	4068	3887	3713
32	5343	5102	4874	4657	4450	4253	4065	3884	3710
33	5339	5098	4870	4653	4447	4250	4062	3881	3708
34	5335	5094	4866	4650	4444	4247	4059	3878	3705
35	5331	5090	4863	4646	4440	4244	4055	3875	3702
36	5326	5086	4859	4643	4437	4240	4052	3872	3699
37	5322	5082	4855	4639	4434	4237	4049	3869	3696
38	5318	5079	4852	4636	4430	4234	4046	3866	3693
39	5314	5075	4848	4632	4427	4231	4043	3863	3691
40	5310	5071	4844	4629	4424	4228	4040	3860	3688
41	5306	5067	4841	4625	4420	4224	4037	3857	3685
42	5302	5063	4837	4622	4417	4221	4034	3855	3682
43	5298	5059	4833	4618	4414	4218	4031	3852	3679
44	5294	5055	4830	4615	4410	4215	4028	3849	3677
45	5290	5051	4826	4611	4407	4212	4025	3846	3674
46	5285	5048	4822	4608	4404	4209	4022	3843	3671
47	5281	5044	4819	4604	4400	4205	4019	3840	3668
48	5277	5040	4815	4601	4397	4202	4016	3837	3665
49	5273	5036	4811	4597	4394	4199	4013	3834	3663
50	5269	5032	4808	4594	4390	4196	4010	3831	3660
51	5265	5028	4804	4590	4387	4193	4007	3828	3657
52	5261	5025	4800	4587	4384	4189	4004	3825	3654
53	5257	5021	4797	4584	4380	4186	4001	3922	3651
54	5253	5017	4793	4580	4377	4183	3998	3820	3649
55	5249	5013	4789	4577	4374	4180	3995	3317	3646
56	5245	5009	4786	4573	4370	4177	3991	3814	3643
57	5241	5005	4782	4570	4367	4174	3988	3811	3640
58	5237	5002	4778	4566	4364	4171	3985	3808	3637
59	5233	4998	4775	4563	4361	4167	3982	3805	3635
60	5229	4994	4771	4559	4357	4164	3979	4802	3632

	26′	27′	28′	29′	30′	31′	32′	33′	34′
0″	3632	3468	3310	3158	3010	2868	2730	2596	2467
1	3629	3465	3307	3155	3008	2866	2728	2594	2465
2	3626	3463	3305	3153	3005	2863	2725	2592	2462
3	3623	3460	3302	3150	3003	2861	2723	2590	2460
4	3621	3457	3300	3148	3001	2859	2721	2588	2458
5	3618	3454	3297	3145	2998	2856	2719	2585	2456
6	3615	3452	3294	3143	2996	2854	2716	2583	2454
7	3612	3449	3292	3140	2993	2852	2714	2581	2452
8	3610	3446	3289	3138	2991	2849	2712	2579	2450
9	3607	3444	3287	3135	2989	2847	2710	2577	2448
10	3604	3441	3284	3133	2986	2845	2707	2574	2445
11	3601	3438	3282	3130	2984	2842	2705	2572	2443
12	3598	3436	3279	3128	2981	2840	2703	2570	2441
13	3596	3433	3276	3125	2979	2838	2701	2568	2439
14	3593	3431	3274	3123	2977	2835	2698	2566	2437
15	3590	3428	3271	3120	2974	2833	2696	2564	2435
16	3587	3425	3269	3118	2972	2831	2694	2561	2433
17	3585	3423	3266	3115	2969	2828	2692	2559	2431
18	3582	3420	3264	3113	2967	2826	2689	2557	2429
19	3579	3417	3261	3110	2965	2824	2687	2555	2426
20	3576	3415	3259	3108	2962	2821	2685	2553	2424
21	3574	3412	3256	3105	2960	2819	2683	2451	2422
22	3571	3409	3253	3103	2958	2817	2681	2548	2420
23	3568	3407	3251	3101	2955	2815	2678	2546	2418
24	3565	3404	3248	3098	2953	2812	2676	2544	2416
25	3563	3401	3246	3096	2950	2810	2674	2542	2414
26	3560	3399	3243	3093	2948	2808	2672	2540	2412
27	3557	3396	3241	3091	2946	2805	2669	2538	2410
28	3555	3393	3238	3088	2943	2803	2667	2535	2408
29	3552	3391	3236	3086	2941	2801	2665	2533	2405
30	3549	3388	3233	3083	2939	2798	2663	2531	2403
31	3546	3386	3231	3081	2936	2796	2060	2529	2401
32	3544	3383	3228	3078	2934	2794	2658	2527	2399
33	3541	3380	3225	3076	2931	2792	2656	2525	2397
34	3538	3378	3223	3073	2929	2789	2654	2522	2395
35	3535	3375	3220	3071	2927	2787	2652	2520	2393
36	3533	3372	3218	3069	2924	2785	2649	2518	2391
37	3530	3370	3215	3066	2922	2782	2647	2516	2389
38	3527	3367	3213	3064	2920	2780	2645	2514	2387
39	3525	3365	3210	3061	2917	2778	2643	2512	2384
40	3522	3362	3208	3059	2915	2775	2640	2510	2382
41	3519	3359	3205	3056	2912	2773	2638	2507	2380
42	3516	3357	3203	3054	2910	2771	2636	2505	2378
43	3514	3354	3200	3052	2908	2769	2634	2503	2376
44	3511	3351	3198	3049	2905	2766	2632	2501	2374
45	3508	3349	3195	3047	2903	2764	2629	2499	2372
46	3506	3346	3193	3044	2901	2762	2627	2497	2370
47	3503	3344	3190	3042	2898	2760	2625	2494	2368
48	3500	3341	3188	3039	2896	2757	2623	2492	2366
49	3497	3338	3185	3037	2894	2755	2621	2490	2364
50	3495	3336	3183	3034	2891	2753	2618	2488	2362
51	3492	3333	3180	3032	2889	2750	2616	2486	2359
52	3489	3331	3178	3030	2887	2748	2614	2484	2357
53	3487	3328	3175	3027	2884	2746	2812	2482	2355
54	3484	3325	3173	3025	2882	2744	2610	2480	2353
55	3481	3323	3170	3022	2880	2741	2607	2477	2351
56	3479	3320	3168	3020	2877	2739	2605	2475	2349
57	3476	3318	3165	3018	2875	2737	2603	2473	2347
58	3473	3315	3163	3015	2873	2735	2601	2471	2345
59	3471	3313	3160	3013	2870	2732	2599	2469	2343
60	3468	3310	3158	3010	2868	2730	2596	2467	2341

	35′	36′	37′	38′	39′	40′	41′	42′	43′
0″	2341	2218	2099	1984	1871	1761	1654	1549	1447
1	2339	2216	2098	1982	1869	1759	1652	1547	1445
2	2337	2214	2096	1980	1867	1757	1650	1546	1443
3	2335	2212	2094	1978	1865	1755	1648	1544	1442
4	2233	2210	2092	1976	1863	1754	1647	1542	1440
5	2331	2208	2090	1974	1862	1752	1645	1540	1438
6	2328	2206	2088	1972	1860	1750	1643	1539	1437
7	2326	2204	2086	1970	1858	1748	1641	1537	1435
8	2324	2202	2084	1968	1856	1746	1640	1535	1433
9	2322	2200	2082	1967	1854	1745	1638	1534	1432
10	2320	2198	2080	1965	1852	1743	1636	1532	1430
11	2318	2196	2078	1963	1850	1741	1634	1530	1428
12	2316	2194	2076	1961	1849	1739	1633	1528	1427
13	2314	2192	2074	1959	1847	1737	1631	1527	1425
14	2312	2190	2072	1957	1845	1736	1629	1525	1423
15	2310	2188	2070	1955	1843	1734	1627	1523	1422
16	2308	2186	2068	1953	1841	1732	1626	1522	1420
17	2306	2184	2066	1951	1839	1730	1624	1520	1418
18	2304	2182	2064	1950	1838	1728	1622	1518	1417
19	2302	2180	2062	1948	1836	1727	1620	1516	1415
20	2300	2178	2061	1946	1834	1725	1619	1515	1413
21	2298	2176	2059	1944	1832	1723	1617	1513	1412
22	2296	2174	2057	1942	1830	1721	1615	1511	1410
23	2294	2172	2055	1940	1828	1719	1613	1510	1408
24	2291	2170	2053	1938	1827	1718	1612	1508	1407
25	2289	2169	2051	1936	1825	1716	1610	1506	1405
26	2287	2167	2049	1934	1823	1714	1608	1504	1403
27	2285	2165	2047	1933	1821	1712	1606	1503	1402
28	2283	2163	2045	1931	1819	1711	1605	1501	1400
29	2281	2161	2043	1929	1817	1709	1603	1499	1398
30	2279	2159	2041	1927	1816	1707	1601	1498	1397
31	2277	2157	2039	1925	1814	1705	1599	1496	1395
32	2275	2155	2037	1923	1812	1703	1598	1494	1393
33	2273	2153	2035	1921	1810	1702	1596	1493	1392
34	2271	2151	2033	1919	1808	1700	1594	1491	1390
35	2269	2149	2032	1918	1806	1698	1592	1489	1388
36	2267	2147	2030	1916	1805	1696	1591	1487	1387
37	2265	2145	2028	1914	1803	1694	1589	1486	1385
38	2263	2143	2026	1912	1801	1893	1587	1484	1383
39	2261	2141	2024	1910	1799	1691	1585	1482	1382
40	2259	2139	2022	1908	1797	1689	1584	1481	1380
41	2257	2137	2020	1906	1795	1687	1582	1479	1378
42	2255	2135	2018	1904	1794	1686	1580	1477	1377
43	2253	2133	2016	1903	1792	1684	1578	1476	1375
44	2251	2131	2014	1901	1790	1682	1577	1474	1373
45	2249	2129	2012	1899	1788	1680	1575	1472	1372
46	2247	2127	2010	1897	1786	1678	1573	1470	1370
47	2245	2125	2009	1895	1785	1677	1571	1469	1368
48	2243	2123	2007	1893	1783	1675	1570	1467	1367
49	2241	2121	2005	1891	1781	1673	1568	1465	1365
50	2239	2119	2003	1889	1779	1671	1566	1464	1363
51	2237	2117	2001	.888	1777	1670	1565	1462	1362
52	2235	2115	1999	1886	1775	1668	1563	1460	1360
53	2233	2113	1997	1884	1774	1666	1561	1459	1359
54	2231	2111	1995	1882	1772	1664	1559	1457	1357
55	2229	2109	1993	1880	1770	1663	1558	1455	1355
56	2227	2107	1991	1878	1768	1661	1556	1454	1354
57	2225	2105	1989	1876	1766	1659	1554	1452	1352
58	2223	2103	1987	1875	1765	1657	1552	1450	1350
59	2220	2101	1986	1873	1763	1655	1551	1449	1349
60	2218	2099	1984	1871	1761	1654	1549	1447	1347

	44′	45′	46′	47′	48′	49′	50′	51′	52′
0″	1347	1249	1154	1061	969	880	792	706	621
1	1345	1248	1152	1059	968	878	790	704	620
2	1344	1246	1151	1057	966	877	789	703	619
3	1342	1245	1149	1056	965	875	787	702	617
4	1340	1243	1148	1054	963	874	786	700	616
5	1339	1241	1146	1053	962	872	785	699	615
6	1337	1240	1145	1051	960	871	783	697	613
7	1335	1238	1143	1050	959	769	782	696	612
8	1334	1237	1141	1048	957	868	780	694	610
9	1332	1235	1140	1047	956	866	779	693	609
10	1331	1233	1138	1045	954	865	777	692	608
11	1329	1232	1137	1044	953	863	776	690	606
12	1327	1230	1135	1042	951	862	774	689	605
13	1326	1229	1134	1041	950	860	773	687	603
14	1324	1227	1132	1039	948	859	772	686	602
15	1322	1225	1130	1037	947	857	770	685	601
16	1321	1224	1129	1036	945	856	769	683	599
17	1319	1222	1127	1034	944	855	767	682	598
18	1317	1221	1126	1033	942	853	766	680	596
19	1316	1219	1124	1031	941	852	764	679	595
20	1314	1217	1123	1030	939	850	763	678	594
21	1313	1216	1121	1028	938	849	762	676	592
22	1311	1214	1119	1027	936	847	760	675	591
23	1309	1213	1118	1025	935	846	759	673	590
24	1308	1211	1116	1024	933	844	757	672	588
25	1306	1209	1115	1022	932	843	756	670	587
26	1304	1208	1113	1021	930	841	754	669	585
27	1303	1206	1112	1019	929	840	753	668	584
28	1301	1205	1110	1018	927	838	751	666	583
29	1300	1203	1109	1016	926	837	750	665	581
30	1298	1201	1107	1015	924	835	749	663	580
31	1296	1200	1105	1013	923	834	747	662	579
32	1295	1198	1104	1012	921	833	746	661	577
33	1293	1197	1102	1010	920	831	744	659	576
34	1291	1195	1101	1008	918	830	743	658	574
35	1290	1193	1099	1007	917	828	741	656	573
36	1288	1192	1098	1005	915	827	740	655	572
37	1287	1190	1096	1004	914	825	739	654	570
38	1285	1189	1095	1002	912	824	737	652	569
39	1283	1187	1093	1001	911	822	736	651	568
40	1282	1186	1091	999	909	821	734	649	566
41	1280	1184	1090	998	908	819	733	648	565
42	1278	1182	1088	996	906	818	731	647	563
43	1277	1181	1087	995	905	816	730	645	562
44	1275	1179	1085	993	903	815	729	644	561
45	1274	1178	1084	992	902	814	727	642	559
46	1272	1176	1082	990	900	812	726	641	558
47	1270	1174	1081	989	899	811	724	640	557
48	1269	1173	1079	987	897	809	723	638	555
49	1267	1171	1078	986	896	808	721	637	554
50	1266	1170	1076	984	894	806	720	635	552
51	1264	1168	1074	983	893	805	719	634	551
52	1262	1167	1073	981	891	803	717	633	550
53	1261	1165	1071	980	890	802	716	631	548
54	1259	1163	1070	978	888	801	714	680	547
55	1257	1162	1068	977	877	799	713	628	546
56	1256	1160	1067	975	885	798	711	627	544
57	1254	1159	1065	974	884	796	710	626	558
58	1253	1157	1064	972	883	795	709	624	541
59	1251	1156	1062	971	881	793	707	623	540
60	1249	1154	1061	969	880	792	706	621	539

	53′	54′	55′	56′	57′	58′	59′
0″	539	458	378	300	223	147	73
1	537	456	377	298	221	146	72
2	536	455	375	297	220	145	71
3	535	454	374	296	219	143	69
4	533	452	373	294	218	142	68
5	532	451	371	293	216	141	67
6	531	450	370	292	215	140	66
7	529	448	369	291	214	139	64
8	528	447	367	289	213	137	63
9	526	446	366	288	211	136	62
10	525	444	365	287	210	135	61
11	524	443	363	285	209	134	60
12	522	442	362	284	208	132	58
13	521	440	361	283	206	131	57
14	520	439	359	282	205	130	56
15	518	438	358	280	204	129	55
16	517	436	357	279	202	127	53
17	516	435	356	278	201	126	52
18	514	434	354	276	200	125	51
19	513	432	353	275	199	124	50
20	512	431	352	274	197	122	49
21	510	430	350	273	196	121	47
22	509	428	349	271	195	120	46
23	507	427	348	270	194	119	45
24	506	426	346	269	192	117	44
25	505	424	345	267	191	116	42
26	503	423	344	266	190	115	41
27	502	422	342	265	189	114	40
28	501	420	341	264	187	112	39
29	499	419	340	262	186	111	38
30	498	418	339	261	185	110	36
31	497	416	337	260	184	109	35
32	495	415	336	258	182	107	34
33	494	414	335	257	181	106	33
34	493	412	333	256	180	105	31
35	491	411	332	255	179	104	30
36	490	410	331	253	177	103	29
37	489	408	329	252	176	101	28
38	487	407	328	251	175	100	27
39	486	406	327	250	174	99	25
40	484	404	326	248	172	98	24
41	483	403	324	247	171	96	23
42	482	402	323	246	170	95	22
43	480	400	322	244	169	94	21
44	479	399	320	243	167	93	19
45	478	398	319	242	166	91	18
46	476	396	318	241	165	90	17
47	475	395	316	239	163	89	16
48	474	394	315	238	162	88	15
49	472	392	314	237	161	87	13
50	471	391	313	235	160	85	12
51	470	390	311	234	158	84	11
52	468	388	310	233	157	83	10
53	467	387	309	232	156	82	8
54	466	386	307	230	155	80	7
55	464	384	306	229	153	79	6
56	463	383	305	228	152	78	5
57	462	382	304	227	151	77	4
58	460	381	302	225	150	75	2
59	459	379	301	224	148	74	1
60	458	378	300	223	147	73	0

SATELLITES OF JUPITER.

Sat.	Mean Distance.	Sidereal Revolution.			Inclination of Orbit to that of Jupiter.			Mass; that of Jupiter being 1000000000
		d.	h.	m.	°	′	″	
1	6.04853	1	18	28	3	5	30	17328
2	9.62347	3	13	14	Variable.			23235
3	15.35024	7	3	43	Variable.			88497
4	26.99835	16	16	32	2	58	48	42659

SATELLITES OF SATURN.

Sat.	Mean Distance.	Sidereal Revolution.			Eccentricities and Inclinations.
		d.	h.	m.	
1	3.351	0	22	38	The orbits of the six interior satellites are nearly circular, and very nearly in the plane of the ring. That of the seventh is considerably inclined to the rest, and approaches nearer to coincidence with the ecliptic.
2	4.300	1	8	53	
3	5.284	1	21	18	
4	6.819	2	17	45	
5	9,524	4	12	25	
6	22.081	15	22	41	
7	64.359	79	7	55	

SATELLITES OF URANUS.

Sat.	Mean Distance.	Sidereal Period.				Inclination to Ecliptic.
		d.	h.	m.	s.	
1?	13.120	5	21	25	0	Their orbits are inclined about 78° 58′ to the ecliptic, and their motion is retrograde. The periods of the 2d and 4th require a trifling correction. The orbits appear to be nearly circles.
2	17.022	8	16	56	5	
3?	19.845	10	23	4	0	
4	22.752	13	11	8	59	
5?	45.507	38	1	48	0	
6?	91.008	107	16	40	0	

www.ingramcontent.com/pod-product-compliance
Lightning Source LLC
LaVergne TN
LVHW010142110826
845151LV00002B/402

* 9 7 8 1 4 2 5 5 3 8 8 6 6 *